Foundations in Signal Processing, Communications and Networking

Volume 12

Series editors

Wolfgang Utschick, Garching, Germany
Holger Boche, München, Germany
Rudolf Mathar, Aachen, Germany

More information about this series at http://www.springer.com/series/7603

Rudolf Ahlswede

Hiding Data – Selected Topics

Rudolf Ahlswede's Lectures on Information Theory 3

Edited by

Alexander Ahlswede
Ingo Althöfer
Christian Deppe
Ulrich Tamm

Springer

Author
Rudolf Ahlswede (1938–2010)
Department of Mathematics
University of Bielefeld
Bielefeld
Germany

Editors
Alexander Ahlswede
Bielefeld
Germany

Ingo Althöfer
Faculty of Mathematics and Computer Science
Friedrich-Schiller-University Jena
Jena
Germany

Christian Deppe
Department of Mathematics
University of Bielefeld
Bielefeld
Germany

Ulrich Tamm
Faculty of Business and Health
Bielefeld University of Applied Sciences
Bielefeld
Germany

ISSN 1863-8538 ISSN 1863-8546 (electronic)
Foundations in Signal Processing, Communications and Networking
ISBN 978-3-319-31513-3 ISBN 978-3-319-31515-7 (eBook)
DOI 10.1007/978-3-319-31515-7

Library of Congress Control Number: 2016935213

Mathematics Subject Classification (2010): 94-XX, 94A60

Printed on acid-free paper

This Springer imprint is published by Springer Nature
The registered company is Springer International Publishing AG Switzerland

Preface

Classical information processing consists of the main tasks of **gaining** knowledge, **storage**, **transmission**, and **hiding** data.

The first named task is the prime goal of statistics and for the next two Shannon presented an impressive mathematical theory, called information theory, which he based on probabilistic models.

Basic in this theory are concepts of codes—lossless and lossy—with small error probabilities in spite of noise in the transmission, which is modeled by channels.

Another way to deal with noise is based on a combinatorial concept of error correcting codes, pioneered by Hamming. This leads to another way to look at information theory, which instead of being looked at by its tasks can be also classified by its mathematical structures and methods: primarily **probabilistic** versus **combinatorial**.

Finally, Shannon also laid foundations of a theory concerning hiding data, called cryptology. Its task is in a sense dual to transmission and we therefore prefer to view it as a subfield of information theory.

Viewed by mathematical structures there is again already in Shannon's work a **probabilistic** and a combinatorial or a **complexity**-theoretical model.

The lectures are suitable for graduate students in mathematics, and also in theoretical computer science, physics, and electrical engineering after some preparations in basic mathematics.

The lectures can be selected for courses or supplements of courses in many ways.

Rudolf Ahlswede

This is the original Preface written by Rudolf Ahlswede for the first 1,000 pages of his lectures. This volume consists of the last third of these pages.

Words and Introduction of the Editors

Rudolf Ahlswede was one of the worldwide accepted experts in nformation theory. Many main developments in this area are due to him. Especially, he made a big progress in multi-user theory. Furthermore, with identification theory and network coding he introduced new research directions. Rudolf Ahlswede died in December 2010.

The topic of this third volume is information hiding. The book starts with a short course on cryptography, which is mainly based on a lecture of Rudolf Ahlswede at the University of Bielefeld in the mid-1990s. It was the second one in his cycle of lectures on information theory which, as usual, started with an introductory course on the basic coding theorems as covered in Volume 1 of this series. In the previous cycles the follow-up lectures were something like "Information Theory II," "Algebraic Coding Theory," "Selected Topics of Information Theory," or "Combinatorial methods in Information Theory" but this time he decided in favor of cryptology.

This turned out to be a very good choice. First, soon after many new areas in cryptology kicked off because of the then new applications in Internet and e-commerce, and, second, Rudolf Ahlswede was about to build up a new group of young students (among them Lars Bäumer, Christian Deppe, Christian Heup, Gohar Khuregyan, Christian Kleinewächter, Rainer Wilmink, and Andreas Winter) who became very much interested in his lectures. Several of them chose information security as a topic of their master or Ph.D. theses.

The short course on cryptography started with a thorough discussion of Shannon's pioneering paper (1949) "Communication Theory of Secrecy Systems" and the presentation of two of Rudolf Ahlswede's own results. After that secret-key and public-key cryptology were introduced. Concerning these standard topics the lecture notes were rather brief and had not been modified since. The reason is that in later lectures he intensively concentrated on the new areas under development those days and the necessary basics were included in some detail in the corresponding lecture notes. This led to the chapters on authentication, the new encryption standard AES, and on elliptic curve cryptosystems. Furthermore, information theoretic aspects as the wiretap channel and oblivious transfers are addressed here which usually are not found in books on cryptology. This lecture about the wiretap channel is written by Holger Boche and Ahmed Mansour. It is an extension of the original text of Rudolf Ahlswede, which was only a one-page summary of the result of Wyner. In this text all new important developments are included. The extension of the original text was a suggestion of one of the reviewers.

So, this volume is rather about selected topics in information hiding and there may be some overlap among the chapters whereas other areas maybe only briefly addressed. The reader is referred to the many excellent books covering the classic stuff in secret-key and public-key cryptography in case he needs a more intensive discussion.

Let us conclude with some related anecdotes. In 1997/1998 the German state Northrhine-Westphalia started a crypto-initiative, which finally led to an institute and several new professor positions at the University of Bochum. Rudolf Ahlswede was included in the preparatory discussions and he and his research assistants, Bernhard Balkenhol and Ulrich Tamm, were regularly invited to the corresponding meetings and conferences. The project was rather important, all leading German experts on information security and also high-ranked officials from the European Union, German government, and the state Northrhine-Westphalia were around at these meetings.

After some time the Ministry of Science of Northrhine-Westphalia asked some of the experts, among them Rudolf Ahlswede, for a statement. As usual, he was quite busy with research and did not answer until the deadline. After several reminders, finally, he was told that everybody else had answered and only his report was missing. Then he decided to write the same day.

One of these meetings took place in the end of November 200 km far from Bielefeld. Because of the bad weather and expected traffic problems in the Ruhr area, we decided to go by train. In spite of the snow Rudolf Ahlswede came to the university without a coat and wearing sandals. Again he concentrated on a research problem, forgetting time and ignoring our reminders. We caught the last possible train only by running through storm and ice and he had not even a minute to stop at his home close by to at least pick up a coat and change the shoes.

The comments for this volume are provided by Rüdiger Reischuk who is Professor for Computer Science at the University of Lübeck. Cryptology is, of course, very close to complexity theory, his area of research. Rüdiger Reischuk obtained his Ph.D. in Bielefeld, where Rudolf Ahlswede had built up a strong group in theoretical computer science at his chair. The situation is described in the preface to the volume "Numbers, Information and Complexity" in honor of Rudolf Ahlswede's 60th birthday:

> Complexity Theory became the main subject in Computer Science. Against all conventions Wolfgang Paul was hired as an Associate Professor at the age of twentyfive and became its prime mover. Among an impressive group of Ph.D.'s we find Ingo Wegener, Friedhelm Meyer auf der Heide and Rüdiger Reischuk, who are now among the leaders in Theoretical Computer Science. Paul and Meyer auf der Heide participated later in two different Leibniz

prizes, the most prestigious award supporting science in Germany. Ingo Wegener is internationally known for his classic on switching circuits. Friedhelm Meyer auf der Heide predominantly contributed to parallel computing. Paul and Reischuk made their famous step towards $P \neq NP$.

Our thanks go to Regine Hollmann, Carsten Petersen, and Christian Wischmann for helping us in typing, typesetting, and proofreading. Furthermore our thanks go to Bernhard Balkenhol who combines the approximately first 2000 pages of lecture scripts in different styles (amstex, latex, etc.) to one big lecture script. He can be seen as one of the pioneers of Ahlswede's lecture notes.

Alexander Ahlswede
Ingo Althöfer
Christian Deppe
Ulrich Tamm

Contents

Chapter 1
A Short Course on Cryptography

Cryptology is the science of information protection. In his pioneering paper "Communication Theory of Secrecy Systems" Claude E. Shannon (1949) investigated the following *secrecy system*.

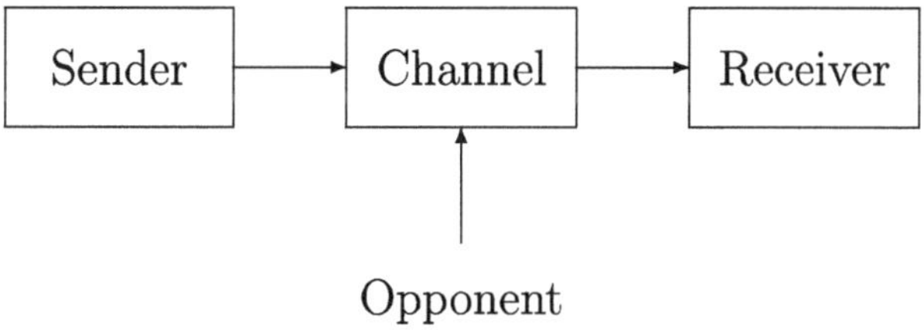

A sender transmits a message to a receiver over a communication channel. A third person, which we shall denote as "opponent", "wiretapper", or "cryptanalyst", has access to the channel and is able to read the message. The aim of the sender and the receiver is to avoid that the opponent can extract any information from the transmitted message. In order to do so, the sender *encrypts* the original message $m \in \mathcal{M}$ (where $\mathcal{M}$ is the set of all possible messages), i.e., he transmits an encoded version $m' = c(m)$, where $c : \mathcal{M} \to \mathcal{M}'$ is a mapping usually denoted as *key* or code.

The receiver simply applies the inverse mapping $c^{-1}(m') = c^{-1}\big(c(m)\big) = m$ to obtain the original message m which is also called *plaintext*. The opponent intercepts the *cryptogram* $c(m)$. The difficulty for him is to apply the inverse function c^{-1}, since sender and receiver have some additional information not contained in the cryptogram.

Shannon introduced the concept of *secret-key* cryptology. Here the key c is chosen from a large set of possible keys $\{c_1, \dots, c_k\}$. The sender transmits the chosen key over a secure channel, which cannot be intercepted by the wiretapper. Hence the opponent's task is to determine the key out of a set of K possible keys. Later we shall learn about *public-key* cryptography. Here we use as key a one-way function c, i.e., it is easy to compute $c(m)$ but it is hard to determine $c^{-1}(m')$ (e.g., the opponent could

A. Ahlswede et al. (eds.), *Hiding Data – Selected Topics*,
Foundations in Signal Processing, Communications
and Networking 12, DOI 10.1007/978-3-319-31515-7_1

have to solve an NP-complete problem in order to decrypt the message, whereas the receiver only has to verify the solution).

Before the presentation of the important results we shall first introduce the notation which we shall use throughout this chapter. The original text, which has to be conveyed to the receiver, is divided into small units—letters over some alphabet. If, e.g., the original text is in English language, these units may be the letters $\{a, \ldots, z\}$ of the Latin alphabet. If the text is a binary string the smallest units are the single bits. It is also convenient to use blocks of a fixed length n, i.e., words of length n over $\{0, 1\}$ or $\{a, \ldots, z\}$ as units. Each of the units is then encrypted subsequently.

The frequency of the letters of the Latin alphabet (cf. Chap. 2) imposes a probability distribution $\mathcal{P}$ on the set of possible messages $\mathcal{M}$. A set with a probability distribution on its elements is denoted as source, hence we have the source $(\mathcal{M}, \mathcal{P})$. We also assume that there is a probability distribution Q on the key space $\mathbb{C} = \{c_1, \ldots, c_k\}$. The pair $(\mathbb{C}, Q)$ is called *cipher*. However, usually Q is the uniform distribution, since this leaves the most amount of uncertainty at the wiretapper. A key is a mapping $c_j : \mathcal{M} \rightarrow \mathcal{M}_j$. In Shannon's model we usually assume that $\mathcal{M}_j = \mathcal{M}'$ for all $j = 1, \ldots, K$, i.e., the range is the same for all keys (often also $\mathcal{M}' = \mathcal{M}$). However, in Simmons' model of authentication in which the opponent can replace the cryptogram by a fraudulent one, it is essential that the ranges do not overlap too much.

The chapter on cryptology will be divided into six sections. The main topics are Secret-Key Cryptology, The Wiretap Channel, Cryptosystems, Homophonic Coding, Spurious Decipherments and Authentication. In Sect. 1.1 we shall consider three measures for the quality of a cipher. Shannon asked for the remaining uncertainty about the plain-text message when the cryptogram is known. We shall denote this as the *entropy criterion*. Hellman later introduced a similar, rather combinatorial, measure, namely *spurious decipherments* are counted. Roughly spoken, these are the possible different interpretations of a cryptogram. Finally, Ahlswede considered the probability of error as criterion on the quality of a code. In the last part of Sect. 1.1, we shall introduce Simmons' model of authentication. In Shannon's model of a cryptosystem, the opponent may intercept the cryptogram and try to decipher it. Simmons introduced a new model of a cryptosystem in which he gave much more power to the opponent. He now is able to replace the cryptogram by a fraudulent one. The receiver's task now is to detect such a deception and the sender has to encrypt in such a way that the receiver can verify the authenticity of the received message.

A rather different cryptological approach will be presented in Sect. 1.2. In Shannon's cryptosystem we did not consider distortions that may occur during the transmission. Now we assume that sender and receiver communicate over some channel W and that the wiretapper receives the cryptogram over a different channel W', which shall be denoted as the *wiretap channel*. The question now is: How can we encode in such a way that the receiver can reconstruct the message with high probability and that, on the other hand, the wiretapper does not gain enough information to decrypt the message? The wiretap channel was introduced by Wyner [37]. Ahlswede independently considered the special case that sender and receiver communicate over a noiseless channel and that the wiretap channel is noisy. In order to leave a maximum amount of uncertainty at the wiretapper, when distortions occur during the transmission, it is necessary to place the codewords as close as possible in the Hamming space (if the channel W is binary). If, e.g., the wiretapper receives the all-zero vector 0^n and all the vectors x^n with weight $w(x^n) = 1$ are possible codewords, then there are already n possible messages from which 0^n may have arisen if only one error occurred.

This is contrasting to Coding Theory, where the codewords are chosen at a certain minimum distance to each other, in order to protect them against distortions during the transmission. In Wyner's model of the wiretap channel, the distortions are used to make the wiretappers life as hard as possible, and hence the codewords are chosen close to each other. So a bad code can be a good cipher.

In all cryptosystems, which we shall discuss in the sequel, we shall concede the cryptanalyst as much information as possible:

(1) He knows about the existence of the message,
(2) there is no special equipment required to recover the message, the cryptanalyst can use the same technical facilities as the receiver.

If (1) is violated, Shannon spoke of a *concealment system*, e.g., the message may be concealed in an innocent text or written with invisible ink.

If (2) is violated, Shannon called this cryptosystem a *privacy system*. He defined a *"true" secrecy system* as a cryptosystem in which the meaning of the message is concealed by a cipher, code, ..., the enemy knows about the existence and has all technical equipment to intercept and record the transmitted signal.

We shall only deal with true secrecy systems. As Shannon pointed out, concealment systems are rather a psychological and privacy systems rather a technical problem, whereas the design of a true secrecy system is a mathematical problem. In Sect. 1.4 we shall consider Shannon's information-theoretic approach to cryptosystems.

Sects. 1.5–1.7 are devoted to Homophonic Coding, Spurious Decipherments and Authentication.

Finally, a remark about the word "cryptology". We use this notion, because it covers both, cryptography and cryptanalysis. In the literature, often the science of information protection is denoted as cryptography. We shall use this notion only for the encryption of messages. Cryptanalysis, the attempt to break a code is a science for itself, which (especially in public-key cryptology) uses quite different methods from those used in cryptography.

1.1 Ahlswede's Immediate Response to Shannon's Work

1.1.1 Introduction

The concept of secret-key cryptology was already presented in the Introduction. The sender chooses a key c_j out of K possible keys in order to encrypt the message m by the prescription $m' = c_j(m)$. The receiver has been informed about the choice

$j \in \{1, \dots, K\}$ of the key index via a secure channel. "Secure" means that the cryptanalyst has no access to this channel. So the cryptanalyst can only intercept the cryptogram m' from which he must conclude the plain-text m. His task hence is to find the key c_j, then he can apply c_j^{-1} to obtain $c_j^{-1}(m') = m$. We shall denote by

m	– the plain-text message
m'	– the cryptogram
$\mathcal{M} = \{1, \dots, M\}$	– the set of all possible plain-text messages
P	– the probability distribution on $\mathcal{M}$
$(\mathcal{M}, P)$	– the message source
$c_j : \mathcal{M} \rightarrow \mathcal{M}_j;\ j \in \{1, \dots, K\}$	– the key
$\mathbb{C} = \{c_1, \dots, c_K\}$	– the key space
Q	– the probability distribution on $\mathbb{C}$
$(\mathbb{C}, Q)$	– the cipher
X	– the random variable for the plain-text
Y	– the random variable for the cryptogram
Z	– the random variable for the key

Deviations from these standard notations will be announced in the respective sections.

Although messages have been encrypted with secret keys already in ancient times, the mathematical foundations of cryptology and especially secret-key cryptology are due to Shannon (1949). For a survey on the history of cryptography until 1945 we refer to Kahn [25].

Shannon introduced a measure for the quality of a cipher, namely he considered the remaining uncertainty $H(X|Y)$ about the plain-text message X when the cryptogram Y is known. He called a secrecy system *perfect* if $H(X|Y) = H(X)$, i.e., the knowledge of the cryptogram does not yield any information about the original message. The mathematical interpretation is that the random variables X and Y for plain-text and cryptogram, respectively, are independent. Shannon demonstrated that in a perfect secrecy system the amount of key space is at least as big as the amount of plain-text $H(Z) \geq H(X)$. He further introduced the key-equivocation function $H(Z|Y^n)$ (the remaining uncertainty about the key when n letters of the cryptogram are known) and the *unicity distance* (the smallest n such that there is exactly one key from which the cryptogram Y^n can have arisen).

Shannon's results will be presented in Sect. 1.4. In Sect. 1.5 we shall discuss how the unicity distance can be augmented by homophonic coding. In homophonic coding a message can be encrypted by several codewords (homophones). Homophonic coding is useful in order to produce an output sequence in which 0's and 1's occur equally often on the average.

After Shannon's pioneering paper there had been little interest in cryptology for almost three decades until Diffie and Hellman introduced public-key cryptology in 1976. Due to its applications in computer networks there is an enormous interest in this branch of cryptology. In secret-key cryptology, however, there are only a few follow-up papers. In 1977 Hellman and Ahlswede presented new criteria for the quality of a code. Hellman considered spurious decipherments which will be discussed in Sect. 1.6. Ahlswede investigated the probability of erroneous decryption (when the cryptanalyst uses the maximum-likelihood decoding rule). He presented a simple cipher which is almost optimal under this criterion as well as under the entropy criterion. A further advantage is that this cipher does not require the asymptotic equipartition property (AEP).

Finally, in this section we shall investigate Simmons' model of authentication. Simmons introduced a cryptosystem in which he gave more power to the opponent. He now is not only able to intercept the cryptogram but he may replace it with a fraudulent one and hence try to cheat the receiver. In order to avoid this, the sender has to encrypt the message in such a way that the receiver can verify its authenticity. This authentication cryptosystem was motivated by economy. Think, for instance, of a casino with a lot of slot machines. Each slot machine is equipped with some software that prints out the sum of money this machine made this day. At the end of each day the manager of the casino collects these billets and reports the total sum to the owner. In order to guarantee that the information really stems from the single slot-machine, the sum of money reported by each machine must be encoded in such a way that the owner can verify the authentication and hence immediately detect when the manager is cheating him.

Simmons investigated two different situations. In an *impersonation* attack the enemy sends the fraudulent cryptogram before intercepting the transmission. So, by chance fraudulent and correct cryptogram might be the same. This is not possible in a *substitution* attack, where the enemy waits until he intercepted the correct cryptogram. It will be shown in Sect. 1.7 that the probability P_I that the fraudulent cryptogram is valid in an impersonation attack is always lower bounded by $P_I \geq 2^{-I(X \wedge Y)}$, where X and Y are the random variables for plain-text and correct cryptogram. Simmons defined a secrecy system to have *perfect authenticity* if this lower bound is assumed for the probability of deception (impersonation or substitution). It will be demonstrated that perfect secrecy and perfect authenticity do not imply each other.

At the end of this introduction we shall present some basic ciphers which we shall use in the sequel. We don't want to discuss the construction of ciphers any further, since we are rather interested in information-theoretic aspects of cryptology.

1. In a *permutation cipher* (or *substitution cipher*) the keys c_j are permutations on the set of possible messages $\mathcal{M}$. If, e.g., $\mathcal{M} = \{a, b, c, \ldots, z\}$ is the Latin alphabet, there are 26! permutations. Caesar already used a special permutation cipher to encrypt messages, namely a cyclic shift of three letters, i.e., $c(m) = m + 3 \pmod{26}$. So the word "CAESAR" was encrypted as "FDHVDU". Augustus later used the key defined by $c'(m) = m + 4 \pmod{26}$.
2. In a *transposition cipher* the smallest units are words of a fixed length n over some alphabet, e.g., $\mathcal{M} = \{a, b, c, \ldots, z\}^n$. Each block of length n is encrypted by applying a fixed permutation to the positions of the letters. For instance, we obtain from "BERLIN" the codeword "RBENLI" if the permutation (123) is applied to blocks of three letters.

Permutation ciphers are easily broken by counting the frequencies of the single letters. So it makes sense to mix permutation and transposition ciphers. We shall later see that for a real secure encryption we have to use a key space at least as big as the set of messages. In the previous examples each unit is encrypted by the same prescription (permutation or transposition). If long texts are encoded this way, the cryptanalyst will sooner or later detect the selected key.

3. In a *one-time pad* the key is changed after each unit (letter, word of fixed length, ...). So each unit is encrypted using a new key. We shall see in Sect. 1.3 that the one-time pad is perfectly secret. A one-time pad was used to encrypt messages over the "hot wire" between Washington and Moscow.

1.1.2 A Simple Cipher for Shannon's Secrecy System

Let $(\mathcal{M}, P)$ be a message source, where w.l.o.g. the probabilities are ordered non-increasingly, hence

$$P(1) \geq P(2) \geq \cdots \geq P(M).$$

We shall now introduce a cipher with K keys $c_1, \ldots, c_K$ which turns out to be very good with respect to the security measures "decrypting error probability" as well as "entropy criterion". If $K \geq M$ one can choose the M keys $c_i(m) = m + i \bmod M$ $(i = 1, \ldots, M)$ with equal probability and then for this cipher $H(X) = H(X|Y)$ (where X and Y are the RVs for the plain-text and the cryptogram, respectively). Therefore we can always assume that there are at most as many keys as messages, $K \leq M$. We further consider only ciphers whose keys are equiprobable.

We now write w.l.o.g. $M = K \cdot \ell, \ell \in \mathbb{N}$ (by assumption, $K \leq M$ and if M is not divisible by K we enlarge the source with messages of probability 0). The K keys $c_0, \ldots, c_{K-1}$ are defined as follows

$$c_i(m) \triangleq K \cdot j + \big((t + i - 1) \bmod K\big) \text{ if } m = K \cdot j + t, 0 \leq j \leq \ell - 1. \tag{1.1.1}$$

So each key c_i yields a cyclic shift (of length i) on each of the ℓ blocks of messages

$$B_j \triangleq \big\{(j-1) \cdot K + 1, \ldots, j \cdot K\big\}. \tag{1.1.2}$$

Obviously the cipher $(\mathbb{C}, Q)$, where $\mathbb{C} = \{c_1, \ldots, c_k\}$ and Q is the uniform distribution, is regular.

The best decoding rule for the cryptanalyst with respect to the error probability criterion is the maximum likelihood decoding rule, i.e., given a cryptogram $m' \in \mathcal{M}$ he votes for an $m \in \mathcal{M}$ maximizing $P(X = m, Y = m')$ (if there are more than one messages with the same joint probability, he votes for the message which is minimal in the order obtained by embedding the set $\mathcal{M}$ into the positive integers [any other decision rule which leaves a unique message is also o.k.]).

For our special key this just means that the cryptanalyst always votes for the first element $K \cdot j + 1$ in the block B_{j+1}. Recall that the messages are ordered with respect to non-increasing probabilities. So the messages $K \cdot j + 1, 0, \leq j \leq \ell - 1$ (which have the greatest probability within the block B_{j+1}) are always decrypted correctly.

Theorem 1 *For the cipher described above the decrypting error probability* λ *satisfies*

$$\frac{K-1}{K}(1 - P_{\max}) \leq \lambda \leq 1 - P_{\max} \tag{1.1.3}$$

(where $P_{\max} = P(1)$ *denotes the maximum probability of a message in the source* $\mathcal{M}$*).*

Proof Since the message $K \cdot j + 1, 0 \leq j \leq \ell - 1$ are always decrypted correctly, the error probability λ of the cipher can be expressed as

$$\begin{aligned}\lambda &= 1 - P(1) - P(K+1) - \cdots - P\big(K \cdot (\ell - 1) + 1\big) \\ &= P(2) + \cdots + P(K) + P(K+2) + \cdots + P(2K) + \cdots \\ &\quad + P\big((\ell-1) \cdot K + 2\big) + \cdots + P(\ell K) \\ &\geq (k-1) \cdot \big(P(K) + P(2K) + \cdots + P(\ell K)\big) \\ &= (k-1)\big(1 - \lambda - P(1) + P(\ell K)\big) \\ &\geq (k-1)\big(1 - \lambda - P(1)\big)\end{aligned}$$

since the probabilities are non-increasing and hence $\sum_{t=2}^{k} P(K \cdot j + t) \geq (k-1) \cdot P\big(K \cdot (j+1)\big)$ and $P(K \cdot j) \geq P(K \cdot j + 1)$ for $j = 0, \ldots, \ell - 1$.

So the left-hand side of (1.1.3) is immediate. Since the cryptanalyst can always vote for message 1, obviously $\lambda \leq 1 - P(1)$ must hold and the theorem is proved.

Remark 1 (1) For large K upper and lower bound in Theorem 1 are close to each other and the probability of error is almost exactly determined.

(2) Observe that the proof of Theorem 1 is constructive, i.e., a cipher with the desired properties is explicitly introduced. When Shannon considered the entropy criterion he derived for a random cipher the bound $H(X|Y) \geq \log K + H(X) - \log M$, which turns out to be tight only if $H(X) = \log M$, i.e., the messages are uniformly distributed. We shall see in Theorem 2 that the cipher introduced above already yields $H(X|Y) \geq \log K - 1$ (under the quite natural condition that $\frac{1}{K} \geq P(m)$ for all $m \in \mathcal{M}$). Indeed, Shannon found the method of random coding in Cryptology, and our cipher demonstrates that the results obtained are, in general, not optimal. On the other hand, random coding turned out to be a central tool in Information Theory, where this method of proof is used to demonstrate the existence of codes reaching channel capacity.

(3) Shannon (and later Hellmann) required the asymptotic equipartition property (AEP) for the source $(\mathcal{M}, P)$, i.e., the set of messages $\mathcal{M}$ can be divided into two groups: one group of high and fairly uniform probability, the second of negligibly small total probability. The AEP does not hold for every source. Observe that the construction of our special cipher does not require any restrictions to the source, especially not the AEP.

Shannon in his pioneering paper considered the entropy criterion as a measure for the quality of a cipher. Given a cryptogram $m' \in \mathcal{M}$, what is the remaining uncertainty about the plain-text m? As usual, we denote by X, Y, and Z the random variables for the plain-text, the cryptogram, and the key, respectively. Obviously for every cipher $H(X|Y) \leq \log K$, since for every cryptogram, there are at most K possible messages in $\mathcal{M}$ (one for each key) from which this cryptogram could have arises (the verification of this last inequality is left as an exercise to the reader).

So from a good code we would require that the conditional entropy $H(X|Y)$ is close to this lower bound. As mentioned before, Shannon showed that

$$H(X|Y) \geq \log K + H(X) - \log M \tag{1.1.4}$$

for a random cipher. However, if $H(X)$ is smaller than $\log M$ (which is always the case when the probability distribution P on the message set $\mathcal{M}$ is not uniform), then this lower bound is far apart from the upper bound.

We shall show that under the rather natural assumption that $P(m) \leq \frac{1}{K}$ for all possible plain-texts $m \in \mathcal{M}$ (no message is "too probable") the conditional entropy $H(X|Y)$ cannot differ by more than one bit from $\log K$ for the special "block-cyclic" cipher introduced above.

First we shall see that the interception of a message $m' \in \mathcal{M}$ doesn't give any further information than the number j of the block B_j in which this message is contained. This information, of course, is inavoidable, because of the definition of the cipher.

We denote by U the RV for the blocks B_j, $j = 1, \dots, \ell$, hence U is distributed according to

$$Pr(U = j) = P(B_j) = \sum_{t=1}^{K} P\big(K(j-1) + t\big).$$

Further for the messages in each block we define RV's X_j, $j = 1, \dots, \ell$ with distribution

$$Pr\big(X_j = K(j-1) + t\big) = \frac{P\big(K(j-1) + t\big)}{Pr(U = j)}, t = 1, \dots, K.$$

Lemma 1 *For the cipher described above*

$$H(X|Y) = \sum_{j=1}^{\ell} Pr(U = j) \cdot H(X_j) = H(X|U). \tag{1.1.5}$$

Proof Recall that (for arbitrary RVs)

$$H(X|Y) + H(Y) = H(Y|X) + H(X).$$

By the grouping axiom for the entropy $H(X) = H(U) + \sum_{j=1}^{\ell} Pr(U = j) \cdot H(X_j)$ and since the cipher is regular, it is $H(Y|X) = \log K$. Hence

$$H(X|Y) + H(Y) = H(U) + \sum_{j=1}^{\ell} Pr(U = j) \cdot H(X_j) + \log K. \tag{1.1.6}$$

Y is equidistributed on the blocks $B_1, \ldots, B_\ell$ and hence by the grouping axiom

$$H(Y) = H(U) + \sum_{j=1}^{\ell} Pr(U = j) \cdot \log K.$$

Replacing now $H(Y)$ in (1.1.6) by the expression of the right-hand side in our last equality, we obtain

$$\begin{aligned} H(X|Y) &= \sum_{j=1}^{\ell} Pr(U = j) \cdot H(X_j) + \log K - \sum_{j=1}^{\ell} Pr(U = j) \cdot \log K \\ &= \sum_{j=1}^{\ell} Pr(U = j) \cdot H(X_j) = H(X|U). \end{aligned}$$

Theorem 2 *Let K be the number of keys and let $P = \big(P(1), P(2), \ldots\big)$ be a probability distribution on satisfying*

$$P(m) \leq \frac{1}{K} \textit{ for all } m \in \mathcal{M},$$

then for our simple cipher

$$H(X|Y) \geq \log K - 1. \tag{1.1.7}$$

Proof By (1.1.5) it suffices to give a lower bound on $H(X|U)$, the remaining uncertainty, when we already know the block B_j in which the plain-text $m \in \mathcal{M}$ is contained. For this we write $Pr(U = j)$ in the form

$$Pr(U = j) = \frac{1}{K^{\epsilon_j}}, \text{ where } 0 \leq \epsilon_1 \leq \epsilon_2 \leq \cdots \leq \epsilon_\ell.$$

Let us look at the first block. Since its total probability equals $Pr(U = 1) = \frac{1}{K^{\epsilon_1}}$ and since the individual probabilities are smaller than $\frac{1}{K}$ by the monotonicity properties of $x \cdot \log x$

$$H(X|U = 1) \cdot Pr(U = 1) \geq \frac{1}{K^{\epsilon_1}} \log K^{1-\epsilon_1}.$$

By the monotonicity of the $P(m)$'s

$$P(K=1) \leq \frac{1}{K^{\epsilon_1}} K^{-1} = \frac{1}{K^{1+\epsilon_1}}$$

and repetition of the previous argument gives

$$H(X|U=2) \cdot Pr(U=2) \geq \frac{1}{K^{\epsilon_2}}\big((1+\epsilon_1-\epsilon_2)\log K - 1\big).$$

By reiteration therefore

$$H(X|Y) \geq \sum_{j=1}^{\infty} \frac{1}{K^{\epsilon_j}}(1+\epsilon_{j-1}-\epsilon_j)\log K$$

(with the convention $\epsilon_0 = 0$)

Of course, also

$$\sum_{j=1}^{\infty} \frac{1}{K^{\epsilon_j}} = 1.$$

These two relations imply

$$H(X|U) \geq \log K - \sum_{j=1}^{\infty} \frac{\epsilon_j - \epsilon_{j-1}}{K^{\epsilon_j}} \log K. \tag{1.1.8}$$

Since for natural logarithms $\log x \leq x - 1$, we have that

$$(\epsilon_j - \epsilon_{j-1}) \cdot \log K \leq K^{\epsilon_j - \epsilon_{j-1}} - 1,$$

which is equivalent to

$$\frac{\epsilon_k - \epsilon_{j-1}}{K^{\epsilon_j}} \leq \frac{1}{\log K}\left(\frac{1}{K^{\epsilon_{j-1}}} - \frac{1}{K^{\epsilon_j}}\right).$$

We can conclude that $\sum_{j=1}^{\infty} \frac{\epsilon_j - \epsilon_{j-1}}{K^{\epsilon_j}} \leq \frac{1}{\log K}$ and the theorem is proved using (1.1.8).

1.1.3 A Robustification of Shannon's Secrecy System

For a regular, canonical cipher, obviously $H(Y|X) = \log K$, since by the definition of regularity, to each message $m \in \mathcal{M}$, every key yields a different cryptogram (and all keys are equiprobable by the definition of canonical). Hence

$$\begin{aligned} H(X|Y) &= H(Y|X) + H(X) - H(Y) \\ &= \log K + H(X) - H(Y) \\ &\geq \log K + H(X) - \log M \end{aligned}$$

for all plain-text variables X, and therefore also

$$H(X|Y) \geq \log K + H_0 - \log M$$

for all plain-text variables X with $H(X) \geq H_0$. We show in the sequel that this bound is essentially best possible for all canonical ciphers (($\mathbb{C}$, Q), where Q is the uniform distribution).

Theorem 3 *For every canonical cipher* ($\mathbb{C}$, Q) *on* $\mathcal{M} = \{1, \ldots, M\}$ *with* K *keys and for every* H_0, $0 \leq H_0 \leq \log M$ *there exists a plain-text variable* X *with values in* $\mathcal{M}$ *and* $H(X) \geq H_0$ *such that*

$$H(X|Y) \leq [\log K + H_0 - \log M]^+ + \log \frac{6}{\epsilon} + \epsilon \cdot \log K, \tag{1.1.9}$$

where $0 < \epsilon < \frac{1}{2}$ *and* $[t]^+ = \max\{t, 0\}$.

In order to prove Theorem 3, we should first point out that to every cipher ($\mathbb{C}$, Q) and every source ($\mathcal{M}$, P) we can associate in a natural way the transmission matrix of a channel $W : \mathcal{M} \to \mathcal{M}$ by

$$W(m'|m) = Pr(Y = m'|X = m) \text{ for all } m, m' \in \mathcal{M}.$$

Theorem 3 is now proved by using methods from Coding Theory. We need Fano's Lemma and Feinstein's maximal coding idea for the construction of a code with codewords from a prescribed subset $\mathcal{A} \subset \mathcal{M}$. In the sequel we denote by ϵ the error probability of a code, i.e., $W(D_i|u_i) \geq 1 - \epsilon$ for all i.

Lemma 2 (Fano's Lemma) *Let* $\{(u_i, D_i) : 1 \leq i \leq N\}$ *be a block code with average error* $\lambda_Q \triangleq \sum_{i=1}^{N} Q(i)w(D_i^c|u_i)$. *Further, let* U *be a random variable with* $P(U = u_i) = Q(i)$ *and let* V *be a random variable induced by the channel, i.e.,* $P(V = y|U = u_i) = w(y|u_i)$ *for all* $i \in \{1, \ldots, N\}$ *and* $y \in \mathcal{Y}$ *and* $P(V = y) = \sum_{i=1}^{N} Q(i) \cdot w(y|u_i)$. *Then*

$$H(U|V) \leq 1 + \lambda_Q \log N.$$

Fano's Lemma states that the conditional entropy is smaller (by a factor λ_Q) than $\log N$, the logarithm of the code length. If, e.g., we would have chosen $Q(i) = \frac{1}{N}$ for $i = 1, \ldots, N$, the uniform distribution, then the uncertainty $H(U) = \log N$ is reduced by a factor at least λ_Q, when we already know the realization of V.

Observe that Fano's Lemma does not make use of the time structure, i.e., the block length n is not important and can be chosen as $n = 1$.

Proof of Lemma 2. Let the decoding function d be given by $d(y) = u_i$ exactly if $y \in D_i$ (we can assume w.l.o.g. that $\bigcup_{i=1}^{N} D_i = \mathcal{Y}$, otherwise the "rest" $\mathcal{Y} \setminus \bigcup_{i=1}^{N} D_i$ is added to some D_i). Then $\lambda_Q = P\big(U \neq d(V)\big) = \sum_{y \in \mathcal{Y}} P\big(U \neq d(y)|V = y\big) \cdot P(V = y)$. Now for $y \in \mathcal{Y}$ let $\lambda(y) \triangleq P\big(U \neq d(v)|V = y\big)$ and think of the random experiment "U given $V = y$" divided into "$U \neq d(y)$" and "$U = d(y)$". "$U \neq d(y)$" will take place with probability $\lambda(y)$ by definition and hence "$U = d(y)$" has probability $1 - \lambda(y)$. So, by the grouping axiom for the entropy function

$$H(U|V = y) \leq h\big(\lambda(y)\big) + \big(1 - \lambda(y)\big) \cdot 0 + \lambda(y) \cdot \log(N - 1),$$

where $h(p) \triangleq H(p, 1 - p)$ for $p \in [0, 1]$.

Multiplication by $P(V = y)$ yields

$$\begin{aligned} H(U|V) &= \sum_{y \in \mathcal{Y}} H(U|V = y) \cdot P(V = y) \\ &\leq \sum_{y \in \mathcal{Y}} P(V = y) \cdot h(\lambda(y)) + \sum_{y \in \mathcal{Y}} P(V = y) \cdot \lambda(y) \cdot \log(N - 1). \end{aligned}$$

Now observe that the second term on the right hand side is just $\lambda_Q \log(N - 1)$. Since the entropy function is concave and $h(p)$ can be at most 1 (for $p = \frac{1}{2}$), we can further conclude that

$$\begin{aligned} H(U|V) &\leq h\left(\sum_{y \in \mathcal{Y}} P(V = y) \cdot \lambda(y)\right) + \lambda_Q \cdot \log(N - 1) \\ &\leq 1 + \lambda_Q \cdot \log(N - 1) \leq 1 + \lambda_Q \cdot \log N. \end{aligned}$$

Lemma 3 *Let W be the transmission matrix associated with a canonical cipher on $\mathcal{M} = \{1, \ldots, M\}$ and let $\mathcal{A} \subset \mathcal{M}$ be a subset of size $|\mathcal{A}| \geq (1 - \delta)M$, $0 < \delta < 1$. Then for any ϵ, $0 < \epsilon < \frac{1}{2}$ there exists an ϵ-code $\big\{(u_i, D_i), i = 1, \ldots, N\big\}$ for W such that $\{u_i, i = 1, \ldots, N\} \subset \mathcal{A}$ and $N = \big\lfloor \frac{\epsilon}{K}(1 - \delta)M \big\rfloor$.*

Proof Let $\big\{(u_i, D_i), i = 1, \ldots, N\big\}$ be an ϵ-code with $\{u_i, i = 1, \ldots, N\} \subset \mathcal{A}$ such that u_i is connected to every element in D_i and such that it is not possible to find a further pair (u, D), $u \in \mathcal{A}$, with $W(D|u) \geq 1 - \epsilon$ (i.e., the code is maximal). Then for all $u \in \mathcal{A}$ it is $W\Big(\bigcup_{i=1}^{N} D_i|u\Big) > \epsilon$ (if u is a codeword $u_i \in \mathcal{A}$, say, then already $W(D_i|u_i) \geq 1-\epsilon > \epsilon$, since $\epsilon < \frac{1}{2}$, if u is not a codeword, then $W\Big(\bigcup_{i=1}^{N} D_i|u\Big) > \epsilon$, since otherwise we could prolong the code by the pair $\Big(u, \overline{\bigcup_{i=1}^{N} D_i}\Big)$.

Therefore $\left|\bigcup_{i=1}^{N} D_i\right| \cdot K \geq \epsilon \cdot |\mathcal{A}|$. Hence $N \cdot K \geq \left|\bigcup_{i=1}^{N} D_i\right| \geq \epsilon \cdot |\mathcal{A}|$ and $N \geq \frac{\epsilon}{K}(1-\delta)M$.

Proof of Theorem 3. By iteratively applying Lemma 3, we can construct ϵ-codes $\{(u_i^{(t)}, D_i^{(t)}), i = 1, \ldots, N\}$ for $t = 1, \ldots, T$ with all codewords $u_i^{(t)}$ distinct provided that

$$T \cdot \frac{\epsilon}{K}(1-\delta)M \leq \delta M.$$

This is satisfied if $T \leq \frac{\delta}{\epsilon(1-\delta)} K$. Define now a random variable X with distribution

$$Pr(X = u_j^{(t)}) = \frac{1}{N \cdot T}$$

(the uniform distribution on all possible codewords).

Let Y be the corresponding output variable with respect to W. By the grouping axiom for the entropy function and by Fano's Lemma (with $\lambda_Q = \epsilon$)

$$H(X|Y) \leq \log T + 1 + \epsilon \cdot \log K. \tag{1.1.10}$$

Actually, Fano's Lemma applied directly would give only a term $\epsilon \cdot \log N$, here we can do better because every $m' \in \mathcal{M}$ is connected with at most K codewords.

Now we choose T as small as possible under the condition $\log(T \cdot N) \geq H_0$. Clearly, $\log T \leq H_0 - \log \frac{\epsilon}{K} - \log(1-\delta) - \log M + 1$ and (1.1.10) yields for $\delta = \frac{1}{2}$

$$H(X|Y) \leq H_0 + \log K - \log M + \log \frac{2}{\epsilon} + 1 + \epsilon \cdot \log K + 1$$

which is (1.1.9) $\left(\log \frac{2}{\epsilon} + 2 = \log \frac{6}{\epsilon}\right)$.

1.2 The Wiretap Channel

Wireless communication nowadays does not only require a reliable data transmission, but it also involve some secrecy requirements. Secure communication over a wireless medium is not an easy task. This because the open nature of the wireless medium makes it easy for non-legitimate receivers to eavesdrop on the transmitted signals. In order to overcome this exposure problem, high-level cryptographic techniques have been used to encrypt the transmitted information. These techniques work under the assumption that limited computational power is available at the eavesdroppers. However, with the rapid improvement in the digital design field, this assumption is no longer valid, which implies that these techniques are becoming less efficient. That is why, physical layer secrecy also known as *information theoretic security* is becoming more attractive because it does not have any constraints on the computational power of the eavesdroppers.[1]

Information theoretic security was first introduced by Shannon in [35], where he proved that secure communication can be achieved by using a secret key shared between the transmitter and the receiver if the entropy of this key is greater than or equal to the entropy of the message to be transmitted. In [37], Wyner ashowed that secure transmission is still achievable in the absence of a secret key by exploiting the noisiness of the channel. He introduced the degraded wiretap channel, in which the channel observation at the eavesdropper is a degraded version from the one at the legitimate receiver. He calculated the maximum rate at which information can be sent to the legitimate receiver, while keeping it secret from the eavesdropper and defined this rate as the secrecy capacity. In [17], Csiszár and Körner extended Wyner's result to the general wiretap channel, where the legitimate receiver has no statistical advantage over the eavesdropper. In [6, 15, 29], secure communication over wiretap channels with more than one legitimate receivers has been investigated. This line of work leads to the introduction of the multi-user wiretap channel which captured a lot of attention recently. Researchers managed to establish the secrecy capacity of many special multi-user wiretap channels. However, despite their tremendous efforts, the secrecy capacity of the general case has remained unknown.

Most of the initial investigation of the wiretap channel was performed under the assumption of the availability of perfect channel state information (CSI) to all users in the network. Although this assumption helped in capturing a better understanding of the wiretap channel, it is not a realistic assumption. This is because in wiretap channels malevolent eavesdroppers will not provide any information about their channels to the transmitter and even if by some means the transmitter managed to gather some information about the CSI, this information will not be perfect. Thus, in order to consider a more realistic and practical CSI assumptions, the compound wiretap channel was introduced [27]. In this channel, instead of knowing the exact channel realization, the users are given an uncertainty set of channels from which the true channel is selected. It is also assumed that the channel state remains constant during the whole transmission. This last assumption was further avoided by considering the principle of arbitrary varying channel [2], where the channel realization may vary from one channel use to another in an unknown and arbitrary manner. This leads to the model of arbitrary varying wiretap channels.

1.2.1 The Classical Wiretap Channel

In this section, we consider the classical wiretap channel that consists of: a transmitter, a legitimate receiver and an eavesdropper. We assume a scenario where perfect

[1]This section was written by Holger Boche and Ahmed Mansour. It is an extension of the original text of Rudolf Ahlswede, which was only a one page summary of the result of Wyner. In this text all new important developements are included. The extension of the original text was a suggestion of one of the reviewers.

channel state information (CSI) is available at all nodes. This implies that the transmitter, the receiver and the eavesdropper know the channel statistics a head of time.

System Model

Let $\mathcal{X}$ be a finite input alphabet at the transmitter, $\mathcal{Y}$ be a finite output alphabet at the legitimate receiver, and $\mathcal{Z}$ be a finite output alphabet at the eavesdropper. We model the channel between the transmitter and the legitimate receiver by the stochastic matrix $\mathrm{W} : \mathcal{X} \rightarrow \mathcal{P}(\mathcal{Y})$. This matrix defines the probability of observing a certain output symbol at the legitimate receiver given that a certain input symbol was transmitted. Similarly we model the channel between the transmitter and the eavesdropper by the stochastic matrix $\mathrm{V} : \mathcal{X} \rightarrow \mathcal{P}(\mathcal{Z})$. Note that since the legitimate receiver and eavesdropper are not supposed to cooperate, there is no loss in representing the wiretap channel by its marginal probability matrices instead of a joint one.

Definition 1 The wiretap channel $\mathfrak{W}$ is given by the pair of channels with common inputs as

$$\mathfrak{W} = \{\mathrm{W}, \mathrm{V}\} \tag{1.2.1}$$

Further, we consider a discrete memoryless channel, such that for a block code of length n, an input sequence $x^n = (x_1, x_2, \ldots, x_n) \in \mathcal{X}^n$, and output sequences $y^n = (y_1, y_2, \ldots, y_n) \in \mathcal{Y}^n$ and $z^n = (z_1, z_2, \ldots, z_n) \in \mathcal{Z}^n$, the transmission matrices are given by

$$\mathrm{W}^n(y^n|x^n) = \prod_{i=1}^{n} \mathrm{W}(y_i|x_i) \quad \text{and} \quad \mathrm{V}^n(z^n|x^n) = \prod_{i=1}^{n} \mathrm{V}(z_i|x_i) \tag{1.2.2}$$

The communication task over the wiretap channel requires the establishment of a reliable communication link between the transmitter and the legitimate receiver, while keeping the eavesdropper ignorant about the information transmitted over this link.

Definition 2 A $(2^{nR}, n)$ code $\mathcal{C}_n$ for the classical wiretap channel consists of: a message set $\mathcal{M} = \left\{1, 2^{nR}\right\}$, a stochastic encoder at the transmitter

$$E : \mathcal{M} \rightarrow \mathcal{P}(\mathcal{X}^n) \tag{1.2.3}$$

which maps a confidential message $m \in \mathcal{M}$ to a codeword $x^n(m) \in \mathcal{X}^n$ according to the conditional probability $E(x^n|m)$, and a deterministic decoder at the legitimate receiver

$$\varphi : \mathcal{Y}^n \rightarrow \mathcal{M} \tag{1.2.4}$$

that maps each channel observation at the receiver node to the corresponding required message.

We assume that the code $\mathcal{C}_n$ is known to the transmitter, legitimate receiver and the eavesdropper. We also assume that the transmitted message is chosen uniformly at random. It is important to point out that the usage of a deterministic encoder in which each confidential message $m \in \mathcal{M}$ is mapped to only one codeword $x^n \in \mathcal{X}^n$ is insufficient for secure communication. On the other hand, there is no need to use a stochastic decoder at the legitimate receiver as a deterministic one is sufficient [10].

Reliability and Secrecy Analysis

In order to judge the performance of the code $\mathcal{C}_n$, we need to evaluate its reliability and secrecy performance. We start by the reliability performance and highlight the fact that a reliable code should ensure the capability of the legitimate receiver to decode the transmitted message correctly. This implies that a code with small decoding error probability is a code with good reliability performance. In order to calculate this probability, we start by assuming that a message $m \in \mathcal{M}$ was transmitted and a sequence $y^n \in \mathcal{Y}^n$ was received at the legitimate receiver. In this case the probability of a decoding error is given by:

$$e(m) = \sum_{x^n \in \mathcal{X}^n} \sum_{y^n : \varphi(y^n) \neq m} \mathcal{W}(y^n|x^n) E(x^n|m) \tag{1.2.5}$$

The previous equation defines the probability of a decoding error for a certain message m. Now in order to measure the reliability performance of the whole code, we can either use the average error probability or the maximum error probability as follows:

$$\bar{e} = \frac{1}{|\mathcal{M}|} \sum_{m \in \mathcal{M}} e(m), \qquad e_{max} = \max_{m \in \mathcal{M}} e(m) \tag{1.2.6}$$

One can notice that the maximum error probability criterion is stronger than the average error probability criterion. However, it was shown that for a wiretap channel, where perfect CSI is available both criteria lead to the same secrecy capacity [17].

On the other hand, a secure coding scheme should make sure that the eavesdropper can not infer any information about the confidential message. In his seminal paper [37], Wyner formulated the previous requirement in terms of equivocation as follows: For a random variable M uniformly distributed over the message set $\mathcal{M}$ and a sequence $\mathrm{Z}^n = (\mathrm{Z}_1, \mathrm{Z}_2, \ldots, \mathrm{Z}_n)$ that represents a random variable for the channel output sequence at the eavesdropper, Wyner required that

$$\frac{1}{n}\mathbb{H}(\mathrm{M}) \leq \frac{1}{n}\mathbb{H}(\mathrm{M}|\mathrm{Z}^n) + \tau_n, \tag{1.2.7}$$

where $\tau_n \to 0$ as $n \to \infty$. This implies that the information available at the eavesdropper represented by the random variable Z^n does not decrease the uncertainty about the confidential message M in terms of rate. This criterion has been known as weak secrecy and is usually written as

$$\frac{1}{n}\mathbb{I}(\text{M}; \text{Z}^n) \leq \tau_n. \tag{1.2.8}$$

The weak secrecy criterion only implies that the rate of information leaked to the eavesdropper vanishes as n approaches infinity. This does not necessarily mean that the term $\mathbb{I}(\text{M}; \text{Z}^n)$ is a decreasing function in n, because as long as $\mathbb{I}(\text{M}; \text{Z}^n)$ grows at most sub-linearly with n, the weak secrecy constraint is fulfilled. Most of the early studies of the classical wiretap channel only considered the weak secrecy criterion. However, recently a stronger secrecy criterion has been introduced to strengthen the weak secrecy constraint by dropping the division by the block length n as follows:

$$\mathbb{I}(\text{M}; \text{Z}^n) \leq \tau_n. \tag{1.2.9}$$

This is criterion is known as strong secrecy, where the total amount of information leaked to the eavesdropper is small. This is achieved by forcing $\mathbb{I}(\text{M}; \text{Z}^n)$ to be a decreasing function in n. The wiretap channel was first studied under the strong secrecy constraint in [16, 31]. Since then different approaches have been proposed to achieve strong secrecy [11, 24].

In order to understand the difference between the previous two criteria, we need to investigate the practical and operational meaning: when for sufficiently large code block length n, the information leakage of the confidential message to the eavesdropper vanishes. This can be understand by considering the following fact: As the information leakage to the eavesdropper approaches zero, the average probability of error of any decoder implemented at the eavesdropper will approach one. This implies that both weak and strong secrecy criteria guarantee a high probability of error at the eavesdropper. However, the difference is in the speed at which the error probability converges to one. Using Fano's inequality, one can show that the speed of convergence for the weak secrecy criterion is $o(1)$. On the other hand, it has been shown in [7], that strong secrecy criterion provides an exponential speed of convergence. This conclusion advocates the fact that strong secrecy is a more conservative criterion compared to the weak one.

Definition 3 A confidential rate $R \in \mathbb{R}_+$ is achievable for the classical wiretap channel, if there exist a sequence of $(2^{nR}, n)$ codes $\mathcal{C}_n$ and two sequences ϵ_n, τ_n, where n is large enough, such that

$$\bar{e} \leq \epsilon_n, \qquad \lim_{n\to\infty} \epsilon_n, \tau_n = 0. \tag{1.2.10}$$

and depending on the selected secrecy criterion, the condition in (1.2.8) or (1.2.9) is fulfilled.

Secrecy Capacity

Secrecy capacity was originally introduced by Wyner in [37] as the maximum rate at which information can be transmitted reliably to the legitimate receiver and secretly from the eavesdropper. In the same paper, Wyner established the secrecy capacity for a special class of wiretap channels known as the degraded wiretap channel. The main characteristic of this channel is that $\mathrm{X} - \mathrm{Y} - \mathrm{Z}$ forms a Markov chain, which implies that the channel observation at the eavesdropper (Z^n) is a degraded version of the channel observation at the legitimate receiver (Y^n).

Theorem 4 ([37]) *The secrecy capacity region of the degraded wiretap channel is the set of all rates* $R \in \mathbb{R}_+$ *that satisfy*

$$R \leq \mathrm{C}(\mathfrak{W}) = \max_{\mathrm{X}-\mathrm{Y}-\mathrm{Z}} \big[\mathbb{I}(\mathrm{X};\mathrm{Y}) - \mathbb{I}(\mathrm{X};\mathrm{Z})\big], \tag{1.2.11}$$

for random variables satisfying the following Markov chain $\mathrm{X} - \mathrm{Y} - \mathrm{Z}$.

The main idea used to establish the previous secrecy capacity region can be explained as follows: Instead of using the full rate of the channel between the transmitter and the legitimate receiver to transmit the confidential message, part of this rate is used to induce a randomization index to confuse the eavesdropper. This technique affects the structure of the code, such that for every confidential message, there exist a set of valid codewords. Now, when a certain message is to be transmitted, the encoder selects one of these codewords uniformly at random and transmit it. The key for this principle to work is to choose for each message roughly 2^{nR_Z} codewords, where $R_\mathrm{Z} = \mathbb{I}(\mathrm{X};\mathrm{Z})$ represents the full rate of the channel between the transmitter and the eavesdropper. This implies that all the available resource at the eavesdropper will be jammed by this randomization index, such that the eavesdropper can not infer any information about the confidential message. Since the legitimate receiver will need to decode both the confidential message and the randomization index and since the maximum rate that allows for a reliable communication between the transmitter and the legitimate receiver is $\mathbb{I}(\mathrm{X};\mathrm{Y})$, the maximum rate available for the confidential message is $\mathbb{I}(\mathrm{X};\mathrm{Y}) - \mathbb{I}(\mathrm{X};\mathrm{Z})$.

Although Theorem 4 was derived for degraded wiretap channels, the established capacity region holds for less noisy and more capable channels as well (cf. for example [19] for a discussion on less noisy and more capable channels). In [17], Csiszár and Körner extended Wyner's result to the general wiretap channel, where the legitimate receiver does not possess any statistical advantage over the eavesdropper.

Theorem 5 ([17]) *The secrecy capacity region of the general wiretap channel is the set of all rates* $R \in \mathbb{R}_+$ *that satisfy*

$$R \leq \mathrm{C}(\mathfrak{W}) = \max_{\mathrm{U}-\mathrm{X}-(\mathrm{Y},\mathrm{Z})} \big[\mathbb{I}(\mathrm{U};\mathrm{Y}) - \mathbb{I}(\mathrm{U};\mathrm{Z})\big], \tag{1.2.12}$$

for random variables satisfying the following Markov chain $\mathrm{U} - \mathrm{X} - (\mathrm{Y},\mathrm{Z})$.

The difference between this capacity region and the one in Theorem 4 is the utilization of an auxiliary random variable U instead of the direct channel input X. U plays the role of a channel prefix, creating new channels $\tilde{\mathrm{W}}(\mathrm{Y}|\mathrm{U})$ and $\tilde{\mathrm{V}}(\mathrm{Z}|\mathrm{U})$. Now applying the same coding strategy used in Theorem 4 to the new channels $\tilde{\mathrm{W}}$ and $\tilde{\mathrm{V}}$

establish the secrecy capacity region in (1.2.12). One might wonder about the necessity of using a channel prefix, specially because according to the data processing inequality pre-coding decreases the mutual information, i.e., $\mathbb{I}(U; Y) \leq \mathbb{I}(X; Y)$ and $\mathbb{I}(U; Z) \leq \mathbb{I}(X; Z)$. However, the target of the channel prefixing is to find a certain U, such that the decrease in the eavesdropper channel quality is bigger than that of the legitimate receiver channel quality leading to an increase in the difference, i.e., $\mathbb{I}(U; Y) - \mathbb{I}(U; Z) \geq \mathbb{I}(X; Y) - \mathbb{I}(X; Z)$. However, Theorem 4 indicates that for wiretap channels with a *stronger* legitimate channel, such U does not exist and channel prefixing can not increase the secrecy capacity. Although the capacity regions in Theorems 4 and 5 were established for the weak secrecy criterion, they are also valid for the strong secrecy one. This because strengthening the secrecy constraint from weak to strong for the classical wiretap channel with perfect CSI comes at no loss in the secrecy capacity.

1.2.2 The Multi-user Wiretap Channel

In this section, we extend the model of the classical wiretap channel with one legitimate receiver to a multi-user scenario, in which we have more than one legitimate receiver in addition to the eavesdropper. For simplicity, we will only consider a two-user wiretap channel. We will also maintain the assumption that, perfect CSI is available all nodes a head of time.

System Model

Let $\mathcal{X}$ be a finite input alphabet at the transmitter, $(\mathcal{Y}_1, \mathcal{Y}_2)$ be two finite output alphabets at the first and second legitimate receiver respectively and $\mathcal{Z}$ be a finite output alphabet at the eavesdropper. We model the channels between the transmitter and the two legitimate receivers by the two stochastic matrices $W_1 : \mathcal{X} \to \mathcal{P}(\mathcal{Y}_1)$ and $W_2 : \mathcal{X} \to \mathcal{P}(\mathcal{Y}_2)$. Simultaneously, we model the channel between the transmitter and the eavesdropper by the stochastic matrix $V : \mathcal{X} \to \mathcal{P}(\mathcal{Z})$.

Definition 4 The two-user wiretap channel $\mathfrak{W}$ is given by the triple of channels with common inputs as

$$\mathfrak{W} = \{W_1, W_2, V\} \tag{1.2.13}$$

For input and output sequences of length n given by: $x^n \in \mathcal{X}^n$, $y_1^n \in \mathcal{Y}_1^n$, $y_2^n \in \mathcal{Y}_2^n$ and $z^n \in \mathcal{Z}^n$, the discrete memoryless two-user wiretap channel is identified by the following transmission matrices

$$W_1^n(y_1^n|x^n) = \prod_{i=1}^{n} W_1(y_{1i}|x_i), \qquad W_2^n(y_2^n|x^n) = \prod_{i=1}^{n} W_2(y_{2i}|x_i), \tag{1.2.14}$$

$$V^n(z^n|x^n) = \prod_{i=1}^{n} V(z_i|x_i) \tag{1.2.15}$$

Definition 5 A $(2^{nR_0}, 2^{nR_1}, 2^{nR_2}, n)$ code $\mathcal{C}_n$ for the two-user wiretap channel consists of: a common confidential message set $\mathcal{M}_0 = \{1, 2^{nR_0}\}$, two individual confidential message sets one for each legitimate receiver $\mathcal{M}_1 = \{1, 2^{nR_1}\}$ and $\mathcal{M}_2 = \{1, 2^{nR_2}\}$, a stochastic encoder at the transmitter

$$E : \mathcal{M}_0 \times \mathcal{M}_1 \times \mathcal{M}_2 \to \mathcal{P}(\mathcal{X}^n) \tag{1.2.16}$$

which maps a confidential messages triple $(m_0, m_1, m_2) \in \mathcal{M}_0 \times \mathcal{M}_1 \times \mathcal{M}_2$ to a codeword $x^n(m_0, m_1, m_2) \in \mathcal{X}^n$ according to the conditional probability $E(x^n|m_0, m_1, m_2)$, and two deterministic decoders: one for each legitimate receiver

$$\varphi_1 : \mathcal{Y}_1^n \to \mathcal{M}_0 \times \mathcal{M}_1$$
$$\varphi_2 : \mathcal{Y}_2^n \to \mathcal{M}_0 \times \mathcal{M}_2$$

that maps the channel observation at each legitimate receiver to the corresponding required messages.

Further, we assume that the random variables that represents the confidential messages M_0, M_1 and M_2 are independent and uniformly distributed.

Reliability and Secrecy Analysis

In order to evaluate the reliability performance of $\mathcal{C}_n$, we need to make sure that the information transmitted over each communication link from the transmitter to the intended legitimate receiver can be decoded correctly. This suggests the usage of the decoding error probability as a measure for the reliability of a certain communication link. Based on Definition 5, we define the following error probabilities:

$$e_{10}(m_0) = \sum_{x^n \in \mathcal{X}^n} \sum_{y_1^n : \varphi(y_1^n) \neq m_0} W_1(y_1^n|x^n) E(x^n|m_0, \cdot, \cdot)$$
$$e_{11}(m_1) = \sum_{x^n \in \mathcal{X}^n} \sum_{y_1^n : \varphi(y_1^n) \neq m_1} W_1(y_1^n|x^n) E(x^n|\cdot, m_1, \cdot)$$
$$e_{20}(m_0) = \sum_{x^n \in \mathcal{X}^n} \sum_{y_2^n : \varphi(y_2^n) \neq m_0} W_2(y_2^n|x^n) E(x^n|m_0, \cdot, \cdot)$$
$$e_{22}(m_2) = \sum_{x^n \in \mathcal{X}^n} \sum_{y_2^n : \varphi(y_2^n) \neq m_2} W_2(y_2^n|x^n) E(x^n|\cdot, m_2, \cdot)$$

Using the previous four error events along with the union bound, we can derive an upper-bound for the average probability of error for the whole code $\mathcal{C}_n$ as follows:

$$\bar{e} \leq \frac{1}{|\mathcal{M}_0|} \sum_{m_0 \in \mathcal{M}_0} \left[e_{10}(m_0) + e_{20}(m_0)\right] + \frac{1}{|\mathcal{M}_1|} \sum_{m_1 \in \mathcal{M}_1} e_{11}(m_1) + \frac{1}{|\mathcal{M}_2|} \sum_{m_2 \in \mathcal{M}_2} e_{22}(m_2), \tag{1.2.17}$$

Similarly, the maximum probability of error for $\mathcal{C}_n$ is given by:

$$e_{max} \leq \max_{m_0 \in \mathcal{M}_0} \left[e_{10}(m_0) + e_{20}(m_0)\right] + \max_{m_1 \in \mathcal{M}_1} e_{11}(m_1) + \max_{m_2 \in \mathcal{M}_2} e_{22}(m_2) \quad (1.2.18)$$

On the other hand, the secrecy performance of $\mathcal{C}_n$ should be evaluated by its ability to protect the two communication links between the transmitter and the two legitimate receivers against eavesdropping. For this requirement, we consider a secrecy constraint known as the joint secrecy criterion, in which these two links are independently protected. For the two-user wiretap channel, the joint secrecy criterion requires the leakage of the confidential messages of one user to the eavesdropper given the individual confidential message of the other user to be small. This can be formulated by the following conditions:

$$\mathbb{I}(\mathrm{M}_0\mathrm{M}_1; \mathrm{Z}^n|\mathrm{M}_2) \leq \tau_{1n} \quad \text{and} \quad \mathbb{I}(\mathrm{M}_0\mathrm{M}_2; \mathrm{Z}^n|\mathrm{M}_1) \leq \tau_{2n} \quad (1.2.19)$$

where $\tau_{1n}, \tau_{2n} \rightarrow 0$ as $n \rightarrow \infty$. These constraints guarantee that the rate of information leaked to the eavesdropper from one user is small even if the individual confidential message of the other user is compromised. This means that the secrecy of the communication link between the transmitter and the first legitimate receiver is not affected even if the link between the transmitter and the second legitimate receiver is compromised. This implies that the joint secrecy criterion does not consider any form of mutual trust between the legitimate receivers. In some literature, the joint secrecy criterion is defined such that, the mutual leakage of all confidential messages to the eavesdropper is small as follows:

$$\mathbb{I}(\mathrm{M}_0\mathrm{M}_1\mathrm{M}_2; \mathrm{Z}^n) \leq \tau_n, \quad (1.2.20)$$

where $\lim_{n\rightarrow\infty} \tau_n = 0$. One can easily show that the definition in (1.2.19) is equivalent to the one in (1.2.20). However, we prefer the definition in (1.2.19), because it provides a better understanding to the relation between the legitimate receivers and allows us to interpret the independence between the secrecy of each confidential communication link.

Definition 6 A confidential rate triple $(R_0, R_1, R_2) \in \mathbb{R}_+^3$ is achievable for the two-user wiretap channel, if there exist a sequence of $(2^{nR_0}, 2^{nR_1}, 2^{nR_2}, n)$ codes $\mathcal{C}_n$ and three sequences $\epsilon_n, \tau_{1n}, \tau_{2n}$, where n is large enough, such that

$$\bar{e} \leq \epsilon_n, \qquad \lim_{n\rightarrow\infty} \epsilon_n, \tau_{1n}, \tau_{2n} = 0. \quad (1.2.21)$$

In the previous definition, we used the average probability of error as our reliability constraint. However, under the assumption of perfect CSI at all nodes, both the maximum and average probability of error lead to the same secrecy capacity. It also worth mentioning, that the joint secrecy constraints in (1.2.19) and (1.2.20) are formulated under the strong secrecy criterion.

Secrecy Capacity: Common Confidential Message

We consider a two-user wiretap channel, where we only have a common confidential message set $\mathcal{M}_0$, i.e., $\mathcal{M}_1 = \mathcal{M}_2 = \emptyset$. In this scenario, the joint secrecy criterion only requires the leakage of M_0 to the eavesdropper to be small, i.e., $\mathbb{I}(\mathrm{M}_0; \mathrm{Z}^n) \leq \tau_n$. This model was investigated by Chia and El Gamal in [15], where they established the secrecy capacity of a special class of degraded two-user wiretap, where $\mathrm{X} - (\mathrm{Y}_1, \mathrm{Y}_2) - \mathrm{Z}$ forms a Markov chain.

Theorem 6 ([15]) *The secrecy capacity region of the degraded two-user wiretap channel with common confidential message is given by the set of all rates $R_0 \in \mathbb{R}_+$ that satisfy*

$$R_0 \leq \mathrm{C}(\mathfrak{W}) = \max_{\mathrm{X}-(\mathrm{Y}_1,\mathrm{Y}_2)-\mathrm{Z}} \Big[\min \big[\mathbb{I}(\mathrm{X}; \mathrm{Y}_1) - \mathbb{I}(\mathrm{X}; \mathrm{Z}), \mathbb{I}(\mathrm{X}; \mathrm{Y}_2) - \mathbb{I}(\mathrm{X}; \mathrm{Z})\big]\Big], \tag{1.2.22}$$

for random variables satisfying the following Markov chain $\mathrm{X} - (\mathrm{Y}_1, \mathrm{Y}_2) - \mathrm{Z}$.

The previous capacity region follows by extending the coding technique used in Theorem 4 to the two-user scenario as follows: The secrecy requirement is achieved by jamming all the resources available at the eavesdropper. This is done by using a randomization index of size equivalent to the full rate of the channel between the transmitter and the eavesdropper, i.e., $\mathbb{I}(\mathrm{X}; \mathrm{Z})$. On the other hand, for a reliable communication, the two legitimate receivers should be able to correctly decode both the confidential message and the randomization index, which implies that the worst channel will control the bound for a reliable transmission, i.e., $\min\big[\mathbb{I}(\mathrm{X}; \mathrm{Y}_1), \mathbb{I}(\mathrm{X}; \mathrm{Y}_2)\big]$. Combining the two bounds leads to the secrecy capacity region in (1.2.22). In [28], it was shown that Theorem 6 holds for the less noisy and more capable two-user wiretap channel as well.

In the previous section, it was shown that an auxiliary random variable that acted as a channel prefix is needed to generalize the secrecy capacity of the degraded wiretap channel to the general one. Many researchers have applied the same technique to the two-user wiretap channel hoping they could generalize the capacity region of the degraded two-user scenario in (1.2.22) to the general one. However, most of these efforts failed suggesting that the straight forward extension of Theorem 5 to the two-user wiretap channel is not optimal. The reason for this is that, in the two-user wiretap channel, we have two independent legitimate channels one for each receiver. This implies that, two independent auxiliary random variables are needed to enhance the bound for each channel. The independence between these two auxiliary random variables makes it hard to find a suitable coding scheme. That is why, the best we have so far is the following achievable region:

Theorem 7 ([15]) *An achievable secrecy rate region for the two-user wiretap channel with common confidential message is given by the set of all rates $R_0 \in \mathbb{R}_+$ that satisfy*

$$R_0 \leq \min\big[\mathbb{I}(\mathrm{V}_1; \mathrm{Y}_1) - \mathbb{I}(\mathrm{V}_1; \mathrm{Z}), \mathbb{I}(\mathrm{V}_2; \mathrm{Y}_2) - \mathbb{I}(\mathrm{V}_2; \mathrm{Z})\big], \tag{1.2.23}$$

for random variables satisfying the following Markov chain $(V_1, V_2) - X - (Y_1, Y_2, Z)$*, such that*

$$\mathbb{I}(V_1V_2; Z) \leq \mathbb{I}(V_1; Z) + \mathbb{I}(V_2; Z) - \mathbb{I}(V_1; V_2).$$

The previous rate region is described by two independent auxiliary random variables V_1 and V_2, where V_1 creates a channel prefix for the channel between the transmitter and the first legitimate receiver, while V_2 creates a channel prefix for the channel between the transmitter and the second legitimate receiver. In order to do so, the Marton coding technique introduced in [30] was used. However, this brought an additional condition on the input distribution.

Secrecy Capacity: Two Individual Confidential Messages

We consider a two-user wiretap channel as described before, but without the common confidential message M_0. This setup was first investigated in [6] under the joint secrecy criterion, where the authors managed to establish the joint secrecy capacity of the class of degraded two-user wiretap, where $X - Y_1 - Y_2 - Z$ forms a Markov chain.

Theorem 8 ([6]) *The joint secrecy capacity region of the degraded two-user wiretap channel is given by the union of all rate pairs* $(R_1, R_2) \in \mathbb{R}_+^2$ *that satisfy*

$$\begin{aligned} R_2 &\leq \mathbb{I}(U; Y_2) - \mathbb{I}(U; Z) \\ R_1 &\leq \mathbb{I}(X; Y_1|U) - \mathbb{I}(X; Z|U) \end{aligned} \tag{1.2.24}$$

where the union is taken over all random variables (U, X)*, such that* $U - X - Y_1 - Y_2 - Z$ *forms a Markov chain.*

The proof of the previous capacity region is based on a combination of the superposition coding principle [26] and wiretap random coding introduced in Theorem 4. The superposition principle is used to establish a reliable communication between the transmitter and the two legitimate receivers, while wiretap random coding is used to assure the ignorance of the eavesdropper about the transmission.

We start by explaining the role of the superposition coding to guarantee a reliable communication. The main idea is to divide the code into two layers: an inner layer known as the cloud centers and an outer layer that contains the satellite codewords. Each layer provides a reliable communication link from the transmitter to one of the legitimate receivers. The inner layer is represented by an auxiliary random variable U, and is used to encode the confidential message of the weaker legitimate receiver Y_2. This creates a channel $\tilde{W}_2 : \mathcal{U} \to \mathcal{P}(\mathcal{Y}_2)$, where the maximum reliable rate that can be transmitted on this channel is bounded by $\mathbb{I}(U; Y_2)$. On the other hand, the confidential message of the stronger legitimate receiver Y_1 is encoded in the outer layer represented by the channel input X. Due to this superposition structure, the channel $W_1 : \mathcal{X} \to \mathcal{P}(\mathcal{Y}_1)$ becomes a conditional channel on the auxiliary random variable U. This implies that the maximum reliable rate available for transmitting M_1 is bounded by $\mathbb{I}(X; Y_1|U)$.

In order to guarantee the ignorance of the eavesdropper about the confidential messages, the full resources of the eavesdropper identified by the channel $V : \mathcal{X} \to \mathcal{Z}$ should be jammed by useless information. This implies that a randomization index of size at least $\mathbb{I}(X; Z)$ is needed. However, due to the superposition structure of the code, this constraint is not enough as we need to make sure that this randomization is distributed in a smart way among the two layers of the code. This distribution should assure that the eavesdropper resources devoted for the virtual channel $\tilde{V} : \mathcal{U} \to \mathcal{Z}$ and main channel conditioned on U are also saturated with useless information. This implies that the part of the randomization index devoted for the inner layer should be roughly $\mathbb{I}(U; Z)$, while the part used in the outer layer should be $\mathbb{I}(X; Z|U)$. This leads to the secrecy capacity region in (1.2.24).

In spite of the tremendous efforts invested in studying secure communication over the two-user wiretap channel, the secrecy capacity region of the general channel is still unknown. Not only that, but also the secrecy capacity of special classes like less noisy and more capable channels has not been established yet.

1.2.3 The Compound Wiretap Channel

In this section, we investigate the model of the compound wiretap channel introduced in [27]. This model aims to simulate a more realistic scenario regarding the availability of perfect CSI at all nodes in the system. Instead of assuming that both the transmitter and all receivers know the exact channel realization, the compound wiretap channel considers a specific uncertainty set from which the actual channel state is selected. Further, it assumes that the channel state does not vary during the entire transmission of each codeword. This is a better model for real life communication systems, where the available CSI is usually imperfect. This imperfection usually originates from inaccurate channel estimation techniques or insufficient feedback schemes. It can also happen due to the presence of active eavesdroppers who are capable of manipulating the channel state.

System Model

In order to modify the model of the classical wiretap channel to account for the uncertainty in the available CSI of the compound wiretap channel, we introduce a state set $\mathcal{S}$. For the finite input and output alphabets $\mathcal{X}$, $\mathcal{Y}$ and $\mathcal{Z}$, the stochastic matrices $W_s : \mathcal{X} \to \mathcal{P}(\mathcal{Y})$ and $V_s : \mathcal{X} \to \mathcal{P}(\mathcal{Z})$ are used to model the channels to the legitimate receiver and the eavesdropper respectively for a given state s. We assume that the transmitter and the legitimate receiver knows the state space $\mathcal{S}$, but have now knowledge regarding the actual state s. We define the compound wiretap channel in terms of families of all possible channel states as follows:

$$\mathcal{W} = \{W_s : s \in \mathcal{S}\} \quad \text{and} \quad \mathcal{V} = \{V_s : s \in \mathcal{S}\} \tag{1.2.25}$$

We further assume a discrete memoryless channel, such that for a block code of length n, an input sequence $x^n \in \mathcal{X}^n$, and output sequences $y^n \in \mathcal{Y}^n$ and $z^n \in \mathcal{Z}^n$, the transmission matrices for a state $s \in \mathcal{S}$ are given by

$$\mathrm{W}_s^n(y^n|x^n) = \prod_{i=1}^{n} \mathrm{W}_s(y_i|x_i) \quad \text{and} \quad \mathrm{V}_s^n(z^n|x^n) = \prod_{i=1}^{n} \mathrm{V}_s(z_i|x_i) \tag{1.2.26}$$

Definition 7 The discrete memoryless compound wiretap channel $\mathfrak{W}$ is given by the families of compound channels with common input as

$$\mathfrak{W} = \{\mathcal{W}, \mathcal{V}\} \tag{1.2.27}$$

We consider a code $\mathcal{C}_n$ as in Definition 2 and assume that the transmitter, the legitimate receiver and the eavesdropper do not possess any information about the actual channel state s. Additionally, we do not impose any prior distribution on the channel state set $\mathcal{S}$ that govern the selection of the channel state. This implies that, the encoder and decoder of the code should be universal in the sense that they work for all possible channel states. This also implies that the code $\mathcal{C}_n$ should fulfill the reliability and secrecy constraints similar to the ones in (1.2.6) and (1.2.9) for all channel states $s \in \mathcal{S}$. For the reliability constraints, we define the average and maximum decoding error probability for the compound wiretap channel as follows:

$$\bar{e} = \max_{s\in\mathcal{S}} \frac{1}{|\mathcal{M}|} \sum_{m\in\mathcal{M}} \sum_{x^n\in\mathcal{X}^n} \sum_{y^n:\varphi(y^n)\neq m} \mathcal{W}_s(y^n|x^n)E(x^n|m) \tag{1.2.28}$$

$$e_{max} = \max_{s\in\mathcal{S}} \max_{m\in\mathcal{M}} \sum_{x^n\in\mathcal{X}^n} \sum_{y^n:\varphi(y^n)\neq m} \mathcal{W}_s(y^n|x^n)E(x^n|m) \tag{1.2.29}$$

Subsequently, the strong secrecy requirement becomes

$$\max_{s\in\mathcal{S}} \mathbb{I}(\mathrm{M}; \mathrm{Z}_s^n) \le \tau_n, \tag{1.2.30}$$

where Z_s^n represents the random variable associated with the output sequence at the eavesdropper for channel state s. It is important to point out that, if the channel state s is selected by an active eavesdropper, this active eavesdropper should be independent from the passive one. This means that it chooses s without possessing any information about channel observation Z_s^n. Now, the target is to formulate the secrecy capacity of the compound wiretap channel, which is the maximal achievable rate that satisfy the following definition:

Definition 8 A confidential rate $R \in \mathbb{R}_+$ is achievable for the compound wiretap channel, if there exist a sequence of $(2^{nR}, n)$ codes $\mathcal{C}_n$ and two sequences ϵ_n, τ_n, where n is large enough, such that $\lim_{n\to\infty} \epsilon_n, \tau_n = 0$ and depending on the selected reliability criterion $\bar{e}$ or e_{max} is smaller than ϵ_n.

Secrecy Capacity

The compound wiretap channel can be visualized as a group of classical wiretap channels, where each channel state $s \in \mathcal{S}$ defines a specific one. This implies that the secrecy capacity of the compound wiretap channel can not exceed the smallest secrecy capacity of the wiretap channels in this group. This bound is known as the worst case secrecy capacity and is given by

Proposition 1 ([27]) *The strong secrecy capacity of the compound wiretap channel is upper-bounded by its worst-case secrecy capacity as follows:*

$$C(\mathfrak{W}) \leq \min_{s\in\mathcal{S}} \max_{U_s - X_s - (Y_s, Z_s)} \mathbb{I}(U_s; Y_s) - \mathbb{I}(U_s; Z_s) \tag{1.2.31}$$

for random variables satisfying the following Markov chain $U_s - X_s - (Y_s, Z_s)$.

This bound is usually a loose upper-bound, because for every state s, there exist a certain capacity achieving input distribution P_s. This distribution may differ from one state to another. This implies that in order to achieve the worst case secrecy capacity, an adaptive encoder depending on the channels state is needed, such that the channel input X_s and the channel prefix U_s depends on the actual channel state s. However, we already pointed out that the encoder for the compound wiretap channel must be universal and independent of the channel state. A universal encoder can have only one input distribution, which should be capable of balancing the rates for all the wiretap channels in the uncertainty set in the most optimum way. This indicates that the actual secrecy capacity of the compound wiretap channel is usually smaller than the worst case capacity.

In order to construct an achievable scheme in which neither the transmitter nor the receiver knows the actual channel state, we need to adapt the coding techniques used in Sect. 1.2.1 for the classical wiretap channel with perfect CSI to the compound setup. This implies that regardless of the channel state s, this coding scheme should provide a reliable communication to the legitimate receiver while keeping the eavesdropper ignorant about the information transmitted. This concept leads to the following rate region:

Theorem 9 ([7, 27]) *An achievable strong secrecy rate region for the compound wiretap channel is given by the set of all rates* $R \in \mathbb{R}_+$ *that satisfy*

$$R \leq \max_{U - X - (Y_s, Z_s)} \Big[\min_{s\in\mathcal{S}} \mathbb{I}(U; Y_s) - \max_{s\in\mathcal{S}} \mathbb{I}(U; Z_s) \Big], \tag{1.2.32}$$

for random variables that satisfy the following Markov chain $U - X - (Y_s, Z_s)$.

Differently from the worst case upper-bound given in Proposition 1, the channel prefix U and the channel input X are chosen independently from the channel state s. This agrees with the fact that, this achievable region is established using universal encoder and decoder independent of the actual channel state s. The previous rate region follows as: In order to guarantee a reliable link between the transmitter and the legitimate receiver for all channel states $s \in \mathcal{S}$, the maximum transmission rate should be bounded by the smallest rate among all the channel states, i.e., $\min_{s\in\mathcal{S}} \mathbb{I}(X; Y_s)$. On the other hand, in order to make sure that the eavesdropper is not capable of inferring any information about the transmitted message, we need to choose the randomization index to be roughly $\max_{s\in\mathcal{S}} \mathbb{I}(X; Z_s)$. This will assure that even the best channel resources available at the eavesdropper will always be jammed by useless information. Combining these two conditions and introducing an auxiliary random variable U that plays the role of additional channel prefixing similarly as in Theorem 5, leads to the previous rate region.

In [27], it was shown that the achievable rate region in (1.2.32) is tight for the class of degraded compound wiretap channel, in which all channel realization to the eavesdropper are degraded with respect to any channel realization to the legitimate receiver. Suppose we have two uncertainty sets $\mathcal{S}$ and $\mathcal{T}$, where $\mathcal{S}$ contains the possible channel states between the transmitter and the legitimate receiver, while $\mathcal{T}$ contains the possible channel states between the transmitter and the eavesdropper. A compound wiretap channel is said to be degraded, if for all $s \in \mathcal{S}$ and $t \in \mathcal{T}$, $X - Y_s - Z_t$ forms a Markov chain. The secrecy capacity for such class of compound channel is established by replacing the auxiliary random variable U in (1.2.32) by the channel input X as follows:

Theorem 10 ([7, 27]) *The strong secrecy capacity region for the degraded compound wiretap channel is given by the set of all rates $R \in \mathbb{R}_+$ that satisfy*

$$R \leq C(\mathfrak{W}) = \max_{X-Y_s-Z_t} \left[\min_{s\in\mathcal{S}} \mathbb{I}(X; Y_s) - \max_{t\in\mathcal{T}} \mathbb{I}(X; Z_t) \right], \tag{1.2.33}$$

for random variables satisfying the following Markov chain $X - Y_s - Z_t$.

Despite the tremendous effort of researchers, finding a single letter characterization for the secrecy capacity of the general compound wiretap channel has remained an unanswered question. However, in [7] a multi-letter upper bound was derived, that matches the achievable rate region in (1.2.32) applied to the n-fold channels $\tilde{W}_s : \mathcal{U} \to \mathcal{P}(\mathcal{Y}^n)$ and $\tilde{V}_s : \mathcal{U} \to \mathcal{P}(\mathcal{Z}^n)$. This leads to a multi-letter description of the secrecy capacity region as follows:

Theorem 11 ([7]) *The strong secrecy capacity region of the general compound wiretap channel is given by the set of all rates $R \in \mathbb{R}_+$ that satisfy the following multi-letter description*

$$R \leq C(\mathfrak{W}) = \lim_{n\to\infty} \frac{1}{n} \max_{U-X^n-(Y_s^n, Z_s^n)} \left[\min_{s\in\mathcal{S}} \mathbb{I}(U; Y_s^n) - \max_{s\in\mathcal{S}} \mathbb{I}(U; Z_s^n) \right], \tag{1.2.34}$$

for random variables satisfying the following Markov chain $U - X^n - (Y_s^n, Z_s^n)$.

Unfortunately, a multi-letter expression depends on the block length n, which makes it not easily computable. However, such a description is still useful as it helps in capturing some insights and deducing some properties regarding the secrecy capacity of the compound wiretap channel.

Continuity and Robustness

The secrecy capacity of the compound wiretap channel depends on the uncertainty set that contains the different channel states. Since active eavesdroppers might be able to control this uncertainty set, it is desirable to have a continuous dependency between the secrecy capacity and the uncertainty set. In other words small variations in the uncertainty set should result only in small variations in the secrecy capacity. This property will assure that an active eavesdropper that can slightly change the uncertainty set, will not be able to cause a dramatic loss in the secrecy capacity.

In order to investigate this property, we need a quantity to measure the distance between two compound wiretap channels. We consider the total variation distance and define the distance between two channels $\mathrm{W}_1, \mathrm{W}_2 : \mathcal{X} \to \mathcal{P}(\mathcal{Y})$ as:

$$d(\mathrm{W}_1, \mathrm{W}_2) = \max_{x \in \mathcal{X}} \sum_{y \in \mathcal{Y}} |\mathrm{W}_1(y|x) - \mathrm{W}_2(y|x)|. \tag{1.2.35}$$

Then the distance $D(\mathfrak{W}_1, \mathfrak{W}_2)$ between two compound wiretap channels $\mathfrak{W}_1$ and $\mathfrak{W}_2$ is given by the largest distance defined by (1.2.35) for all possible channel realizations for the legitimate and eavesdropper channels.

Theorem 12 ([14]) *Let $\epsilon \in (0, 1)$ be arbitrary and let $\mathfrak{W}_1$ and $\mathfrak{W}_2$ be two compound wiretap channels. If $D(\mathfrak{W}_1, \mathfrak{W}_2) < \epsilon$, then it holds that*

$$|\mathrm{C}(\mathfrak{W}_1) - \mathrm{C}(\mathfrak{W}_2)| \leq \delta(\epsilon, |\mathcal{Y}|, |\mathcal{Z}|), \tag{1.2.36}$$

where $\delta(\epsilon, |\mathcal{Y}|, |\mathcal{Z}|)$ is a constant that depends only on the distance ϵ and the output alphabet sizes $|\mathcal{Y}|$ and $|\mathcal{Z}|$.

This theorem implies that the strong secrecy capacity of the compound wiretap channel is a continuous function in the uncertainty set. It also bounds the difference in the secrecy capacities with respect to the distance between the uncertainty sets. Theorem 12 also ensures that: If there is a "good" (i.e., capacity-achieving) code for $\mathfrak{W}_1$, then there exists another "good" code that achieves a similar rate over $\mathfrak{W}_2$ as long as $D(\mathfrak{W}_1, \mathfrak{W}_2) < \epsilon$.

Another important property is the robustness of the code $\mathcal{C}_n$. A code is robust if its reliability and secrecy performance depend continuously on the underlying uncertainty set. In [9], it was shown that a code $\mathcal{C}_n$ for the classical compound wiretap channel is robust, such that a "good" code in the sense of small decoding error probability will also perform well for other compound channels within a small distance. This implies that the reliability performance of a code $\mathcal{C}_n$ for the compound wiretap channel is robust. On the other hand, it was shown that the weak secrecy criterion is also robust against small changes in the uncertainty set.

Theorem 13 ([14]) *Let $\mathcal{V}_1$ be a compound channel to the eavesdropper with uncertainty set $\mathcal{S}_1$. Then for any code that achieves weak secrecy criterion*

$$\max_{s_1 \in \mathcal{S}_1} \frac{1}{n} \mathbb{I}(\mathrm{M}; \mathrm{Z}^n_{s_1}) \leq \tau_n, \tag{1.2.37}$$

it holds that for all compound channels $\mathcal{V}_2$ with uncertainty set $\mathcal{S}_2$ and $D(\mathcal{V}_1, \mathcal{V}_2) < \epsilon$, that

$$\max_{s_2 \in \mathcal{S}_2} \frac{1}{n} \mathbb{I}(\mathrm{M}; \mathrm{Z}^n_{s_2}) \leq \tau_n + \delta(\epsilon, |\mathcal{Z}|), \tag{1.2.38}$$

where $\delta(\epsilon, |\mathcal{Z}|)$ is a constant that depends only on the distance ϵ and the output alphabet size $|\mathcal{Z}|$.

This theorem implies that any code for the compound wiretap channel is robust with respect to the weak secrecy criterion as follows: If the information leakage rate over the eavesdropper compound channel $\mathcal{V}_1$ is small, then the information leakage rate over a compound channels $\mathcal{V}_2$, where $D(\mathcal{V}_1, \mathcal{V}_2) < \epsilon$ will also be small and bounded by (1.2.38).

1.2.4 The Arbitrary Varying Wiretap Channel

In this section, we investigate the model of the arbitrary varying wiretap channel [8]. This channel is similar to the compound wiretap channel, where the main difference is that the channel state may vary from one channel use to the other. This implies that symbols of the same codeword are transmitted over different channel realizations. This model captures some real life communication scenarios such as fast fading channels and wiretap channels with active eavesdroppers capable of maliciously manipulating the channel state for each channel use.

System Model

We consider an input alphabet $\mathcal{X}$, two output alphabets $(\mathcal{Y}, \mathcal{Z})$ and a finite set $\mathcal{S}$ that contains all the possible channel realizations. We use the stochastic matrices $\mathrm{W} : \mathcal{X} \times \mathcal{S} \rightarrow \mathcal{P}(\mathcal{Y})$ and $\mathrm{V} : \mathcal{X} \times \mathcal{S} \rightarrow \mathcal{P}(\mathcal{Z})$ to model the channels from the transmitter to the legitimate receiver and the eavesdropper respectively. Now for a discrete memoryless channel and a block code of length n, the transmission matrices are given by

$$\mathrm{W}^n_{s^n}(y^n|x^n) = \prod_{i=1}^{n} \mathrm{W}(y_i|x_i, s_i) \quad \text{and} \quad \mathrm{V}^n_{s^n}(z^n|x^n) = \prod_{i=1}^{n} \mathrm{V}(z_i|x_i, s_i), \tag{1.2.39}$$

where $x^n \in \mathcal{X}^n$, $y^n \in \mathcal{Y}^n$ and $z^n \in \mathcal{Z}^n$ represent the channel input and output sequences, while $s^n = (s_1, s_2, \ldots, s_n) \in \mathcal{S}^n$ represents the channel state sequence.

We consider the scenario in which the channel state sequence s^n is produced independently from the transmitted message m without any presumed a priori distribution. We also assume that the transmitter and the legitimate receiver knows the state space $\mathcal{S}$, but have no knowledge regarding the actual state sequence s^n.

Definition 9 The discrete memoryless arbitrary varying wiretap channel $\mathfrak{W}$ is given by the families of marginal AVCs with common input as

$$\mathfrak{W} = \{\mathcal{W}, \mathcal{V}\} = \{\mathrm{W}^n_{s^n}, \mathrm{V}^n_{s^n} : s^n \in \mathcal{S}^n\} \tag{1.2.40}$$

Since the channel is memoryless, the behavior of the channel should depend on the number of times each channel state s is imposed, and not on the order of these states. This observation motivates the introduction of the average channel notation. For any probability distribution $q \in \mathcal{P}(\mathcal{S})$, the average channel is given by:

$$\mathrm{W}_q(y|x) = \sum_{s \in \mathcal{S}} \mathrm{W}(y|x, s)q(s) \quad \text{and} \quad \mathrm{V}_q(z|x) = \sum_{s \in \mathcal{S}} \mathrm{V}(z|x, s)q(s). \tag{1.2.41}$$

Another important concept for AVCs is the symmetrizability property. In order to capture the implications of this property, let us consider an AVC, where $\mathcal{S} = \mathcal{X}$, and assume that $\mathrm{W}(y|x, s) = \mathrm{W}(y|s, x)$ is true for all $x \in \mathcal{X}$, $y \in \mathcal{Y}$ and $s \in \mathcal{S}$. In this case the receiver cannot distinguish between the scenario in which the transmitter sent the symbol x over a channel with state s and the scenario in which the symbol s is transmitted over a channel with state x. The previous condition is an example of a symmetrizable AVC, which is generally defined as:

Definition 10 An AVC defined by the stochastic matrix $\mathrm{W} : \mathcal{X} \times \mathcal{S} \to \mathcal{P}(\mathcal{Y})$ is called symmetrizable if there exists an auxiliary channel $\sigma : \mathcal{X} \to \mathcal{P}(\mathcal{S})$ such that

$$\sum_{s \in \mathcal{S}} \mathrm{W}(y|x, s)\sigma(s|\tilde{x}) = \sum_{s \in \mathcal{S}} \mathrm{W}(y|\tilde{x}, s)\sigma(s|x) \tag{1.2.42}$$

holds for all $x, \tilde{x} \in \mathcal{X}$ and $y \in \mathcal{Y}$.

The symmetrizability property plays an important role for AVWC, where the state sequence s^n does not originate solely from channel uncertainty but might be controlled by an active eavesdropper. In that case the eavesdropper can change it's attacking strategy from trying to infer information about the transmitted message to symmetrizing the channel between the transmitter and the legitimate receiver. This might lead to the incapability of the legitimate receiver to detect the transmitted message correctly.

Coding Techniques

Two different coding techniques are usually used to support communication over the AVWC. The first technique is known as the unassisted coding scheme and is an extension to the class of common deterministic codes used for the AVC. The second technique is known as the common randomness (CR) assisted codes. The CR-assisted codes are simply a collection of unassisted codes among them one is selected for communication based on some random experiment. The CR-assisted codes usually outperform the unassisted ones, however, it is harder to implement CR-assisted codes compared to the unassisted ones. The two coding schemes are defined as follows:

1. Unassisted Codes: We consider a code $\mathcal{C}_n$ as in Definition 2, where the term "unassisted" is used to highlight the fact that the encoder (1.2.3) and the decoder (1.2.4) are universal for the whole transmission and their choice cannot be coordinated in any way. This implies that the code $\mathcal{C}_n$ should fulfill the reliability and security constraints for all state sequences $s^n \in \mathcal{S}^n$. We start by the reliability requirement and define the average probability of error as follows:

$$\bar{e} = \max_{s^n \in \mathcal{S}^n} \frac{1}{|\mathcal{M}|} \sum_{m \in \mathcal{M}} \sum_{x^n \in \mathcal{X}^n} \sum_{y^n : \varphi(y^n) \neq m} \mathrm{W}^n_{s^n}(y^n|x^n) E(x^n|m). \tag{1.2.43}$$

Although the secrecy capacity for the wiretap channel with perfect CSI and the compound wiretap channel turned out to be the same for the average and maximum error probability, the situation is different for the AVWC. It has been shown that even for the classical AVC without any secrecy constraints, the average and maximum error probability have different capacities, where the maximum error capacity is still unknown. That is why, we will only consider the average error probability as our reliability constraint. On the other hand, the strong secrecy criterion is given by:

$$\max_{s^n \in \mathcal{S}^n} \mathbb{I}(\mathrm{M}; \mathrm{Z}^n_{s^n}) \leq \tau_n, \tag{1.2.44}$$

where $\tau_n \to 0$ as $n \to \infty$. Thus, the maximal achievable rate for the unassisted code $\mathcal{C}_n$, that satisfy the requirements in (1.2.43) and (1.2.44) is the unassisted strong secrecy capacity of the AVWC under the average error probability.

2. CR-Assisted Codes: In order to construct a CR-assisted code, we need a random experiment, whose output is available to the transmitter and the legitimate receiver. This random experiment can be realized by some common satellite signal or a common synchronization procedure. We will model the CR by a random variable Γ that takes values in a finite set $\mathcal{G}_n$ according to a distribution $\mathrm{P}_\Gamma \in \mathcal{P}(\mathcal{G}_n)$. Now, as the transmitter and the legitimate receiver observes $\gamma \in \mathcal{G}_n$, they choose a certain encoder and a corresponding decoder.

Definition 11 A $(2^{nR}, \mathcal{G}_n, \mathrm{P}_\Gamma, n)$ CR-assisted code $\mathcal{C}_{CR}$ is given by a family of unassisted codes as

$$\mathcal{C}_{CR} = \{\mathcal{C}(\gamma) : \gamma \in \mathcal{G}_n\} \tag{1.2.45}$$

with Γ is a random variable taking values in the finite set $\mathcal{G}_n$ according to the probability distribution $\mathrm{P}_\Gamma \in \mathcal{P}(\mathcal{G}_n)$.

Since a CR-assisted code consists of a family of unassisted ones, a given message may be assigned to different codewords in different communication instances. For such codes, we extend the reliability constraint in (1.2.43) to include the expectation over the underlying family of unassisted codes as follows:

$$\bar{e}_{CR} = \max_{s^n \in \mathcal{S}^n} \frac{1}{|\mathcal{M}|} \sum_{m \in \mathcal{M}} \sum_{\gamma \in \mathcal{G}_n} \sum_{x^n \in \mathcal{X}^n} \sum_{y^n : \varphi(y^n) \neq m} \mathrm{W}^n_{s^n}(y^n|x^n) E_\gamma(x^n|m,) \mathrm{P}_\Gamma(\gamma). \quad (1.2.46)$$

Similarly, the strong secrecy criterion becomes

$$\max_{s^n \in \mathcal{S}^n} \sum_{\gamma \in \mathcal{G}_n} \mathbb{I}(\mathrm{M}; \mathrm{Z}^n_{s^n,\gamma}) \mathrm{P}_\Gamma(\gamma) \leq \tau_n, \quad (1.2.47)$$

where $\tau_n \to 0$ as $n \to \infty$ and $\mathrm{Z}^n_{s^n,\gamma}$ represents the output sequence at the eavesdropper for state sequence s^n and CR realization γ. It is important to note that the previous criterion only implies that the average leakage over all realizations of the CR is small. This requirement is sufficient if we assumed that the eavesdropper has no knowledge about the instantaneous CR realization γ. However, this assumption is not practical, because if the eavesdropper has no access to the CR resources, the CR resources can be used to generate a secret key between the transmitter and the legitimate receiver. That is why, it is better to strengthen the previous criterion by replacing the average over all CR realizations by the maximum as follows:

$$\max_{s^n \in \mathcal{S}^n} \max_{\gamma \in \mathcal{G}_n} \mathbb{I}(\mathrm{M}; \mathrm{Z}^n_{s^n,\gamma}) \leq \tau_n, \quad (1.2.48)$$

Surprisingly, it was shown that strengthening the secrecy criterion from (1.2.47) to (1.2.48) comes at no cost in terms of secrecy capacity [32]. Finally, we highlight the fact that, the maximal achievable rate for a CR-assisted code of Definition 11 that guarantee the reliability constraint in (1.2.46) and the secrecy constraint in (1.2.48) is the CR-assisted secrecy capacity of the AVWC.

Secrecy Capacity

We present some of the main bounds that highlights the secrecy capacity of the AVWC.

Theorem 14 ([8, 32]) *The unassisted strong secrecy capacity of the AVWC is characterized by the following:*

1. $\mathrm{C}(\mathfrak{W}) = 0$, *if $\mathcal{W}$ is symmetrizable.*
2. *Otherwise,* $\mathrm{C}(\mathfrak{W}) = \mathrm{C}_{CR}(\mathfrak{W})$.

This theorem reflects the same behavior of the AVC without secrecy constraint, where the unassisted capacity is either equivalent to the CR-assisted capacity or zero. It is important to note that the vanishing behavior of the unassisted secrecy capacity depends only on the symmetrizability of the legitimate channel $\mathcal{W}$ and does not depend on the eavesdropper channel $\mathcal{V}$. This result is due to the failure of the unassisted codes to provide a reliable communication over a symmetric channel. On the other hand, the previous theorem suggests that if the legitimate receiver channel is not symmetrizable, using a code with a complicated structure, i.e., a CR-assisted code does not provide any gain in terms of the secrecy capacity over a code with simpler structure, i.e., an unassisted code.

Instead of using the entropic relations between the input and output distributions to bound the unassisted secrecy capacity of the AVWC, Theorem 14 used the CR-assisted secrecy capacity. This implies that, we still need to bound the CR-assisted secrecy capacity in terms of those entropic quantities. Unfortunately a single-letter characterization of the CR-assisted secrecy capacity remains unknown, where only a multi-letter description has been established.

Theorem 15 ([33]) *The CR-assisted strong secrecy capacity region of the AVWC is given by the set of all rates $R_{CR} \in R^+$ that satisfy the following multi-letter description*

$$R_{CR} \leq C_{CR}(\mathfrak{W}) = \lim_{n\to\infty} \frac{1}{n} \max_{U-X^n-(Y_q^n, Z_{s^n}^n)} \Big[\min_{q\in\mathcal{P}(\mathcal{S})} \mathbb{I}(U; Y_q^n) - \max_{s^n\in\mathcal{S}^n} \mathbb{I}(U; Z_{s^n}^n) \Big], \tag{1.2.49}$$

where Y_q^n is the random variable associated with the output sequence of the averaged channel W_q^n.

The previous multi-letter description followed from a multi-letter achievable secrecy rate instead of a single-letter one, because establishing a single-letter secrecy rate which is achievable for the general AVWC remains unsolved. The single-letter achieveability scheme that has been established is only valid for a special class of AVWC, where a best channel to the eavesdropper exists. An AVWC is said to have a best channel to the eavesdropper if there exist a channel $V_{q^*} \in \{V_q : q \in \mathcal{P}(\mathcal{S})\}$, all other channels in this set are degraded versions of V_{q^*}. In other words, V_{q^*} is called a best channel to the eavesdropper if the following Markov chain:

$$X - Z_{q^*} - Z_q \tag{1.2.50}$$

holds for all $q \in \mathcal{P}(\mathcal{S})$, where Z_{q^*} and Z_q are the random variables associated with the output sequences of the averaged channels V_{q^*} and V_q respectively.

Theorem 16 ([8]) *If there exists a best channel to the eavesdropper, an achievable CR-assisted strong secrecy rate region for the AVWC is given by the set of all rates $R \in R^+$ that satisfy*

$$R \leq \max_{X-(Y_q, Z_q)} \Big[\min_{q\in\mathcal{P}(\mathcal{S})} \mathbb{I}(X; Y_q) - \max_{q\in\mathcal{P}(\mathcal{S})} \mathbb{I}(X; Z_q) \Big], \tag{1.2.51}$$

where Y_q and Z_q represents the random variables associated with the output sequences of the averaged channels W_q and V_q respectively.

Continuity and Robustness

Since the secrecy capacity of the AVWC depends on the state set $\mathcal{S}$, it is important to investigate whether this dependence is continuous or not. We start by the CR-assisted secrecy capacity and present the following theorem.

Theorem 17 ([14]) *Let $\epsilon \in (0, 1)$ be arbitrary and let $\mathfrak{W}_1$ and $\mathfrak{W}_2$ be two AVWCs, such that $D(\mathfrak{W}_1, \mathfrak{W}_2) < \epsilon$, then the following holds*

$$|C_{CR}(\mathfrak{W}_1) - C_{CR}(\mathfrak{W}_2)| \leq \delta(\epsilon, |\mathcal{Y}|, |\mathcal{Z}|), \tag{1.2.52}$$

where $\delta(\epsilon, |\mathcal{Y}|, |\mathcal{Z}|)$ is a constant that depends only on the distance ϵ and the output alphabet sizes $|\mathcal{Y}|$ and $|\mathcal{Z}|$.

The previous theorem indicates that the CR-assisted secrecy capacity is continuous with respect to the uncertainty set, such that small changes in the uncertainty set will only result in small changes in the CR- assisted secrecy capacity. On the other hand, Theorem 14 raises some doubts about the continuity of the unassisted secrecy capacity. In order to investigate these doubts, we will need the following function:

$$F(\mathcal{W}) = \min_{\sigma:\mathcal{X}\rightarrow\mathcal{P}(\mathcal{S})} \left[\max_{x\neq\tilde{x}} \sum_{y\in\mathcal{Y}} \left| \sum_{s\in\mathcal{S}} \mathrm{W}(y|\tilde{x}, s)\sigma(s|x) - \mathrm{W}(y|x, s)\sigma(s|\tilde{x}) \right| \right] \tag{1.2.53}$$

This function is related to the symmetrizability property of the AVC $\mathcal{W}$ between the transmitter and the legitimate receiver as follows: $\mathcal{W}$ is symmetrizable if and only if $F(\mathcal{W}) = 0$. One can easily show that the function $F(\mathcal{W})$ is a continuous function in $\mathcal{W}$. Now regarding the continuity of the unassisted secrecy capacity, we present the following result:

Theorem 18 ([32]) *The unassisted secrecy capacity of the AVWC is discontinuous if and only if the following holds:*

1. $C_{CR}(\mathfrak{W}) > 0$.
2. $F(\mathcal{W}) = 0$ *and for every* $\epsilon > 0$, *there is a finite* $\mathcal{W}_\epsilon$ *with* $D(\mathcal{W}, \mathcal{W}_\epsilon) \leq \epsilon$ *and* $F(\mathcal{W}_\epsilon) > 0$.

The previous theorem interestingly characterizes the discontinuity behavior of the unassisted secrecy capacity in terms of two continuous functions: CR-assisted secrecy capacity and the function $F(\mathcal{W})$. The previous two condition defines the scenario where a discontinuity point occurs as follows: First, $\mathcal{W}$ must be symmetrizable. Second the CR-assisted secrecy capacity must be greater than zero to make sure that unassisted secrecy capacity is not a zero function. Finally, there should exist another non-symmetrizable AVC $\mathcal{W}_\epsilon$, such that the distance between $\mathcal{W}_\epsilon$ and $\mathcal{W}$ is small. The discontinuity behavior established in Theorem 18 implies that small changes in the uncertainty set of the AVWC can lead to a dramatic loss in the unassisted secrecy capacity $C(\mathfrak{W})$. It is important to highlight the fact that $C(\mathfrak{W})$ is a continuous function in the eavesdropper channel $\mathcal{V}$, where the discontinuity only originates from the legitimate channel $\mathcal{W}$.

In addition to the continuity of the secrecy capacity, we need to investigate the robustness of the unassisted and CR-assisted codes against small changes in the uncertainty set. We start by the CR-assisted codes and present the following result:

Theorem 19 ([14]) *Let $\mathcal{V}_1$ be an AVC to the eavesdropper with uncertainty set $\mathcal{S}_1$. Then for any CR-assisted code that achieves weak secrecy criterion*

$$\max_{s_1^n \in \mathcal{S}_1^n} \sum_{\gamma \in \mathcal{G}_n} \frac{1}{n} \mathbb{I}(\mathrm{M}; \mathrm{Z}^n_{s_1^n,\gamma}) P_\Gamma(\gamma) \leq \tau_n \tag{1.2.54}$$

it holds that for all AVCs $\mathcal{V}_2$ with finite state set $\mathcal{S}_2$ and $D(\mathcal{V}_1, \mathcal{V}_2) < \epsilon$ that

$$\max_{s_2^n \in \mathcal{S}_2^n} \sum_{\gamma \in \mathcal{G}_n} \frac{1}{n} \mathbb{I}(\mathrm{M}; \mathrm{Z}^n_{s_2^n,\gamma}) P_\Gamma(\gamma) \leq \tau_n + \delta(\epsilon, |\mathcal{Z}|), \tag{1.2.55}$$

where $\delta(\epsilon, |\mathcal{Z}|)$ is a constant that depends only on the distance ϵ and the output alphabet size $|\mathcal{Z}|$.

This theorem indicates that a "good" CR-assisted code with small information leakage rate over the eavesdropper AVC will also has small information leakage rate for all AVCs in the neighborhood. In [39] it has been shown that, not only the CR-assisted codes are robust under the weak secrecy criterion, but the unassisted codes are also robust. This result agrees with the previous observation that the discontinuity in the unassisted secrecy capacity originates for the legitimate link and has nothing to do with the eavesdropper link.

1.2.4.1 Super-Activation

Medium access control and in particular resource allocation plays an important role in determining the overall performance of a wireless communication system. Consider an OFDM system, the overall capacity of such system is given by the sum of the capacities of all orthogonal sub-channels. This implies that given a system that consists of two orthogonal channels, where both have zero capacity, the overall capacity of the system should be zero as well. This result is known as the classical additivity of basic resources, i.e., "$0 + 0 = 0$". On the other hand, this result does not hold in quantum information theory, where there exist some scenarios in which a system with two orthogonal zero capacity channels has a non-zero capacity, i.e., "$0+0 > 0$". This phenomena is known as super-activation and has been investigated in the field of quantum information theory in [20].

Super-activation remained a distinct phenomena for quantum information theory, until it has been shown that it can also happen in the classical non-quantum world. In [13], it has been demonstrated that: two orthogonal AVWCs $\mathfrak{W}_1$ and $\mathfrak{W}_2$ with zero secrecy capacity $\mathrm{C}(\mathfrak{W}_1) = \mathrm{C}(\mathfrak{W}_2) = 0$, can be super-activated to provide a non-zero secrecy rates, i.e., $\mathrm{C}(\mathfrak{W}_1 \otimes \mathfrak{W}_2) > \mathrm{C}(\mathfrak{W}_1) + \mathrm{C}(\mathfrak{W}_2) = 0$, where $\mathfrak{W}_1 \otimes \mathfrak{W}_2$ represents the joint usage of both orthogonal AVWCs. The joint usage of both orthogonal channels implies that instead of designing two individual encoder-decoder pairs (one for each channel), only one encoder-decoder pair is jointly designed for the combined channel. The phenomena of super-activation has been completely characterized for AVWCs [32] as follows:

Theorem 20 ([32]) *Let $\mathfrak{W}_1$ and $\mathfrak{W}_2$ be two AVWCs. Then the following properties hold:*

- *If $\mathrm{C}(\mathfrak{W}_1) = \mathrm{C}(\mathfrak{W}_2) = 0$, then $\mathrm{C}(\mathfrak{W}_1 \otimes \mathfrak{W}_2) > 0$, if and only if $(\mathcal{W}_1 \otimes \mathcal{W}_2)$ is non-symmetrizable and $\mathrm{C}_{CR}(\mathfrak{W}_1 \otimes \mathfrak{W}_2) > 0$. Further, if $\mathfrak{W}_1$ and $\mathfrak{W}_2$ can be super-activated, it holds that*

$$\mathrm{C}(\mathfrak{W}_1 \otimes \mathfrak{W}_2) = \mathrm{C}_{CR}(\mathfrak{W}_1 \otimes \mathfrak{W}_2) \tag{1.2.56}$$

- *If C_{CR} shows no super-activation for $\mathfrak{W}_1$ and $\mathfrak{W}_2$, then super-activation of C can only happen if $\mathcal{W}_1$ is non-symmetrizable and $\mathcal{W}_2$ is symmetrizable, in addition $\mathrm{C}_{CR}(\mathfrak{W}_1) = 0$ and $\mathrm{C}_{CR}(\mathfrak{W}_2) > 0$. The statement is independent of the specified labeling.*
- *There exist AVWCs that exhibit the behavior according to the second property.*

In order to understand how super-activation is possible for AVWCs, let us consider the following example: Assume $\mathfrak{W}_1$ and $\mathfrak{W}_2$ are two orthogonal AVWCs, such that the legitimate AVC $\mathcal{W}_1$ is symmetrizable, while the eavesdropper AVC $\mathcal{V}_2$ is less noisy than the legitimate channel $\mathcal{W}_2$. This implies that the unassisted secrecy capacity of both AVWCs is zero, i.e., $\mathrm{C}(\mathfrak{W}_1) = \mathrm{C}(\mathfrak{W}_2) = 0$. Now, it is important to note that the legitimate AVC $\mathcal{W}_2$ can support a reliable non-secure communication between the transmitter and the legitimate receiver. This implies that $\mathcal{W}_2$ can be used to generate a sort of common randomness, then CR-assisted codes can be used between the transmitter and the legitimate receiver on $\mathfrak{W}_1$ to achieve a non-zero CR-assisted secrecy rate.

Since Theorem 20 describes the conditions needed for two orthogonal AVWCs to be super-activated, it remains to investigate how super-activation as a property depends on the channels. In other words, it is important to find out how super-activation behaves when the channels are slightly changed. In [34], it was shown that super-activation is a generic property, such that if two orthogonal AVWCs can be super-activated, then all AVWCs in a certain neighborhood can be super-activated as well. This result was further strengthened as follows:

Theorem 21 ([34]) *Let $\mathfrak{W}_1$ and $\mathfrak{W}_2$ be two orthogonal AVWCs with zero secrecy capacity that can be super-activated. Then there exists an $\epsilon > 0$, such that all orthogonal AVWCs $\tilde{\mathfrak{W}}_1$ and $\tilde{\mathfrak{W}}_2$ that satisfy:*

$$D(\mathcal{W}_1, \tilde{\mathcal{W}}_1) < \epsilon, \qquad D(\mathcal{W}_2, \tilde{\mathcal{W}}_2) < \epsilon \qquad \textit{and} \qquad C_{CR}(\tilde{\mathfrak{W}}_1 \otimes \tilde{\mathfrak{W}}_2) > 0, \quad (1.2.57)$$

can be super-activated as well.

The previous result indicates that only the legitimate AVCs need to be within a certain neighborhood, while no explicit condition is required for the distance between the eavesdropper AVCs. This implies that the legitimate AVC plays a much more important role in controlling the super-activation phenomena compared to the eavesdropper AVC. This result was further extended to show that super-activation leads to a more robust and continuous system as follows:

Theorem 22 ([34]) *Let $\mathfrak{W}_1$ and $\mathfrak{W}_2$ be two orthogonal AVWCs with zero secrecy capacity that can be super-activated. Then the unassisted strong secrecy capacity* $C(\mathfrak{W}_1 \otimes \mathfrak{W}_2)$ *depends in a continuous way on the channels $\tilde{\mathfrak{W}}_1$ and $\tilde{\mathfrak{W}}_2$ with $D(\mathcal{W}_1, \tilde{\mathcal{W}}_1) < \epsilon$ and $D(\mathcal{W}_2, \tilde{\mathcal{W}}_2) < \epsilon$.*

The previous theorem demonstrates a very interesting observation which is: Although the unassisted secrecy capacity of a single AVWC cannot be guaranteed to be continuous in general, bonding of orthogonal AVWCs can lead to a more robust system which is continuous. Another important observation which also is a consequence of Theorem 21, is that super-activation mostly depends on the legitimate AVC and is robust in the eavesdropper AVC. This result inspired the investigation of super-activation for AVCs without secrecy requirement in [34].

The problem of reliable communication over orthogonal AVCs has been indirectly highlighted by Shannon's question of the additivity of the zero error capacity [36]. This indirect relation was discovered by Ahlswede in [1], where he showed that the capacity of the AVC under the maximum error probability criterion includes the characterization of the zero error capacity as a special case. Although Shannon predicted that the zero error capacity is additive, Alon constructed a counter-example in [5], where he showed that the capacity of reliable communication over orthogonal AVCs under the maximum error probability criterion is super-additive. In [34], the capacity of reliable communication over orthogonal AVCs was investigated under the average error probability criterion. It was shown that the capacity is super-additive under certain circumstances. However, super-activation was proved to be not possible for reliable communication over orthogonal AVC, making it a unique feature of AVWCs in the classical information theory world.

1.2.5 Discussion and Open Questions

The usage of information theoretic techniques to achieve secure communication has captured a lot of attention in the last few years. The main target is to establish the secrecy capacity for a given channel and to develop a coding scheme that achieves it. A single-letter characterization for the secrecy capacity of the classical wiretap channel has been established in terms of mutual information. This result has been extended to some special cases for the compound and arbitrary varying wiretap channels. However, a general single-letter formula for the secrecy capacity of the compound and arbitrary varying wiretap channels remains unknown, where only multi-letter formula has been established. The usage of multi-letter descriptions to establish secrecy capacity has raised many doubts in the information theory community, because they are not efficiently computable. Yet, it has been shown that multi-letter descriptions can be used to prove some important characteristics of the secrecy capacity like continuity and super-activation. Further, there are some speculations that multi-letter formulas might be able to provide other useful insights.

Consider a classical-quantum channel (CQC), where the channel input is a classical random variable, while the channel output is a quantum state. It was shown for some classes of CQC that despite a single-letter characterization of the capacity in terms of mutual information is unknown, a multi-letter description is possible. It was Holevo, who suggested to tackle the capacity characterization problem using a different information quantity other than the mutual information. He introduced the Holevo quantity in [22] and used it to establish a single-letter description of the CQC capacity in [23]. This result raises two questions: The first is whether other information quantities rather than the mutual information is capable of establishing a single-letter description of the secrecy capacity of the general compound and arbitrary varying wiretap channels. The second question is can the existence of a multi-letter description for the capacity of some channels, when a single-letter is not known be an indicator of using other information quantities instead of the mutual information. More discussion regarding this point can be found in [12].

Another important question that we need to address is related to the relation between the compound and the arbitrary varying wiretap channels. Consider an AVWC $\mathfrak{W}_{\text{AV}} = \{\text{W}_{s^n}, \text{V}_{s^n} : s^n \in \mathcal{S}^n\}$ with a CR-assisted secrecy capacity $\text{C}(\mathfrak{W}_{\text{AV}})$ and a corresponding compound wiretap channel $\mathfrak{W}_{\text{C}} = \{\text{W}_q, \text{V}_q : q \in \mathcal{P}(\mathcal{S})\}$ with a secrecy capacity $\text{C}(\mathfrak{W}_{\text{C}})$. It is known that if $\mathfrak{W}_{\text{AV}}$ is strongly degraded, which implies that $\mathfrak{W}_{\text{C}}$ is a degraded compound wiretap channel, then $\text{C}(\mathfrak{W}_{\text{AV}}) = \text{C}(\mathfrak{W}_{\text{C}})$. However, the relation between the two capacities is not known in general. This relation is very important because if one can prove that there exists an AVWC where $\text{C}(\mathfrak{W}_{\text{AV}}) < \text{C}(\mathfrak{W}_{\text{C}})$, this will imply that $\text{C}(\mathfrak{W}_{\text{AV}})$ can not be expressed as a single-letter expression using mutual information. This will support the previous speculation about the role played by multi-letter descriptions.

It was shown in [34], that although super-activation is possible for the unassisted secrecy capacity of two orthogonal AVWCs, it is not possible for two orthogonal AVCs. This result raises some questions about whether some of the well established concepts in the non-secrecy domain are valid for their corresponding secrecy scenarios or not. For example in [34], it was shown that the CR-assisted capacity of two orthogonal AVCs is additive, however we do not know if this additivity also holds for the CR-assisted secrecy capacity of two orthogonal AVWCs or not.

Investigating the validity of the established results for non-secure communication to secrecy scenarios is not restricted to the additivity and super-activation of orthogonal channels. Another example where this phenomena occurs is given as follows: Consider a compound channel $\mathfrak{W}_{\text{C}}$ with a channel state set $\mathcal{S}$. It has been shown that if $C(\mathfrak{W}_C) = 0$, then there must be a state $s \in \mathcal{S}$ where the channel is useless, i.e., $C(W_s) = 0$. It was also shown that the reverse result holds, that If for all states $s \in \mathcal{S}$ $C(W_s) > 0$, then the capacity of the compound channel $\mathfrak{W}_C$ is also greater than zero. This result does not hold for the compound wiretap channel $\mathfrak{W}_{CW}$. It was shown in [7], that there exists some compound wiretap channels, where although for every state $s \in \mathcal{S}$, the secrecy capacity $C_S(W_s, V_s) > 0$, the secrecy capacity of the whole compound wiretap channel $C_S(\mathfrak{W}_{CW})$ can actually be zero.

1.3 Worst Codes for the BSC

The result presented below is for binary symmetric channels with transmission matrix $W = \begin{pmatrix} 1-\epsilon & \epsilon \\ \epsilon & 1-\epsilon \end{pmatrix}$, $0 \leq \epsilon \leq \frac{1}{2}$, its extension to general DMC's seems to be an interesting mathematical problem.

Coding theory has been concerned with the problem to find (n, R)-codes, i.e., codes of block length n and rate R, for which to average error probability is small. Nobody found for arbitrary n and positive rate codes which are optimal in the sense that error probability assumes its minimum. This is a very hard combinatorial extremal problem and has led to numerous investigations in probabilistic and algebraic coding theory.

We study here the dual problem: find (n, R) codes with *distinct* code words for which the decoding error probability is maximal. More generally we also permit an arbitrary message statistic rather than just the equidistribution.

The problem then takes the following form: Given a probability distribution $P = (P_1, \dots, P_{2^n})$ on 2^n elements, find a bijective map $U : \{1, \dots, 2^n\} \to \{0, 1\}^n$ such that

$$\lambda_c(P) = \max_{\mathcal{D}} \sum_{i=1}^{2n} P_i \cdot W^n(D_i|u_i) \tag{1}$$

is minimal. Here $u_i = U(i)$; $W^n(\cdot|\cdot)$ denotes the n-fold product of the transmission probability function of the BSC, and $\mathcal{D} = \{D_1, \dots, D_{2^n}\}$ is a decoding rule.

We describe now an explicit solution to the problem. W.l.o.g. we can assume that

$$P_1 \geq P_2 \geq \cdots \geq P_{2^n}.$$

Let us order the vectors v in $\{0, 1\}^n$ primarily according to the number of components with value 0 and secondarily lexicographically, where 1 precedes 0. Thus

$$v_1 \geq v_2 \geq \cdots \geq v_{n+1} \geq v_{n+2} \geq \cdots \geq v_{\binom{n}{2}+n+1} \geq \cdots \geq v_{2^n}.$$

Theorem 23 ([3]) *Let $P = (P_1, \dots, P_{2^n})$ be a probability distribution on the messages, $P_i \geq P_{i+1}$, then the encoding $U(i) = v_i$ for $i = 1, \dots, 2^n$ minimizes the probability of correct decoding $\lambda_c(P)$ (as defined in (1)).*

For (n, R) codes one gets the solution to the above problem by choosing $F_i = \frac{1}{N}$ for $i = 1, \dots, N = [e^{nR}]$.

For the proof of Theorem 23 we need an extension of a result of Harper ([21]). Let us denote by $S_r(x^n)$ the Hamming sphere in $\{0, 1\}^n$ with center $x^n \in \{0, 1\}^n$ and radius r. Then we have:

Theorem 24 (General isoperimetry theorem of Harper and Ahlswede) *Let $\{r_i\}_{i=1}^N$ be a decreasing sequence of integers, then for any distinct $x_1^n, \dots, x_N^n \in \{0, 1\}^n$:*

$$\left| \bigcup_{i=1}^{N} S_{r_i}(x_i^n) \right| \geq \left| \bigcup_{i=1}^{N} S_{r_i}(v_i) \right| .$$

Harper proved this in the case $r_i = r, i = 1, \dots, N$. We show here that the given general case easily follows from his result.

Proof Fix any $j \in \{0, \dots, N-1\}$. Then for any $i \in \{1, \dots, N-j\}$ $i \leq N-j$ holds and by the monotonicity of the radii we have for those i

$$r_i \geq r_{N-j} \text{ and } |S_{r_i}(x_i^n)| \geq |S_{r_{s-i}}(x_i^n)|.$$

Hence,

$$\left| \bigcup_{i=1}^{N} S_{r_i}(x_i^n) \right| \geq \max_{j \in \{0, \dots, N-1\}} \left| \bigcup_{i=1}^{N-j} S_{r_{N-j}}(x_i^n) \right| .$$

By Harper's theorem the expression on the right-hand side is minimal if $x_i^n = c_i$ for $i = 1, \dots, N$. Furthermore it can easily be verified that $\bigcup_{i=1}^{N-j} S_{r_{N-j}}(v_i)$ equals $\{v, \dots, v_{t_j}\}$ for a suitable t_j.

Therefore, there is a $j' \in \{0, \dots, N-1\}$ such that $\bigcup_{i=1}^{N-j'} S_{r_{N-j'}}(v_i)$ contains all the sets $\bigcup_{i=2}^{N_j} S_{r_{N-j}}(v_i)$, $j \in \{0, \dots, N-1\}$. We conclude that

$$\left| \bigcup_{i=1}^{N} S_{r_i}(v_i) \right| = \left| \bigcup_{j=0}^{N-j} \left(\bigcup_{i=1}^{N-j} S_{r_{N-j}}(v_i) \right) \right| = \max_{j \in \{0, \dots, N-1\}} \left| \bigcup_{i=1}^{N-j} S_{r_{N-j}}(v_i) \right|$$

which proves the theorem.

Proof of Theorem 23. For a map $U : \{1, \dots, 2^n\} \to \{0, 1\}^n \big(U(i) := u_i\big)$, a decoding rule is optimal iff

$$D_i \subset \left\{ y^n \in \{0, 1\}^n | P_i \beta^{d(y^n, u_i)} \geq P_j \beta^{d(y^n, u_j)} \text{ for all } j \right\} \tag{2}$$

and $\bigcup_{i=1}^{2^n} D_i = \{0,1\}^n$, where $\beta = \frac{\epsilon}{1-\epsilon} \le 1$ and where $d(\cdot)$ denotes the Hamming distance. Note that in (2) we have formulated just the concept of maximum likelihood decoding for the special case of the BSC. It should be clear intuitively that best decoding sets for the code word u_i are "like spheres around u_i", the diameter of which depends on P_i. We make this heuristic precise and apply the general isoperimetry theorem. For $y^n \in \{0,1\}^n$ define

$$m(y^n, U) = \max_i P_i \cdot \beta^{d(y^n, u_i)}.$$

Then our problem is equivalent to the problem of minimizing

$$\sum_{y^n \in \{0,1\}^n} m(y^n, U),$$

as a function of U. Order now the elements of $\{P_i \beta^j | 1 \le i \le N; 0 \le j \le n\}$ in increasing order and denote them by $\alpha_1, \dots, \alpha_{(n+1)N}$, $N = 2^n$.

We can write

$$\sum_{y^n \in \{1,0\}^n} m(y^n, U) = \sum_{\ell=1}^{(n+1)N} \alpha_\ell |\delta_\ell(U)|,$$

where $\delta_\ell(U) = \{y^n | m(y^n, U) = \alpha_\ell\}$. Further, set $\delta_\ell^*(U) = \delta_\ell(U) \cup \delta_{\ell+1}(U) \cup \dots \cup \delta_{n \cdot N}(U)$. Then with $\alpha_0 := 0$

$$\sum_{\ell=1}^{(n+1)N} \alpha_\ell |\delta_\ell(U)| = \sum_{\ell=1}^{(n+1)N} (\alpha_\ell - \alpha_{\ell-1}) |\delta_\ell^*(U)|.$$

Since $\alpha_\ell \ge 0$ and $\alpha_\ell - \alpha_{\ell-1} \ge 0$ for $\ell = 2, \dots, (n+1)N$ we are done if the *same* U minimizes all $\delta_\ell^*(U)$; $\ell = 1, \dots, (n+1)N$. We write now $\delta_\ell^*(U)$ as a union of spheres. Define radii

$$r_{\ell i} = \begin{cases} -1 & \text{if } P_i < \alpha_\ell \\ \max\{t | t \text{ integer with } P_i \beta^t \ge \alpha_\ell\} & \text{else} \end{cases}$$

and observe that with the convention $S_{-1}(x^n) = \emptyset$

$$\delta_\ell^*(U) = \bigcup_{i=1}^{N} S_{r_{\ell i}}(u_i).$$

Since $r_{\ell 1} \ge r_{\ell 2} \ge \dots \ge r_{\ell N}$ for $\ell = 1, \dots, nN$ the general isoperimetry theorem gives the result.

1.4 Shannon's Information-Theoretic Approach to Cryptosystems

Definition 12 A secrecy system is called **perfect** if the random variables X and Y for the plain-text and the cryptogram are independent.

This is a quite natural definition, since X and Y are independent exactly if $H(X|Y) = H(X)$, i.e., the knowledge of the cryptogram does not reduce the uncertainty about the plain-text, or, in other words the cryptogram does not yield any information about the plain-text (recall that X and Y are independent exactly if $I(X \wedge Y) =)$).

As an example, Shannon demonstrated that the one-time pad is perfectly secret. Recall that in a one-time pad, there are M possible keys defined by $c_k(m) = m + k(\text{mod } M)$ for $k, m = 1, \dots, M$, each occurring with equal probability $\frac{1}{M}$. To see this, observe that for given $X = m$ all possible cryptograms $m' \in \mathcal{M}$ can occur with the same probability $\frac{1}{M}$ and hence

$$\begin{aligned} H(Y|X) &= \sum_{m \in \mathcal{M}} P(X=m) \cdot H(Y|X=x) = \sum_{m \in \mathcal{M}} P(X=m) \cdot \log M \\ &= \log M = H(Y), \end{aligned}$$

which means that X and Y are independent.

The disadvantage of this one-time pad is that the amount of secret key (in bits) is as large as the number of plain-text bits which have to be encrypted.

However, when we require perfect secrecy, this cannot be avoided, as the following theorem shows.

Theorem 25 *In a perfect secrecy system*

$$H(Z) \geq H(X).$$

Proof By elementary properties of the entropy (cf. Chapter on Data Compression) for a perfect secrecy system

$$\begin{aligned} H(X) &= H(X|Y) \leq H(X, Z|Y) = H(Z|Y) + H(X|Y, Z) \\ &= H(Z|Y) \leq H(Z). \end{aligned}$$

Remark 2 Central in the previous proof is the easy but useful observation that in a secrecy system always

$$H(X|Y, Z) = 0.$$

This is clear, since in knowledge of the cryptogram Y and the secret key Z the cryptanalyst can, of course, reconstruct the plain-text.

Definition 13 A secrecy system is **robustly perfect** (for the set $\mathcal{M} = \{1, \ldots, M\}$) if it is perfect for all possible sources $(\mathcal{M}, P)$, i.e., for an arbitrary choice of the probability distribution P on $\mathcal{M}$.

Since a robustly perfect secrecy system is perfect particularly for the source $(\mathcal{M}, P)$, where P is the uniform distribution on $\mathcal{M}$, by Theorem 25 it follows immediately that there are at least as many keys as possible plain-texts, i.e., $K \geq M$.

Observe that the one-time pad is robustly perfect with a minimal number of keys, since here $K = M$.

Definition 14 The **key-equivocation** $H(Z|Y)$ is the remaining uncertainty about the key when the cryptogram is known. Accordingly, the **message equivocation** $H(X|Y)$ is defined as the remaining uncertainly about the plain-text, when the cryptogram is known.

Remark 3 From the proof of Theorem 25 it is immediate that for all secrecy systems

$$H(Z|Y) \geq H(X|Y),$$

i.e., the key equivocation is always larger then the message equivocation.

In the following, we assume that the encoder uses the same key to encipher n messages represented by the RV $X^n = (X_1, \ldots, X_n)$. For the sequence of cryptograms we use the RV $Y^n = (Y_1, \ldots, Y_n)$.

Definition 15 The **unicity distance**

$$u \triangleq \min_n \{H(Z|Y^n) = 0\}$$

is the smallest n such that there is exactly one key from which the sequence of cryptograms $Y_1, \ldots, Y_n$ could have arisen.

Let us now assume that

(a) a natural language is such that $H(X^n) \tilde{=} n \cdot H(X)$,
(b) the cryptosystem is such that all sequences of length n are equally likely as cryptogram, i.e., $H(Y^n) = n \cdot \log M$.

Under these two assumptions we can express the unicity distance in terms of the redundancy $R = 1 - \frac{H(X)}{\log M}$ of the language, namely in this case

$$\begin{aligned} H(Z|Y^n) &= H(Z, Y^n) - H(Y^n) \\ &= H(X^n, Z, Y^n) - H(Y^n) \\ &= H(X^n, Z) - H(Y^n) \\ &= H(X^n) + H(Z) - H(Y^n) \end{aligned}$$

(since X^n and Z are independent).

Hence for the unicity distance

$$O = H(Z|Y^n) = H(X^n) + H(Z) - H(Y^n)$$
$$= u \cdot H(X) + H(Z) - u \cdot \log M$$

from which follows that

$$u = \frac{H(Z)}{\log M - H(X)}$$

and if all the keys are equiprobable

$$u = \frac{\log K}{\log M - H(X)} = \frac{\log K}{R \cdot \log M}.$$

For a substitution cipher used to encrypt a text in English language this would yield a unicity distance $u = \dfrac{\log(26!)}{\log 26 - 2} = \dfrac{88.4}{2.7} \approx 32.$

This result is compatible with Shannon's empirical observations. He conjectured that in this case the unicity distance is between 20 and 30.

Remark 4 (1) The unicity distance is the amount of ciphertext needed (in theory) to break the cipher (in case the cryptanalyst doesn't have any information about the plain-text).

(2) For the Date Encryption Standard (DES) it can be shown that the unicity distance is about 70 bits.

1.5 Homophonic Coding

Definition 16 If it is possible to use several codewords for one message, these codewords are called **homophones**. A coding procedure in which homophones are possible is called **homophinic coding**.

We shall use homophonic coding in order to realize a binary symmetric source (BSS), i.e., a source which produces each bits with equal probability $P(0) = P(1) = \frac{1}{2}$, from an arbitrary binary memoryless source (where 0 and 1 are not necessarily equiprobable). Homophonic coding has applications in Cryptology, since a binary symmetric source has useful properties, e.g., the unicity distance is ∞ under certain additional assumptions.

Consider first the following example: We are given a binary memoryless source, where each bit occurs with probability $P(0) = \frac{1}{4}$, $P(1) = \frac{3}{4}$. To achieve a BSS we use the following homophonic coding procedure. 0 will always be represented by ∞ and 1 will be represented by either 01, 10, or 11, where each homophone will be assumed with probability $\frac{1}{3}$. Obviously, the source obtained this way is symmetric, since each of the possible 2-blocks 00, 01, 10, 11 takes place with probability $\frac{1}{4}$ and hence $P(0) = P(1) = \frac{1}{2}$ in the source obtained this way.

The homophonic coding in the above example is a fixed-length encoding all possible codewords have (fixed) length 2. Günther introduced variable-length homophonic coding, e.g., in our example we could use the coding $0 \to 00$ and $1 \to 1$ (with probability $\frac{2}{3}$) or $1 \to 01$ (with probability $\frac{1}{3}$). It is an easy exercise to verify that this encoding procedure also yields a binary symmetric source.

We can represent the encoding procedure by a so called *homophonic channel* with input alphabet U and output alphabet V (in our last example $U = \{0, 1\}$, $V = \{00, 1, 01\}$) where the transition probabilities are defined according to the encoding procedure here $P(V = 00|U = 0) = 1$, $P(V = 1|U = 1) = \frac{2}{3}$, $P(V = 01|U = 1) = \frac{1}{3}$ and $P(V = v|U = u) = 0$, else.

Theorem 26 *There exists a binary prefix-free encoding of V such that the output sequence is a BSS sequence, exactly if all the probabilities $P(V = v)$ are negative-integer powers of 2. Moreover, when such a coding exists, the codeword for v has length $-\log P(V = v)$.*

Proof Let L denote the RV for the codeword length, hence

$\mathbb{E}L = \sum_{v \in V} P(V = v) \cdot \ell(v)$ (where $\ell(v)$ denotes the length of the encoding of v) is the expected codeword length.

The output sequence is a BSS sequence, exactly if the redundancy $r \triangleq \mathbb{E}L - H(V) = 0$, so

$$\begin{aligned} r &= \mathbb{E}L - H(V) \\ &= \sum_v P(V = v)\ell(v) + \sum_v P(V = v) \cdot \log P(V = v) \\ &= \sum_v P(v = v) \log \frac{P(V = v)}{2^{-\ell(v)}} \\ &= D(P\|Q) \end{aligned}$$

where P is the probability distribution on V and Q is the probability distribution defined by $Q(v) = 2^{-\ell(v)}$.

Now $D(P\|Q) \geq 0$ with equality exactly if $P = Q$. Hence $P(V = v) = 2^{-\ell(v)}$ and the theorem is proved.

Theorem 27 *For the homophonic coding described above*

$$H(U) \leq H(V) < H(U) + 2.$$

Proof Obviously $H(V) \geq H(U)$, since V determines U, or in other words $H(U|V) = 0$. This last identity is also useful in order to prove the inequality on the right-hand side. Observe that

$$H(V) = H(V) + H(U|V) = H(U) + H(V|U).$$

We are done if we can show that $H(V|U) < 2$. From Theorem 26 we can conclude that $P(U = u) = \sum i \in I2^{-u_i}$ is a sum of negative-integer powers of 2. Hence

$$H(V|U = u) = -\sum_{i \in I} \frac{2^{-u_i}}{P(U = u)} \cdot \log \frac{2^{-u_i}}{P(U = u)}.$$

Since $P(U = u) < 1$, we have

$$H(V|U = u) < -\sum_{i \in I} 2^{-u_i} \log 2^{-u_i} = \sum_{i \in I} u_i \cdot 2^{-u_i} < \sum_{n=1}^{\infty} n2^{-n} = 2$$

(expected value of the geometric distribution).

So $H(V|U) = \sum_{u \in U} P(U = u) \cdot H(V|U = u) < 2$.

The following example demonstrate that the upper bound in Theorem 27 cannot be improved. Let $P(u_1) = 1 - \frac{1}{2^m} = \frac{1}{2} + \frac{1}{4} + \cdots + \frac{1}{2^m}$ and $P(u_2) = \frac{1}{2^m}$ be the probability distribution of the RV U on a 2-elementary source.

Now $H(V) = 2 \cdot \left(1 - \frac{1}{2^m}\right)$. So for intending to infinity we have

$$H(V) \underset{m \to \infty}{\to} \longrightarrow 2 \text{ and } H(U) \underset{m \to \infty}{\to} \longrightarrow 0.$$

1.6 Spurious Decipherments

We now assume that the message source has the AEP, i.e., the messages are divided into two groups $\mathcal{M}_1$ and $\mathcal{M}_2$, one group $\mathcal{M}_1$ of high and fairly uniform probability ($\sim 2^{-H(P)}$, when $(\mathcal{M}, P)$ is the message source), the second group $\mathcal{M}_2$ of negligibly small total probability. We assign probability 0 to all messages in $\mathcal{M}_2$ and probability exactly $2^{-H(P)}$ to the messages in the first group $\mathcal{M}_1$.

A random cipher $\mathbb{C}$ is considered consisting of K keys $c_j : \mathcal{M} \to \mathcal{M}$, $j = 1, \ldots, K$, each occurring with equal probability. For each cryptogram $m' \in \mathcal{M}$ we count the number $a(m')$ of possible ways in which this cryptogram may arise, hence

$$a(m') = |(m, c_j); m \in \mathcal{M}_1, j \in \{1, \ldots, K\} : c_j(m) = m'\}|. \tag{1.6.1}$$

Homophonic coding now is not allowed, so, for each key there is only one cryptogram assigned to each message; also, the cryptogram m' can occur at most once when a fixed key c_j is used.

Definition 17 If $a(m') > 1$, the cryptogram m' is said to have a *spurious key decipherment*, i.e., m' can occur under more than one key as a cryptogram.

We are interested in the expected number of spurious key decipherments, which will be denoted by $\overline{s}$.

Theorem 28

$$\overline{s} \geq \frac{2^{H(P)} \cdot K}{|\mathcal{M}|} - 1. \tag{1.6.2}$$

Proof Obviously the expected number of spurious key decipherments is

$$\overline{s} = \sum_{m' \in \mathcal{M}} s(m') \cdot P^*(m'), \tag{1.6.3}$$

where

$$s(m') = \max\{a(m') - 1, 0\} \tag{1.6.4}$$

and P^* is the probability distribution imposed on the set $\mathcal{M}$ of cryptograms by the random cipher, i.e.,

$$P^*(m') = \frac{a(m')}{2^{H(P)} \cdot K} \tag{1.6.5}$$

($2^{H(P)} \cdot K$ is the total number of cryptograms and $a(m')$ the frequency of m' as a cryptogram).

Now observe that with (1.6.3) and (1.6.5)

$$\begin{aligned} \overline{s} &= \sum_{m' \in \mathcal{M}} a(m') \cdot P^*(m') - 1 \\ &= \sum_{m' \in \mathcal{M}} \frac{a(m')^2}{2^{H(P)} \cdot K} - 1 \end{aligned} \tag{1.6.6}$$

Since no cryptogram can occur twice under the same key, and hence P^* is a probability distribution

$$\sum_{m' \in \mathcal{M}} a(m') = 2^{H(P)} \cdot K$$

and hence

$$\sum_{m' \in \mathcal{M}} a(m')^2 \geq \frac{(2^{H(P)} \cdot K)^2}{|\mathcal{M}|} \tag{1.6.7}$$

(since for any $(x_1, \ldots, x_n)$ with $\sum_{i=1}^n x_i = a$ it is $\sum_{i=1}^n x_i^2 \geq \frac{a^2}{n}$). Combining (1.6.6) and (1.6.7) we obtain

$$\overline{s} \geq \sum_{m' \in \mathcal{M}} \frac{(2^{H(P)} \cdot K)^2}{|\mathcal{M}| \cdot 2^{H(P)} \cdot K} - 1$$

$$= \sum_{m' \in \mathcal{M}} \frac{2^{H(P)} \cdot K}{|\mathcal{M}|} - 1,$$

which is the desired result (1.6.2).

Remark 5 With $\mathcal{M} = \{1, \ldots, M\}$ and hence $|\mathcal{M}| = M$ (1.6.2) is equivalent to

$$\overline{s} \geq 2^{\log K + H(P) - \log M} - 1$$

and the exponent $\log K + H(P) - \log M$ is exactly the term occurring in Shannon's lower bound for the conditional entropy

$$H(X|Y) \geq \log K + H(X) - \log M.$$

Again, the expected number of spurious decipherments is maximized if $H(P) = H(X) = \log M$, i.e., the messages are uniformly distributed.

1.7 Authentication

In Shannon's model of a secrecy system, the enemy (cryptanalyst) had the possibility to intercept a cryptogram and he could try to decipher it. Simmons introduced the model of an *authenticity attack*. Here the enemy is much more powerful. He has the possibility to replace the cryptogram by a fraudulent cryptogram, which then will be sent to the decrypter.

The purpose of the key, in this model, is to guarantee the authenticity of a message, i.e., encrypt in such a way that the decrypter recognizes that a fraudulent cryptogram cannot have been sent by the encrypter and must hence have been replaced by the enemy. As in Shannon's model, encrypter and decrypter communicate over a secure channel in order to agree upon a secret key $c_k : \mathcal{M} \to \mathcal{M}'$.

When does the decrypter realize that the cryptogram Y' he receives must have been replaced by the enemy? This is clearly the case, when Y' is not a valid cryptogram under the key c_k, i.e., Y' is not contained in the range of c_k.

There are two basic options for the enemy to replace the correct cryptogram Y by the fraudulent cryptogram Y' depending on the time of the replacement.

Definition 18 In an **impersonation attack** the enemy sends the fraudulent cryptogram Y' before he intercepts the correct cryptogram Y. In a **substitution attack** the enemy sends the fraudulent cryptogram $Y^{\mathcal{P}}$ after having intercepted the correct cryptogram Y.

So in a substitution attack, the enemy always knows the correct cryptogram Y and will, of course, replace it with a $Y' \neq Y$. In an impersonation attack, the enemy has no information about the correct cryptogram, when he sends Y'. So it may happen that $Y^{\mathcal{P}}$ and Y are the same.

Definition 19 We denote by P_I and P_S, respectively, the probability that the fraudulent cryptogram $Y^{\mathcal{P}}$ is valid under the key Z in the best possible impersonation (P_I) or substitution (P_S) attack. The probability of **deception** P_D is defined as $P_D = \max\{P_I, P_S\}$.

Let us slightly modify the notation we used so far. We denote by

$(\mathcal{M}, P_X)$	– the message source,
$\mathcal{M} = \{1, \dots, M\}$	– the set of possible messages (plain-texts)
P_X	– a probability distribution on $\mathcal{M}$
$\mathbb{C} = (c_1, \dots, c_k)$	– the key space
$c_j : \mathcal{M} \to \mathcal{M}'$	– the possible keys
P_Z	– a probability distribution on $\mathbb{C}$
$\mathcal{M}'$	– the set of all possible cryptograms when key c_j is used
$\mathcal{M}' = \bigcup_{z=1}^{k} \mathcal{M}'_z$	– the set of all possible cryptograms
X, Y, Y', Z	– RVs for plain-text, correct cryptogram, fraudulent cryptogram, and key
P_Y	– the probability distribution on $\mathcal{M}'$

Observe that in Shannon's model we usually assumed that $\mathcal{M}_j = \mathcal{M}$, i.e., the space for plain-text and cryptograms were identical.

In Simmons' authenticity model this assumption is nonsense, since in this case all cryptograms will be valid under each key and the enemy, hence, will always replace the correct cryptogram Y by a valid cryptogram Y'.

Theorem 29

$$P_I \geq \frac{M}{|\mathcal{M}'|}$$

with equality possible, only if the cipher is not randomized.

Proof Since the keys c_j cannot map two different plain-texts onto the same cryptogram, obviously $|\mathcal{M}'_j| \geq |\mathcal{M}| = M$ for all $z = 1, \dots, K$. Further $|\mathcal{M}'_j|$ can only be larger than M if for some plain-text $m \in \mathcal{M}$ there are several possible cryptograms, e.g., in homophonic coding. This is only possible if the cipher is randomized.

We can now lower bound

$$P_I \geq \sum_{z=1}^{K} P_Z(z) \cdot \frac{|\mathcal{M}'_z|}{\mathcal{M}'|} \geq \sum_{z=1}^{K} P_Z(z) \cdot \frac{M}{|\mathcal{M}'|} = \frac{M}{|\mathcal{M}'|}.$$

Remark 6 (1) By the previous theorem, $P_I = 0$ is impossible. This is somehow plausible, since there is always the possibility for the enemy to transmit the correct cryptogram.

(2) $P_I = 1$ can be achieved if $M = |\mathcal{M}'|$, i.e., if every cryptogram $y \in \mathcal{M}'$ is in the range $\mathcal{M}_z$ of every key c_z.

(3) The one-time pad had been proved to be perfectly secret. Obviously, it provides no authenticity at all, since the enemy can choose any cryptogram, all of them are valid. This example shows that perfect secrecy with no authenticity is possible.

The following lower bound for P_I is due to Simmons. Observe that P_I can be made small only if the cryptogram yields much information about the secret key.

Theorem 30

$$P_I \geq 2^{-I(X \wedge Y)}. \tag{1.7.1}$$

Proof The authentication function ϕ is defined by

$$\phi(y, z) = \begin{cases} 1 & \text{if } Y \text{ is a valid cryptogram for key } Z \\ 0 & \text{else.} \end{cases}$$

So $Pr(y \text{ valid}) = \sum_z \phi(y, z) \cdot Q(z)$.

The best impersonation attack for the enemy is of course to choose a cryptogram y' with maximum likelihood of validity. Hence

$$P_I = \max_{y'} Pr(y' \text{ valid}).$$

From this follows immediately

$$P_I \geq \sum_{y} P_y(y) \cdot Pr(y \text{ valid}) = \sum_{y,z} P_y(y) \cdot Q(z) \cdot \phi(y, z)$$

with equality only if $Pr(y \text{ valid})$ is constant for all y.

The key observation now is that it is possible to write the right-hand side of this last inequality as an expected value, namely

$$P_I \geq \mathbb{E}\frac{P)Y(y) \cdot P_Z(z)}{P_{YZ}(y, z)}.$$

This is possible, since the pair (y, z) has joint probability $P_{YZ}(y, z) \neq 0$ exactly if $\phi(y, z) = 1$ and $P_Z(z) \neq 0$. The last inequality is equivalent to

$$\log P_I \geq \log \mathbb{E} \frac{P_Y(y) \cdot P_Z(z)}{P_{YZ}(y, z)}.$$

From Jensen's Inequality

$$\log \mathbb{E} \frac{P_Y(y) \cdot P_Z(z)}{P_{YZ}(y, z)} \geq \mathbb{E} \log \frac{P_Y(y) \cdot P_Z(z)}{P_{YZ}(y, z)} = -I(Y \wedge Z)$$

from which Theorem 30 is immediate.

The necessary and sufficient conditions for equality in (1.7.1) are

(i) $\frac{P_Y(y) \cdot P_Z(z)}{P_{YZ}(y, z)}$ is constant for all pairs (y, z) with $P_{YZ}(y, z) \neq 0$,
(ii) $P(y \text{ valid})$ is constant for all Y.

Condition (i) is necessary and sufficient for equality in Jensens Inequality, (ii) was mentioned before.

Remark 7 Observe that in the proof of Theorem 30 the probability distribution P_X for the source had no influence. However, generally the mutual information $I(Y \wedge Z)$ depends on P_X.

Hence we proved indeed

$$P_I \geq 2^{-\inf I(Y \wedge Z)}$$

where the infimum is taken over all probability distributions P_X on $\mathcal{M}$ that leave the authentification function $\phi(y, z)$ unchanged.

Definition 20 A secrecy system has **perfect authenticity**, if the probability of deception

$$P_D = 2^{-I(Y \wedge Z)}.$$

This definition of perfect authenticity is due to Simmons. It is somehow problematic. We only considered the probability of "successful" impersonation P_I and found that this is lower bounded by $2^{-I(Y \wedge Z)}$. A system now is called perfectly authentically if this lower bound is assumed. Observe that every system with $I(Y \wedge Z) = 0$ provides trivially perfect authenticity. Further observe that we didn't investigate the probability for successful substitution P_S up to now.

In the rest of this paragraph we shall demonstrate that the concepts of perfect secrecy and perfect authenticity are generally not comparable. We already saw that the one-time pad is a perfectly secret system with no authenticity. We shall now give an example of a system with perfect authenticity and no secrecy at all.

Example 1 For the plain-text we have only two possibilities, hence $\mathcal{M} = \{0, 1\}$. The key space consists of all possible binary sequences of even length T, say, each sequence occurring with equal probability $\frac{1}{2^T}$. The cryptogram is now obtained by appending the bits of the key

$Z = \left(Z_1, \ldots, Z_{\frac{T}{2}}, Z_{\frac{T}{2+1}}, \ldots, Z_T\right)$ to the message, the first $\frac{T}{2}$ bits of the message in 0, and the last $\frac{T}{2}$ bits if the message is 1. So if $X = 0$, then $Y = \left(0, Z_1, \ldots, Z_{\frac{T}{2}}\right)$, if $X = 1$, then $Y = \left(1, Z_{\frac{T}{2+1}}, \ldots, Z_t\right)$.

Obviously the system is not secret, since the first bit is the plain-text.

However, the system has perfect authenticity. To see this, observe that $P_I = P_S = 2^{-\frac{T}{2}}$ and hence $P_D = 2^{-\frac{T}{2}}$.

On the other hand $I(Y \wedge Z) = H(Z) - H(Z|Y) = T - \frac{T}{2} = \frac{T}{2} = -\log P_D$.

In the last example the key was used rather as a signature than as encrypting function.

The next example is a secrecy system with perfect authenticity and perfect secrecy. It is quite similar to the previous one. However, we now use an additional key bit to manipulate the plain-text.

Example 2 Again, we have two possible messages, each occurring with probability $\frac{1}{2}$. The key space now consists of all possible binary sequences of odd length T, so each key $Z = (Z_1, \ldots, Z_T)$. The first bit of the key is now added to the message bit, the other bit serve as a signature as in Example 1.

If $X = 0$ then $Y = (X + Z_1, Z_2, \ldots, Z_{(T+1)/2})$,
if $X = 1$ then $Y = (X + Z_1, Z_{T+1)/2+1}, \ldots, Z_T)$.

The system is perfectly secret, since $H(X|Y) = H(X)$. As in the previous example $P_D = P_I = P_S = 2^{-\frac{(T-1)}{2}}$ and $I(Y \wedge Z) = H(Z) - H(Z|Y) = T - \frac{T+1}{2} = \frac{T-1}{2}$ and hence we have perfect authenticity.

References

1. R. Ahlswede, A note on the existence of the weak capacity for channels with arbitrarily varying channel probability functions and its relation to Shannon's zero error capacity. Ann. Math. Stat. **41**(3), 1027–1033 (1970)
2. R. Ahlswede, Elimination of correlation in random codes for arbitrarily varying channels. Zeitschrift für Wahrscheinlichkeitstheorie und verwandte Gebiete **44**(2), 159–175 (1978)
3. R. Ahlswede, Remarks on Shannon's secrecy systems. Probl. Control Inf. Theory **11**(4), 301–318 (1982)
4. R. Ahlswede, G. Dueck, Bad codes are good ciphers. Probl. Control Inf. Theory **11**(5), 337–351 (1982)
5. N. Alon, The Shannon capacity of a union. Combinatorica **18**(3), 301–310 (1998)
6. G. Bagherikaram, A.S. Motahari, A.K. Khandani, Secrecy rate region of the broadcast channel with an eavesdropper, in *Proceedings of the Forty-Sixth Annual Allerton Conference* (2009), pp. 834–841

7. I. Bjelaković, H. Boche, J. Sommerfeld, Secrecy results for compound wiretap channels. Probl. Inf. Transm. **49**(1), 73–98 (2013)
8. I. Bjelaković, H. Boche, J. Sommerfeld, Capacity results for arbitrarily varying wiretap channels, *Information Theory, Combinatorics, and Search Theory* (Springer, New York, 2013), pp. 123–144
9. D. Blackwell, L. Breiman, A.J. Thomasian, The capacity of a class of channels. Ann. Math. Stat. **30**(4), 1229–1241 (1959)
10. M. Bloch, J. Barros, *Physical-Layer Security: From Information Theory to Security Engineering* (Cambridge University Press, Cambridge, 2011)
11. M.R. Bloch, J.N. Laneman, Strong secrecy from channel resolvability. IEEE Trans. Inf. Theory **59**(12), 8077–8098 (2013)
12. H. Boche, N. Cai, J. Nötzel, *The classical-quantum channel with random state parameters known to the sender, CoRR* (2015). arXiv:abs/1506.06479
13. H. Boche, R.F. Schaefer, Capacity results and super-activation for wiretap channels with active wiretappers. IEEE Trans. Inf. Forensics Secur. **8**(9), 1482–1496 (2013)
14. H. Boche, R.F. Schaefer, H.V. Poor, On the continuity of the secrecy capacity of compound and arbitrarily varying wiretap channels. IEEE Trans. Inf. Forensics Secur. **10**(12), 2531–2546 (2015)
15. Y.-K. Chia, A. El Gamal, Three-receiver broadcast channels with common and confidential messages. IEEE Trans. Inf. Theory **58**(5), 2748–2765 (2012)
16. I. Csiszár, Almost independence and secrecy capacity. Probl. Peredachi Inf. **32**(1), 48–57 (1996)
17. I. Csiszár, J. Körner, Broadcast channels with confidential messages. IEEE Trans. Inf. Theory **24**(3), 339–348 (1978)
18. I. Csiszár, J. Körner, *Information Theory: Coding Theorems for Discrete Memoryless Channels* (Academic, New York, 1981)
19. A. El Gamal, Y.-H. Kim, *Network Information Theory* (Cambridge University Press, New York, 2012)
20. G. Giedke, M.M. Wolf, Quantum communication: super-activated hannels. Nat. Photonics **5**(10), 578–580 (2011)
21. L.H. Harper, Optimal assignments of numbers to vertices. J. Soc. Ind. Appl. Math. **12**, 131–135 (1964)
22. A.S. Holevo, Bounds for the quantity of information transmitted by a quantum communication channel. Probl. Inf. Transm. **9**(3), 177–183 (1973)
23. A.S. Holevo, The capacity of the quantum channel with general signal states. IEEE Trans. Inf. Theory **44**(1), 269–273 (1998)
24. J. Hou, G. Kramer, Effective secrecy: reliability, confusion and stealth, in *Proceedings of the IEEE International Symposium on Information Theory (ISIT)*, Honolulu, HI, USA (2014), pp. 601–605
25. D. Kahn, *The Codebreakers - The Story of Secret Writing* (MacMillan Publishing Co, New York, 1979). 9th printing
26. J. Körner, K. Marton, General broadcast channels with degraded message sets. IEEE Trans. Inf. Theory **23**(1), 60–64 (1977)
27. Y. Liang, G. Kramer, H.V. Poor, S.S. (Shitz), Compound wiretap channels. EURASIP J. Wirel. Commun. Netw. **2009**(1), 1–12 (2009)
28. A.S. Mansour, R.F. Schaefer, H. Boche, Joint and individual secrecy in broadcast channels with receiver side information, in *IEEE 15th International Workshop on Signal Processing Advances in Wireless Communications (SPAWC)*, Toronto, Canada (2014), pp. 369–373
29. A.S. Mansour, R.F. Schaefer, H. Boche, The individual secrecy capacity of degraded multi-receiver wiretap broadcast channels, in *Proceedings of the 2015 IEEE International Conference on Communications (ICC)*, London, United Kingdom (2015)
30. K. Marton, A coding theorem for the discrete memoryless broadcast channel. IEEE Trans. Inf. Theory **25**(3), 306–311 (1979)
31. U. Maurer, S. Wolf, Information-theoretic key agreement: from weak to strong secrecy for free, *Advances in Cryptology — EUROCRYPT 2000*, Lecture Notes in Computer Science (Springer, Berlin, 2000), pp. 351–368

32. J. Nötzel, M. Wiese, H. Boche, The arbitrarily varying wiretap channel - secret randomness, stability and super-activation, in *Proceedings of the 2015 IEEE International Symposium on Information Theory (ISIT)* (2015), pp. 2151–2155
33. J. Nötzel, M. Wiese, H. Boche, The arbitrarily varying wiretap channel - deterministic and correlated random coding capacities under the strong secrecy criterion, in *Proceedings of the 2015 IEEE International Symposium on Information Theory (ISIT)* (2015)
34. R.F. Schaefer, H. Boche, H.V. Poor, Super-activation as a unique feature of secure communication in malicious environments (2015)
35. C.E. Shannon, Communication theory of secrecy systems. Bell Syst. Tech. J. **28**, 656–715 (1949)
36. C.E. Shannon, The zero error capacity of a noisy channel. IRE Trans. Inf. Theory **2**(3), 8–19 (1956)
37. A.D. Wyner, The wire-tap channel. Bell Syst. Tech. J. **54**(8), 1355–1387 (1975)

Chapter 2
Authentication and Secret-Key Cryptology

2.1 Introduction

The transmission of information in a communication process faces various threats. These threats arise if during the transmission, the messages are at the mercy of unauthorized actions of an adversary, that is, if the channel used for the communication is insecure. Basically there are three attacks the communicants have to be aware of when using an information transmission system. An adversary might observe the communication and gain information about it, he might insert false messages or he might replace legally sent messages by false messages. The protection against the first attack is a question of *secrecy* and the protection against the latter two attacks is a question of *authenticity*.

The need to protect communication has been appreciated for thousands of years. It is not surprising that most of the historical examples arise from the battleground, where secrecy and authenticity of messages is directly related to a potential loss of life. But apart from those military applications, the fast development of information technology has led to a number of economical applications in our days. From electronic fund transfer in international banking networks to the transmission of private electronic mail, there are vast amounts of sensitive information routinely exchanged in computer networks that demand for protection.

From ancient times on up to now, the authenticity of documents or letters has been guaranteed by the usage of seals and handwritten signatures, which are difficult to imitate. In order to guarantee secrecy, people have used methods in which the very existence of a message is hidden. Those techniques are known as concealment systems, including, for instance, the usage of invisible ink or the microscopical reduction of messages to hide them in meaningless text. An historical example of such a concealment goes back to the Greeks. Learning that the Persian king Darius was about to attack Greece, a Greek living in Persia scratched a warning message on a wooden writing tablet, then covered the tablet with wax so that it looked like a fresh writing surface. He sent it to Sparta, where Gorgo, the wife of the Spartan king

A. Ahlswede et al. (eds.), *Hiding Data – Selected Topics*,
Foundations in Signal Processing, Communications
and Networking 12, DOI 10.1007/978-3-319-31515-7_2

Leonidas, guessed that the blank wax writing surface covered something important, scraped it off and discovered the message that enabled the Greeks to prepare for Darius' attack and to defeat him ([16], pp. 38).

We will not deal with such physical devices for information protection but discuss a different method known as encryption or cryptographic coding, which allows a mathematical treatment. The idea is to transform the messages before transmission in order to make them unintelligible and difficult to forge for an adversary. Perhaps one of the first who employed such a method was Julius Caesar when replacing in his correspondence each letter by its third successor (cyclically) in the Latin alphabet ([15], pp. 83). The general usage of such a cryptosystem can be imagined as follows. Sender and receiver agree upon one of several possible methods to transform the messages. Using this method the sender transforms an actual message and transmits the result over the insecure channel. The receiver, knowing which method was used by the sender, can invert the transformation and resolve the original message. The possible transformations are usually referred to as *keys* and the transformed messages sent over the insecure channel are referred to as *cryptograms*. Further, the transformation of the original message into the cryptogram done by the sender is called *encryption* and the opposite action by the receiver is called *decryption*.

The mathematical model to analyze secrecy systems of this type was introduced by Shannon [24] in 1949. His work on this subject is generally accepted as the starting point of the scientific era of cryptology. As indicated, cryptosystems have been used for more than 2000 years and they were thought to be secure if no one who had tried to break them, had succeeded. Shannon's theory made it possible to prove the security of cryptosystems and to give bounds on the amount of information, which has to be securely transmitted to achieve this provable security.

The problem of authenticity, when a cryptosystem is used, was treated much later than Shannon's development of a theory for secrecy systems. The systematic study of authentication problems is the work of G.J. Simmons [28]. Although he is not among the originators of the earliest publication [12] from 1974 on this subject, the authors of this paper already mentioned that Simmons drew their attention to the model considered ([12], pp. 406).

The successful usage of a cryptosystem of the described form is primarily based on the ability of the sender and the receiver to agree upon a key to be used for the encryption and to keep this key secret. Therefore one has to assume that they can use a secure channel to exchange the identity of that key. Systems of this type are called *secret-key cryptosystems*. One might object that if sender and receiver have a secure channel at their disposal, they could use it directly for the transmission of the messages, but it might be possible that the secure channel is only available at some time instance before the transmission of the messages. Furthermore the secure channel might be unsuitable for the transmission of the messages, for instance, if it has a capacity that is too small. Hence, the assumption that a secure channel is available can be justified in a lot of cases and, in particular, systems with a small number of keys compared to the number of messages are of practical interest.

An example of a secret-key cryptosystem is the DES (data encryption standard), which was developed at IBM around 1974 and adopted as a national standard for the USA in 1977. It uses keys specified by binary strings of length 56 and encrypts using these keys messages given as binary strings of length 64 [7].

We will analyze both the authentication and the secrecy problem on a theoretical level, where we assume that the adversary has infinite computing power. In 1976 Diffie and Hellman [9] invented a new type of cryptosystems where a secure channel to exchange the key is no longer needed. Each participant has a publically available key and a secret private key. These so called *public-key cryptosystems* are mostly based on an intractability assumption on the adversaries ability to solve a certain computational problem, like the factorization of large composite integers or the evaluation of the discrete logarithm, and are in this way based on a bound on the computational power of the adversary. Those systems are beyond the scope of this section.

The present chapter is organized as follows. In Sect. 2.1 the models of secret-key cryptology and authentication are introduced. We start with the classical model of a secrecy system formulated by Shannon [24]. As a measure for the secrecy provided by such a system the entropy criterion and the opponent's error probability when decrypting will be introduced and a relation between these criteria will be derived. In order to analyze the authentication problem we extend the so far discussed model in such a way that the adversary is allowed to become an active wiretapper, which means that he has more influence on the communication channel. We introduce the two different actions an opponent can try in order to deceive the receiver, namely, the so called impersonation attack and the substitution attack and we define the corresponding success probabilities P_I and P_S, respectively.

Although the model of the classical secrecy system is extended, it is still possible to analyze the introduced criteria for secrecy. Especially the class of authentication systems with no secrecy at all is of interest for some applications.

Section 2.3 is concerned with the authentication problem. We begin with deriving some general bounds on P_I and P_S. The derivation of Simmons' bound for P_I leads to the definition of perfect authenticity. We will see that, in general, authenticity and secrecy are two independent attributes of a cryptosystem (Sect. 2.3.1).

Then we will analyze the special class of authentication systems without secrecy. We derive the bound on P_S in such a case, which was originally proved in [12] and in a more general form in [2]. We show that a certain generalization to a larger class of message sources is not possible and we derive from the proof given in [2] necessary and sufficient conditions that an authentication system achieves the lower bound on P_S (Sect. 2.3.2).

The problem of the maximal number of messages in an authentication system under certain constraints on the success probabilities of the opponent will be treated in the next section. We study the behavior of the maximal number of messages for large values of Kp^2, where K is the number of keys and p is an upper bound on the opponent's success probability. The problem is still not completely solved and we derive the known upper and lower bounds. A typical result is that $M \sim \exp(K \cdot f(p))$ where M is the number of messages and f is some positive function. The special shape of f is up to now not exactly known. The difference between the upper and lower bounds for M consists (for small p) essentially of a factor of order $\log \frac{1}{p}$ in the exponent of the bounds (Sect. 2.3.3).

The observation that the receiver's decision problem to accept a received message or not, can be viewed as an hypothesis testing problem will lead to a simpler derivation of information-theoretic lower bounds on the opponent's success probability. This approach, which was made in [19], allows also to generalize the model in several directions (Sect. 2.3.4).

In Sect. 2.4 we start the analyzation of secrecy systems with the derivation of some upper bounds on the secrecy measured by the entropy criterion. This leads to Shannon's result that a necessary condition for perfect secrecy is that the number of keys is at least as big as the number of messages. Afterwards we introduce the notions of regular and canonical ciphers and derive a lower bound on the secrecy for every locally regular cipher (Sects. 2.4.1 and 2.4.2). Furthermore we give an explicit construction of a good locally regular cipher and derive various bounds for the secrecy of this cipher (Sect. 2.4.3). Finally we present an approach to extend the model with a source coder and a (private) randomizer (Sects. 2.4.4 and 2.4.5).

In Sect. 2.4 we shall take a closer look at public-key cryptology. In Shannon's original model of a cryptosystem it is assumed that the cryptanalyst has unlimited computational power and hence is able to decipher the cryptogram immediately, once he knows the key. Shannon already remarked that this assumption often is not realistic. In their pioneering paper "New Directions in Cryptography" Diffie and Hellman [9] introduced public-key cryptology. They presented a protocol using only one key, which is a one-way function. In order to encrypt and decrypt the message, sender i and receiver j have to rise a special value to the power a_i (resp. a_j). This can be done very fast by repeated squaring. In principle a_i and a_j are known to the cryptanalyst, since they are stored in a public directory. However, they are published in the form $b_i = w^{a_i}$ and $b_j = w^{a_j}$, where w is a primitive element in a finite field. In order to conclude from b_i to a_i, the cryptanalyst has to take the *discrete logarithm* $a_i = \log_w b_i$ and for this task up to now no efficient algorithm is known. So, the cryptanalyst has all the necessary information to obtain the original message, but he cannot do this in a reasonable amount of time. There are several advantages of public-key cryptology compared to secret-key cryptology:

(1) the existence of a secure channel is no longer required;
(2) communication is faster, since the key has not to be transmitted;
(3) most public-key protocols are extendable to multi-user systems;
(4) public-key protocols also can be designed for further purposes, such as verification of identity, digital signatures, etc.

Whereas in secret-key cryptology the mathematical tools mostly stem from Information Theory, in public-key cryptology we need some background in Complexity Theory (one-way functions, zero-knowledge proofs) and in Number Theory, since most of the protocols we shall present are based on the hardness of integer factorization. We shall only present the ideas and facts which are important to understand the protocols presented and refer the reader to standard literature in the respective sections.

2.2 Models and Notation

2.2.1 *Secret-Key Cryptology*

In this paragraph the models of secret-key cryptology and authentication will be introduced. In both models we have three actors, a *sender*, a *receiver* and an *opponent*. Sender and receiver act together against the opponent. The sender has to inform the receiver about the state of a message source, in presence of the opponent who has access to the communication channel. The two models differ mainly in the abilities and actions of the opponent.

The opponent reads what is transmitted by the sender. The aim of sender and receiver is to avoid that the opponent can obtain any information from the transmitted message. To this aim sender and receiver share a *secret key* which is not known to the opponent. The sender uses this key to encrypt the original message into a different message, the so called cryptogram. This cryptogram is transmitted over the insecure channel to the receiver who can reconstruct the original message using the key. As the opponent does not know the secret key he hopefully can do nothing useful with the cryptogram. Such a secrecy system is depicted in Fig. 2.1, later this model will be extended with a randomizer and a source coder.

For the components of this model the following notation is used:

- Message Source $(\mathcal{M}, P)$
  where $\mathcal{M} \triangleq \{1, \dots, M\}$ is a set of M messages and P is a probability distribution on $\mathcal{M}$.

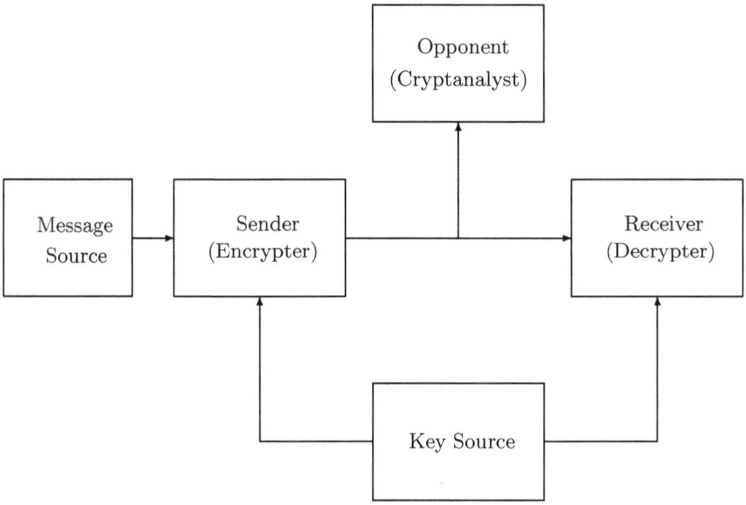

Fig. 2.1 A secret-key crypto system

- Key Source $(\mathcal{C}, Q)$
  where $\mathcal{C} \triangleq \{c_1, \ldots, c_K\}$ is a set of K keys and Q is a probability distribution on $\mathcal{C}$. Every key c_z is a mapping $c_z : \mathcal{M} \to \mathcal{M}'$ from the set of messages $\mathcal{M}$ to the set of cryptograms $\mathcal{M}'$, i.e., the sender encrypts the message $m \in \mathcal{M}$ into the cryptogram $c_z(m) \in \mathcal{M}'$, if the key with index z is used. In order to enable the receiver to reconstruct the original message we have to require

$$c_z(m_1) \neq c_z(m_2) \tag{2.2.1}$$

  for all $m_1, m_2 \in \mathcal{M},\ m_1 \neq m_2,\ z \in \{1, \ldots, K\}$.
  This implies that $|c_z(\mathcal{M})| = M$ for all $z \in \{1, \ldots, K\}$. When considering secrecy systems it is usually also assumed that $c_i(\mathcal{M}) = c_j(\mathcal{M})$ for all $i, j \in \{1, \ldots, K\}$ and therefore one can identify $\mathcal{M}$ and $\mathcal{M}'$ via isomorphy and regard the keys c_z as permutations on $\mathcal{M}$.
  The pair $(\mathcal{C}, Q)$ is also referred to as *cipher*.
- Random variables X, Y, Z
  It is often convenient to work with random variables for message, cryptogram and key rather then with the probability distributions P and Q itself, i.e.:
  Let X be a random variable with values in $\mathcal{M}$ and distribution $P_X = P$.
  Let Z be a random variable with values in $\{1, \ldots, K\}$ and distribution P_Z with $P_Z(z) = Q(c_z)$ for all $z \in \{1, \ldots, K\}$.
  Let Y be a random variable with values in $\mathcal{M}'(=\mathcal{M})$ and distribution P_Y, which is determined by the common distribution P_{XZ}. If not explicitly stated in another way, then we assume that the message and the key are generated by independent random experiments, i.e., $P_{XZ} = P_X P_Z$ and therefore

$$P_Y(m') = \sum_{m \in \mathcal{M}} P_X(m) \sum_{z : c_z(m) = m'} P_Z(z) \qquad \text{for all } m' \in \mathcal{M}. \tag{2.2.2}$$

In order to avoid trivialities we assume that we have more than one message ($M \geq 2$) and we will only deal with messages and keys that occur with strictly positive probability, otherwise they are irrelevant at all. We therefore assume that

$$P_X(m) > 0 \quad \text{and} \quad P_Z(z) > 0 \quad \text{for all } m \in \mathcal{M},\ z \in \{1, \ldots, K\}.$$

The triple $(X, Z, \mathcal{C})$ is referred to as *secrecy system*.

The Opponent's Knowledge

The secrecy provided by such a cryptosystem should be measured according to the fact that the value of the secret key can be kept unknown to the opponent but nothing more. This means it should not be assumed that one can prevent the opponent from getting information about other elements of the secrecy system. This is known as *Kerckhoffs' Principle*[1] in cryptology, which means that the opponent is assumed to

[1] First enunciated by A. Kerckhoffs (1835–1903) ([15], pp. 235).

know all details of the cryptosystem except for the value of the secret key, especially we also assume that the opponent has full knowledge about the probability distributions of messages and keys. Of course this worst-case assumption is quite pessimistic. Nevertheless in the long run it might not be too difficult for an opponent to get information about the design of the cryptosystem.

Measurements for Secrecy

We will introduce two measures for the secrecy provided by a cryptosystem of this type.

Entropy Criterion As the opponent reads the cryptogram $m' \in \mathcal{M}$ which is a realization of the random variable Y and tries to draw conclusions about the original message $m \in \mathcal{M}$ which is a realization of the random variable X, it is natural to use the average uncertainty about the state of the message source given the observation of the cryptogram. This is expressed by the conditional entropy

$$H(X|Y).$$

A 'very good' secrecy system will not decrease the uncertainty about X if Y is observed, i.e., $H(X|Y) = H(X)$. This leads to the following definition.

Definition 21 A secrecy system is perfect if X and Y are independent.

Cryptanalyst's Error Probability Beside the entropy criterion, already studied by Shannon [24], Ahlswede [1] considered as a measure for secrecy the cryptanalyst's error probability in deciding which message was sent.

Given a secrecy system by X, Z and $\mathcal{C}$ the probability of decrypting correctly is

$$\lambda_c(X, Z, \mathcal{C}) = \sum_{m' \in \mathcal{M}} \max_{m \in \mathcal{M}} P_{XY}(m, m'),$$

assuming that the cryptanalyst is using the maximum-likelihood decision rule, which is best possible. Therefore the opponent's error probability is

$$\lambda(X, Z, \mathcal{C}) = 1 - \lambda_c(X, Z, \mathcal{C}).$$

Lemma 4 *The two criteria for secrecy are not unrelated, namely for every secrecy system*

$$\lambda_c \geq 2^{-H(X|Y)}.$$

Proof

$$\begin{aligned} -\log \lambda_c &= -\log \sum_{m' \in \mathcal{M}} \max_{m \in \mathcal{M}} P_{XY}(m, m') \\ &\leq -\log \sum_{m' \in \mathcal{M}} \sum_{m \in \mathcal{M}} P_{X|Y}(m|m') P_{XY}(m, m') \end{aligned}$$

$$\leq - \sum_{m' \in \mathcal{M}} \sum_{m \in \mathcal{M}} P_{XY}(m, m') \log P_{X|Y}(m|m')$$

$$- H(X|Y),$$

where the first inequality is due to the fact that the maximum is greater than the average of terms and the second one follows by application of Jensen's inequality for the ∪-convex function $-\log$. □

This lemma can be used to convert lower bounds on λ into lower bounds on $H(X|Y)$ and upper bounds on $H(X|Y)$ into upper bounds on λ.

Apart from the two measurements introduced so far, as a further criterion for secrecy Hellman [13] considered the average number of *spurious decipherments*.

2.2.2 Authentication

In general, authentication theory is concerned with providing evidence to the receiver of a message that it was sent by a specified and legitimate sender, even in presence of an opponent who can send fraudulent messages to the receiver or intercepts legally sent messages and replaces them by fraudulent ones.

In the model of secret-key cryptology the encryption with a secret key was done in order to guarantee secrecy, i.e., an opponent cannot decipher the cryptogram. In the model of authentication the encryption with a secret key is used to guarantee the authenticity of a transmitted message, which means that the encryption is done in such a way that the receiver recognizes if a fraudulent cryptogram was inserted by an opponent. So in this model the opponent is considered to be more powerful in the sense that he has more influence on the communication channel than before. The opponent can try two types of attacks:

- He can intercept a legally sent cryptogram and replace it by a different one.
  This is the so called *substitution attack*.
- He can send a fraudulent cryptogram to the receiver, even when no cryptogram was transmitted by the sender.
  This is the so called *impersonation attack*.

The opponent tries to deceive the receiver about the actual value of the random variable X. In the case of a successful substitution attack the receiver believes the random variable X to attain a value different from the true one. In the case of a successful impersonation attack the receiver believes the random variable X to attain some value but actually the message source has not generated a message. In both cases the aim of the opponent is to misinform the receiver about the state of the message source. (In fact this is the basic aim. For instance, it would be not very useful for a cheater to make his bank believe that on his account is a less amount of money than there actually is. Therefore one might think about more ambitious aims for the opponent. This will be treated in Sect. 2.3.4).

Such an authentication system is depicted in Figs. 2.2 and 2.3. In Fig. 2.2 a substitution attack is shown. In case of an impersonation attack the opponent simply sends a cryptogram to the receiver, sender and message source are thought to be inactive. Such a situation is shown in Fig. 2.3.

We will use the same notation for the components of this model as before:

- Message Source $(\mathcal{M}, P)$.
- Key Source $(\mathcal{C}, Q)$.
- Keys $c_z : \mathcal{M} \rightarrow \mathcal{M}'$, $z \in \{1, \dots, K\}$.
- Random Variables X, Y, Z for messages,cryptograms and keys, respectively.

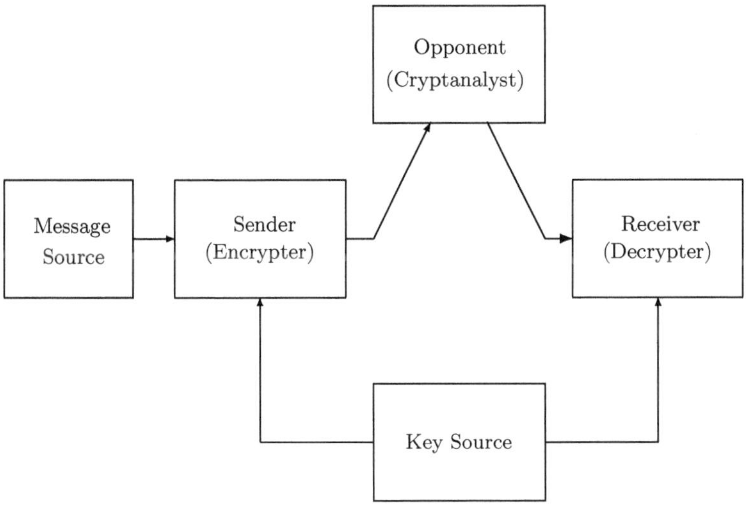

Fig. 2.2 A substitution attack

Fig. 2.3 An impersonation attack

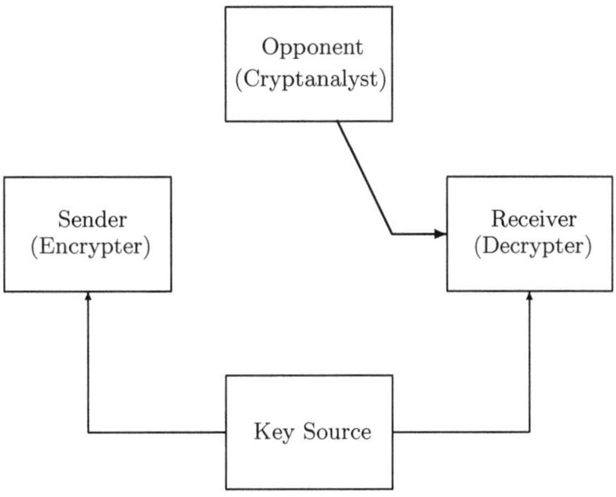

In addition to this we need a random variable Y' for the cryptogram the opponent inserts. We use Y' for both cases of impersonation- and substitution attacks. To specify when the opponent is successful, we need the following definition.

Definition 22 A cryptogram $y \in \mathcal{M}'$ is valid under the key $c_z \in \mathcal{C}$ if y is in the range of c_z, i.e., $y \in c_z(\mathcal{M})$.

If the opponent inserts a cryptogram y', then the receiver does not detect the deception, if the cryptogram y' is valid under the secret key used by sender and receiver. On the other hand if y' is not valid under the secret key, then the receiver is sure that the cryptogram does not come from the sender and must have been inserted by the opponent.

Definition 23 The opponent is considered to be successful in each case if the receiver accepts the inserted y' as a valid cryptogram.

We call a probability distribution $P_{Y'}$ on $\mathcal{M}'$ an impersonation strategy and a family $\{P_{Y'|Y}(\cdot|y) : y \in \mathcal{M}'\}$ of conditional distributions on $\mathcal{M}'$ with $P_{Y'|Y}(y|y) = 0$ for all $y \in \mathcal{M}'$ a substitution strategy.

Let P_I and P_S denote the probabilities for the opponent using his optimal strategy to be successful in an impersonation attack and in a substitution attack, respectively.

Remark 8 1. Note that in a substitution attack we force the opponent to replace the intercepted cryptogram y by a *different* cryptogram y' because otherwise he would not misinform the receiver about the state of the message source.
2. In the model of secret-key cryptology it was assumed $\mathcal{M} = \mathcal{M}'$. Now this does not make sense any longer because it would imply that every cryptogram is valid under every key, therefore $P_I = P_S = 1$ and one cannot guarantee any authenticity of messages. Therefore we will allow in this context that $\mathcal{M}$ and $\mathcal{M}'$ are different sets with $|\mathcal{M}'| \geq |\mathcal{M}|$.

The triple $(X, Z, \mathcal{C})$ is referred to as *authentication system* or *authentication code*.

Such an authentication system can either provide no secrecy, i.e., $H(X|Y) = 0$, or it can provide some degree of secrecy, i.e., $H(X|Y) > 0$. Sometimes authentication codes without secrecy are called *cartesian* or *systematic* in the literature.

For this model of authentication we will keep the assumption of Section "The Opponent's Knowledge" that the opponent knows all details of the elements of the system except for the value of the secret key. In fact, Simmons [26, 27], who introduced this model, had a different notion. He thought of a game-theoretic authentication model. This means sender and receiver play against the opponent. In a game one needs to define the strategy sets of the players. Clearly the strategies for the opponent are the distributions introduced in Definition 23. The strategies of sender and receiver Simmons then defined as the possible distributions P_Z of the keys. Therefore he had to assume that the opponent does not know the key statistics. This approach has not further been developed in literature and we will keep Kerckhoffs' assumption, which means that also P_Z is fixed and known to the opponent.

Remark 9 In order to avoid confusion it should be noted that in a lot of papers concerning authentication theory (for example those of Simmons) a different notation is used. Messages are called source states, cryptograms are called messages and keys are called encoding rules.

2.3 Authentication

2.3.1 General Bounds and Perfectness

In Shannon's model of secret-key cryptology it was clear how to define the perfectness of the system. In the authentication model it is no longer obvious, when one can say that a system provides perfect authenticity. We will see that a complete protection against deception is impossible. Therefore we have to start with the analysis to what degree the opponent is able to deceive the receiver.

Hence, we try to give lower bounds on the probabilities P_I and P_S. It should be noted that there is no general relationship of the form $P_S \geq P_I$, as one might think at first sight because in a substitution attack the opponent has the additional information about a valid cryptogram. Recall that in a substitution attack the opponent is restricted to choose a cryptogram different from the original one, as he wants to misinform the receiver. The next example shows that this can lead to a situation with $P_S < P_I$.

Example 3 Let us define an authentication system as follows:

- Two messages, $\mathcal{M} \triangleq \{1, 2\}$, which occur each with probability $\frac{1}{2}$, i.e., $P_X(1) \triangleq P_X(2) \triangleq \frac{1}{2}$.
- 3 keys, $\mathcal{C} \triangleq \{c_1, c_2, c_3\}$, with $P_Z(z) \triangleq \frac{1}{3}$ for all $z \in \{1, 2, 3\}$.
- 3 possible cryptograms, $\mathcal{M}' \triangleq \{y_1, y_2, y_3\}$ and the encryption is done according to the following table.

	y_1	y_2	y_3
c_1	1	2	
c_2		1	2
c_3	2		1

For instance, the message 2 is encrypted using the key c_3 to the cryptogram y_1 or formally $c_3(2) = y_1$.

Clearly $P_I = \frac{2}{3}$, as $Pr(y_i \text{ valid}) = \frac{2}{3}$ for all $i \in \{1, 2, 3\}$.

But after having observed any valid cryptogram, the probability that a different one is also valid under the used key is always $\frac{1}{2}$.

Therefore $P_S = \frac{1}{2} < \frac{2}{3} = P_I$.

Combinatorial Bounds

Theorem 31 *For every authentication system*

$$P_I \geq \frac{M}{|\mathcal{M}'|} \quad \textit{and} \quad P_S \geq \frac{M-1}{|\mathcal{M}'|-1}.$$

Proof The statement immediately follows by consideration of the following impersonation strategy and substitution strategy, respectively.

- *Impersonation*: The opponent chooses $y \in \mathcal{M}'$ according to the uniform distribution, i.e., $P_{Y'}(y) = \frac{1}{|\mathcal{M}'|}$ for all $y \in \mathcal{M}'$.
- *Substitution*: Observing $y \in \mathcal{M}'$ the opponent chooses $y' \neq y$ according to the uniform distribution from $\mathcal{M}' \backslash \{y\}$, i.e., $P_{Y'|Y}(y'|y) = \frac{1}{|\mathcal{M}'|-1}$ for all $y' \neq y$.

As these strategies are not necessarily optimal, by calculation of the corresponding success probabilities we obtain lower bounds on P_I and P_S, namely

$$P_I \geq \sum_{z=1}^{K} P_Z(z) \frac{|c_z(\mathcal{M})|}{|\mathcal{M}'|} = \frac{M}{|\mathcal{M}'|}$$

and similarly

$$P_S \geq \sum_{z=1}^{K} P_Z(z) \frac{|c_z(\mathcal{M})|-1}{|\mathcal{M}'|-1} = \frac{M-1}{|\mathcal{M}'|-1},$$

where we used that $|c_z(\mathcal{M})| = M$, as c_z is injective. □

Remark 10 1. Note that in Example 3 the bounds hold with equality.
2. If we consider also randomized ciphers (i.e., some messages may be mapped to different cryptograms under the same key according to some probability distribution), then we have $|c_z(\mathcal{M})| \geq M$ and therefore equality in the bounds is only possible if the cipher is not randomized.
3. $P_I = 0$ or $P_S = 0$ is impossible (recall that $M \geq 2$).

Simmons' Bound

In this section we present the basic information-theoretic lower bound on P_I, first given by Simmons [26, 27].

Before this, note that one can get two rough bounds on P_I and P_S in terms of entropy simply by bounding the probabilities of guessing the key correctly (in case of a substitution attack after observing the cryptogram y). Doing this we get:

$$P_I \geq 2^{-H(Z)} \quad \text{and} \quad P_S \geq 2^{-H(Z|Y)}.$$

The derivation of this type of bounds is done in Sect. 2.3.4, where we will treat the bound on P_S in a more general context. The next theorem shows that it is possible to add $H(Z|Y)$ in the exponent of the bound for P_I.

Theorem 32 (Simmons) *For every authentication system*

$$P_I \geq 2^{-I(Y \wedge Z)}.$$

At first sight this bound may look somewhat strange, as it tells us that P_I can be made small only if the cryptogram gives away much information about the key. But recall that in an impersonation attack the opponent does not have access to a legally sent cryptogram. Furthermore one could interprete the bound from the receivers viewpoint. The receiver can only hope for a small P_I if his knowledge of the key gives him a lot information about the cryptogram.

The proof for Simmons' bound presented below was taken from Johannesson and Sgarro [14]. It is simpler than Simmons' original derivation and one easily sees how the bound can be strengthened.

Proof of the theorem. The best impersonation attack for the opponent is to choose a cryptogram $y \in \mathcal{M}'$ with maximal probability of validity, i.e.,

$$P_I = \max_{y \in \mathcal{M}'} \Pr(y \text{ valid}) = \max_{y \in \mathcal{M}'} \sum_{z:\phi(y,z)=1} P_Z(z), \tag{2.3.1}$$

where the function ϕ is defined as follows

$$\phi(y, z) \triangleq \begin{cases} 1, & \text{if } P_{YZ}(y, z) > 0 \\ 0, & \text{otherwise,} \end{cases}$$

i.e., $\phi(y, z)$ is equal to one exactly if y is a valid cryptogram under the key c_z.

Now we calculate $I(Y \wedge Z)$ and apply the log-sum inequality.

$$I(Y \wedge Z) = \sum_y P_Y(y) \sum_z P_{Z|Y}(z|y) \log \frac{P_{Z|Y}(z|y)}{P_Z(z)}.$$

We can restrict the summation to terms with $\phi(y, z) = 1$ (because only for these we have $P_{Z|Y}(z|y) > 0$) and apply the log-sum inequality. In this way we obtain

$$I(Y \wedge Z) = \sum_y P_Y(y) \sum_{z:\phi(y,z)=1} \phi(y,z) P_{Z|Y}(z|y) \log \frac{\phi(y,z)P_{Z|Y}(z|y)}{\phi(y,z)P_Z(z)}.$$

$$\geq \sum_y P_Y(y) \underbrace{\Big(\sum_{z:\phi(y,z)=1} \phi(y,z) P_{Z|Y}(z|y) \Big)}_{=1} \log \frac{\overbrace{\sum_{z:\phi(y,z)=1} \phi(y,z) P_{Z|Y}(z|y)}^{=1}}{\underbrace{\sum_{z:\phi(y,z)=1} \phi(y,z) P_Z(z)}_{\Pr(y \text{ valid})}}$$

$$= - \sum_y P_Y(y) \log \Pr(y \text{ valid}) \geq - \log \max_y \Pr(y \text{ valid}) = - \log P_I. \square$$

Corollary 1 *Necessary and sufficient conditions for equality in Simmons' bound are:*

1. $\Pr(y \text{ valid})$ *is constant in* y.
2. $\frac{P_Z(z)P_Y(y)}{P_{YZ}(y,z)}$ *is constant for all* (y, z) *with* $P_{YZ}(y,z) > 0$.

Proof The first condition follows from the last inequality in the proof and the condition for equality in the log-sum inequality is in our case:

$$P_{Z|Y}(z|y) \Pr(y \text{ valid}) = P_Z(z) \quad \text{for all } (y, z) \text{ with } \phi(y,z) = 1,$$

which is equivalent to condition 2. as we saw already that $\Pr(y \text{ valid})$ must be constant in y. □

Strengthening of Simmons' Bound

The first strengthening by Johannesson and Sgarro [14] is easily derived by the following observation. From Eq. (2.3.1) it is clear that $\Pr(y \text{ valid})$ and therefore also P_I is independent of the distribution P_X of messages, but the mutual information $I(Y \wedge Z)$ is not, in general. This implies that if we change our distribution P_X of messages to some $P_{\tilde{X}}$ in such a way that the function ϕ is kept unchanged, then we get a new value $2^{-I(\tilde{Y} \wedge Z)}$ which is also a bound for P_I in our original authentication system. Therefore we obtain a stronger bound in the following way.

Proposition 2 (Johannesson, Sgarro)

$$P_I \geq 2^{-\inf I(Y \wedge Z)},$$

where the infimum is taken over all distributions P_X *which leave* ϕ *unchanged.*

In the next example we show that this new bound can return values, which are strictly better than those of the former bound.

Example 4 Let us define an authentication system in the following way.

- Two messages, $\mathcal{M} \triangleq \{1, 2\}$ with $P_X(1) \triangleq p \leq \frac{1}{2}$ (w.l.o.g.).
- Four equiprobable keys, $\mathcal{C} \triangleq \{c_1, \ldots, c_4\}$ with $P_Z(z) \triangleq \frac{1}{4}$ for all $z \in \{1, \ldots, 4\}$.
- Four cryptograms, $\mathcal{M}' \triangleq \{y_1, \ldots, y_4\}$.

The encryption is shown in the table below.

	y_1	y_2	y_3	y_4
c_1	1	2		
c_2	2		1	
c_3		1		2
c_4			2	1

For this authentication system we have $P_I = \frac{1}{2}$ and $P_S = 1 - p \geq \frac{1}{2}$, which implies $P_D = 1 - p$.

$I(Y \wedge Z) = H(Y) - H(Y|Z) = \log 4 - h(p) = 2 - h(p)$, where h is the binary entropy, i.e., $h(p) \triangleq -p \log p - (1 - p) \log(1 - p)$. Therefore $2^{-I(Y \wedge Z)} = \frac{2^{h(p)}}{4} \leq \frac{1}{2}$ with equality exactly if $p = \frac{1}{2}$.

Hence, the strengthened bound for P_I is sharp and the old bound is not sharp for $p \neq \frac{1}{2}$.

We could strengthen the bound by observing that P_I is independent of P_X (if ϕ is kept unchanged). We can obtain a further strengthening by analyzing on what P_I depends. Again from Eq. (2.3.1) it is clear that P_I depends only on the (marginal) distribution of Z and on the function ϕ. Thus, given that these are kept fixed, both the message distribution and *any correlation of X and Z* are totally irrelevant. Therefore we get a new bound:

Theorem 33 (Johannesson, Sgarro)

$$P_I \geq 2^{-\inf I(Y \wedge Z)},$$

where now the infimum is taken over all (possibly dependent) random couples (X, Z) such that

1. *Z has the same marginal distribution as for the given system*
2. *the resulting function ϕ is the same as for the given system.*

Again this new bound can return values that are strictly better than those of the previously considered bounds, which is shown in the next example.

Example 5 Let us define an authentication system in the following way:

- Two messages, $\mathcal{M} \triangleq \{1, 2\}$ with $P_X(1) \triangleq p$.
- Two equiprobable keys, $\mathcal{C} \triangleq \{c_1, c_2\}$ with $P_Z(1) \triangleq P_Z(2) \triangleq \frac{1}{2}$.

- Three cryptograms $\mathcal{M}' \triangleq \{y_1, y_2, y_3\}$.

The encryption is done according to the following table.

	y_1	y_2	y_3
c_1	1	2	
c_2	2		1

For this authentication code we have $P_I = 1$, because $\Pr(y_1 \text{ valid}) = 1$ and $I(Y \wedge Z) = H(Y) - H(Y|Z) = 1 + \frac{1}{2}h(p) - h(p) = 1 - \frac{1}{2}h(p)$.

If we take $p = \frac{1}{2}$, then $I(Y \wedge Z)$ is minimized and we obtain the (old) bound

$P_I \geq 2^{-\frac{1}{2}} = \frac{1}{\sqrt{2}}$, which is not sharp. Suppose now that X and Z are no longer independent and assume that X and Z return the same values with probability close to one (we cannot say with probability equal to 1 because this would change ϕ). Then with probability close to one $Y = y_1$ and therefore $I(Y \wedge Z) = H(Y) - H(Y|Z) \leq H(Y) \approx 0$. So the new bound gives the correct estimate $P_I = 1$ for the original system where X and Z are independent.

There are also nondegenerate examples ($P_I < 1$) with this effect (see [14]).

Perfectness

Up to now we derived lower bounds on P_I. With each of these lower bounds we obtain also a lower bound on the probability of deception P_D, which we define as $P_D \triangleq \max\{P_I, P_S\}$. For instance,

$$P_D \geq 2^{-I(Y \wedge Z)} \tag{2.3.2}$$

Simmons [26, 27] defined perfect authenticity to mean that equality holds in (2.3.2). In this case, he noted that *the information capacity of the transmitted cryptogram is used either to inform the receiver as to the state of the message source or else to confound the opponent.*

Definition 24 An authentication system is perfect if

$$P_D = 2^{-I(Y \wedge Z)}.$$

One could also think of perfect authenticity to mean that equality holds in (2.3.2), where instead of Simmons' bound the stronger bound on P_I from Theorem 33 is used on the right-hand side. However we will keep the original definition by Simmons. This was also done by Massey [18] who noted that the information that Y gives about Z, $I(Y \wedge Z)$, is a measure of how much of the secret key is used to provide authenticity. Therefore, if the stronger bound $2^{-\inf I(Y \wedge Z)}$ is greater than $2^{-I(Y \wedge Z)}$, then this indicates that the authentication system is wasting part of the information $I(Y \wedge Z)$ and therefore should not be called 'perfect'.

Remark 11 1. Note that we may have to call a system perfect although it provides no authenticity at all, i.e., $P_D = 1$. For instance, the "One-Time Pad" described in Example 7 provides perfect secrecy and Y and Z are independent. Therefore $P_D = 2^{-I(Y \wedge Z)} = 1$.

2. The authentication system of Example 4 provides for $p = \frac{1}{2}$ both perfect secrecy and perfect authenticity with $P_D = \frac{1}{2}$. For $p \neq \frac{1}{2}$ it still provides perfect secrecy but has no longer perfect authenticity. The next example shows an authentication system with perfect authenticity but without perfect secrecy. Therefore we can say that *in general* authenticity and secrecy are two independent attributes of a cryptographic system. Massey [18] says that this is a lesson that is too often forgotten in practice.

Example 6 Let us define an authentication system in the following way:

- Two messages $\mathcal{M} \triangleq \{1, 2\}$, with $P_X(1) \triangleq P_X(2) \triangleq \frac{1}{2}$.
- Four keys, $\mathcal{C} \triangleq \{c_1, \dots, c_4\}$, which are chosen according to the uniform distribution.
- Four cryptograms $\mathcal{M}' \triangleq y_1, \dots, y_4\}$.

The encryption is shown in the following table.

	y_1	y_2	y_3	y_4
c_1	1		2	
c_2	1			2
c_3		1	2	
c_4		1		2

For this authentication system we have $P_I = P_S = \frac{1}{2}$, $I(Y \wedge Z) = H(Y) - H(Y|Z) = \log 4 - \log 2 = 1$ and therefore $P_I = P_S = 2^{-I(Y \wedge Z)}$, which means that the system provides perfect authenticity but it is clearly not perfectly secret as $H(X|Y) = 0 \neq 1 = H(X)$.

A Bound on P_S

In this section we derive a bound on P_S presented in [23] which is based on Simmons' bound for P_I.

Definition 25 For every cryptogram $y \in \mathcal{M}'$ let $\mathcal{K}(y) \triangleq \{z \in \{1, \dots, K\} : P_{Y,Z}(y, z) > 0\}$ be the set of key-indices such that y is a valid cryptogram under the corresponding keys.

Let $P_S(y)$ denote the probability of successful substitution after observing that $Y = y$.

If the opponent intercepts y and substitutes y' then his probability of success is $P_{Z|Y}(\mathcal{K}(y')|y)$. Therefore $P_S(y)$ can be written as

$$P_S(y) = \max_{y' \neq y} P_{Z|Y}(\mathcal{K}(y')|y). \tag{2.3.3}$$

We will now obtain a lower bound on $P_S = \sum_y P_Y(y)\, P_S(y)$ by bounding $P_S(y)$ below. Therefore let us define for every $y \in \mathcal{M}'$ random variables Y_y, with values in $\mathcal{M}'\backslash\{y\}$, and Z_y, with values in $\{1, \dots, K\}$, as follows

$$P_{Z_y}(z) \triangleq P_{Z|Y}(z|y) \text{ and } P_{Y_y|Z_y}(y'|z) \triangleq \frac{P_{Y|Z}(y'|z)}{a_y(z)} \quad \text{for all } y' \neq y, \tag{2.3.4}$$

where $a_y(z) \triangleq \sum_{y' \neq y} P_{Y|Z}(y'|z)$ is the normalization constant such that $P_{Y_y|Z_y}(\cdot\,|z)$ is a probability distribution. Note that $a_y(z)$ is always greater 0 because $M \geq 2$ and there are M valid cryptograms for every key.

Although one cannot assure that there always exists an authentication system which induces this random couple (Y_y, Z_y), we can (formally) look at the corresponding probability of successful impersonation, since this only depends on the joint distribution of Y_y and Z_y (recall (2.3.1) and the definition of ϕ). We denote this probability by $P_I(y)$. Then from (2.3.1) it follows

$$P_I(y) = \max_{y' \neq y} P_{Z_y}(\mathcal{K}(y')) = \max_{y' \neq y} P_{Z|Y}(\mathcal{K}(y')|y) = P_S(y).$$

Hence, we can apply to $P_S(y)$ the lower bound from Theorem 32 and get

$$P_S(y) \geq 2^{-I(Y_y \wedge Z_y)}.$$

Therefore the next theorem is immediate.

Theorem 34 (Sgarro) *For every authentication code*

$$P_S \geq \sum_y P_Y(y)\, 2^{-I(Y_y \wedge Z_y)},$$

where Y_y and Z_y are defined in (2.3.4).

Remark 12 As already mentioned we can bound P_S by $2^{-H(Z|Y)}$ and given some value $y \in \mathcal{M}'$ we have $P_S(y) \geq 2^{-H(Z|Y=y)}$ (compare also Sect. 2.3.4). The bound just derived returns always values at least as good as this bound because by definition of Z_y we obtain

$$\begin{aligned} -I(Y_y \wedge Z_y) &= -H(Z_y) + H(Z_y|Y_y) \\ &= -H(Z|Y = y) + H(Z_y|Y_y) \geq -H(Z|Y = y). \end{aligned}$$

2.3.2 *Authentication Codes Without Secrecy*

Now we discuss authentication codes without secrecy, which means that the opponent knows the state of the message source after observing the correct cryptogram, i.e.,

$H(X|Y) = 0$. This applies to situations where secrecy is not required or can not be guaranteed (for instance if the opponent has full access to the message source) but the authenticity of messages is still desired.

Preliminaries

In those cases a convenient method of enciphering is the following. We consider only keys c_z which produce cryptograms y of the form

$$c_z(m) = y = (m, n),$$

where n is an extra symbol (string) dependent on m and z which is simply added to the clear message m. We can restrict ourselves, w.l.o.g., to this class of keys because if we are given an arbitrary set of K keys $\{c'_1, \dots, c'_K\}$, we can define $c_z(m) \triangleq (m, c'_z(m))$ for all $z \in \{1, \dots, K\}$, $m \in \mathcal{M}$. This modification leads to a set of K keys $\{c_1, \dots, c_K\}$ of the desired form and for the opponent the situation is as before since m was already uniquely determined by $c'_z(m)$.

Keys of this form have the property that for different messages the sets of possible cryptograms are always disjoint, i.e.,

$$c_i(m) \neq c_j(m') \quad \text{for all } i, j \in \{1, \dots, K\},\ m, m' \in \mathcal{M},\ m \neq m'.$$

The second part n of such a cryptogram $y = (m, n)$ is the so called *authenticator* [12]. It is used by the receiver to check if he can accept the cryptogram as an authentic one. If the opponent is successful in an impersonation attack or in a substitution attack, respectively, he knows in addition to the general case also exactly to which message the receiver decrypts the fraudulent cryptogram.

For instance, in a substitution attack the opponent replaces the original cryptogram (m, n) by a fraudulent one (m', n') with $m' \neq m$. He will be successful if the secret key is also consistent with (m', n'), i.e., if $z \in \mathcal{K}((m', n'))$ (recall Definition 25) and $Z = z$. For ease of notation we will omit sometimes the brackets of (m, n). For instance, we write $\mathcal{K}(m, n) = \mathcal{K}((m, n))$ and for the success probability after observing the cryptogram $y = (m, n)$ we write $P_S(m, n)$ instead of $P_S((m, n))$ (recall Definition 25).

Note that for every message m the sets $\mathcal{K}(m, n)$ form a partition of $\{1, \dots, K\}$, i.e., $\bigcup_n \mathcal{K}(m, n) = \{1, \dots, K\}$ and the sets are disjoint.

We denote as $P_S(m', n', m, n)$ the probability of successful substitution of (m, n) with (m', n').

$$P_S(m', n', m, n) \triangleq \begin{cases} \frac{P_Z(\mathcal{K}(m,n) \cap \mathcal{K}(m',n'))}{P_Z(\mathcal{K}(m,n))}, & m' \neq m \\ 0, & m' = m. \end{cases} \tag{2.3.5}$$

For a chosen substitution strategy of the opponent $\{P_{Y'|Y}(\cdot|m, n) : (m, n) \in \mathcal{M}'\}$ (recall Definition 23) his success probability $P_{S,Y'}$ is given by

$$P_{S,Y'} \triangleq \sum_{m,n,m',n'} P_{Y'Y}(m',n',m,n)\, P_S(m',n',m,n). \tag{2.3.6}$$

From (2.3.5) and (2.3.6) it follows that an optimal strategy for the opponent is to select $(m^\star, n^\star)$ for given (m, n) such that

$$P_Z(\mathcal{K}(m,n) \cap \mathcal{K}(m^\star,n^\star)) = \max_{m' \neq m,\, n'} P_Z(\mathcal{K}(m,n) \cap \mathcal{K}(m',n')), \tag{2.3.7}$$

i.e., an optimal strategy for the opponent is given by

$$P_{Y'|Y}(m',n'|m,n) = \begin{cases} 1, & \text{if } (m',n') = (m^\star,n^\star) \\ 0, & \text{otherwise,} \end{cases} \tag{2.3.8}$$

where $(m^\star, n^\star)$ is in each case the maximizer in (2.3.7) dependent on (m, n) (if $(m^\star, n^\star)$ is not unique, one can choose any of the maximizers).

We denote as $P_S(m)$ the probability of successful substitution if the message m occurs. Then with (2.3.5) and (2.3.8) it follows

$$\begin{aligned} P_S(m) &= \sum_n P_{Y|X}(m,n|m)\, P_S(m^\star,n^\star,m,n) \\ &= \sum_n P_Z(\mathcal{K}(m,n))\, P_S(m^\star,n^\star,m,n) = \sum_n P_Z(\mathcal{K}(m,n) \cap \mathcal{K}(m^\star,n^\star)), \end{aligned} \tag{2.3.9}$$

where $(m^\star, n^\star)$ is in each case the maximizer in (2.3.7) dependent on (m, n).

The Lower Bound on P_S in the Case of No Secrecy

The bound on P_S presented in Theorem 35 was first given by Gilbert, MacWilliams and Sloane and proved in [12] for the case of an equiprobable message distribution. It can be generalized to arbitrary distributions P_X with the property $P_X(m) \leq \frac{1}{2}$ for all $m \in \mathcal{M}$ as it was done by Bassalygo in [2]. We will present this derivation.

In order to get a lower estimate on P_S one can consider the following two strategies, which are not optimal in general. The strategies are described as follows. If the original cryptogram is (m, n) then in both strategies the message m', which shall be substituted for m, is chosen at random from the $M - 1$ messages different from m (according to the uniform distribution). The two strategies differ only in the choice of n' given (m, n) and m'. In the first strategy n' is chosen with probability $\frac{P_S(m',n',m,n)}{\sum_{n''} P_S(m',n'',m,n)}$, i.e., the opponent uses as weights for the authenticators their success probabilities. In the second strategy n' is chosen optimal given (m, n) and m'.

To describe the strategies formally let Y'_1 and Y'_2 be the corresponding random variables for strategy 1 and 2, respectively. Then we define

$$P_{Y'_1|Y}(m',n'|m,n) \triangleq \frac{1}{M-1} \frac{P_S(m',n',m,n)}{\sum_{n''} P_S(m',n'',m,n)}$$

$$\text{and } P_{Y_2'|Y}(m', n'|m, n) \triangleq \begin{cases} \frac{1}{M-1}, & n' = n^\star \\ 0, & n' \neq n^\star, \end{cases}$$

where $n^\star$ is chosen for given m, n, m' in such a way that

$$P_Z(\mathcal{K}(m,n) \cap \mathcal{K}(m', n^\star)) = \max_{n'} P_Z(\mathcal{K}(m,n) \cap \mathcal{K}(m', n'))$$

(if $n^\star$ is not unique we choose any of the maximizers).

We denote as $P_{S,Y_1'}$ and $P_{S,Y_2'}$ the success probabilities for these strategies. It was shown in [12] that for equiprobable messages $P_S \geq P_{S,Y_2'} \geq 2^{-\frac{H(Z)}{2}} \geq \frac{1}{\sqrt{K}}$. To generalize this result for other distributions on $\mathcal{M}$ a lower bound on the sum of the probabilities of successful substitution of m with m' and m' with m for the first strategy, which is presented in the next lemma, is essential.

Definition 26 For any substitution strategy of the opponent and any two messages m and m' let $P_{S,Y'}(m', m)$ be the probability of successful substitution of message m with message m'.

Lemma 5 *For any two messages* $m, m' \in \mathcal{M},\ m \neq m'$

$$P_{S,Y_1'}(m', m) + P_{S,Y_1'}(m, m') \geq 2^{1-\frac{H(Z)}{2}}.$$

Proof Let $m, m' \in \mathcal{M},\ m \neq m'$. By (2.3.5) and the choice of Y_1' it follows that

$$\begin{aligned} P_{S,Y_1'}(m', m) &= \sum_n P_{Y|X}(m, n|m) \sum_{n'} \frac{P_S(m', n', m, n)}{\sum_{n''} P_S(m', n'', m, n)} P_S(m', n', m, n) \\ &= \sum_n P_Z(\mathcal{K}(m,n)) \sum_{n'} \frac{P_Z(\mathcal{K}(m,n) \cap \mathcal{K}(m', n'))^2}{P_Z(\mathcal{K}(m,n)) \sum_{n''} P_Z(\mathcal{K}(m,n) \cap \mathcal{K}(m', n''))} \\ &= \sum_{n,n'} \frac{P_Z(\mathcal{K}(m,n) \cap \mathcal{K}(m', n'))^2}{P_Z(\mathcal{K}(m,n))}, \end{aligned} \tag{2.3.10}$$

where we used in the last step that $\bigcup_{n''} \mathcal{K}(m', n'') = \{1, \ldots, K\}$ and the sets are disjoint.

Therefore,

$$P_{S,Y_1'}(m', m) + P_{S,Y_1'}(m, m')$$

$$= \sum_{n,n'} P_Z(\mathcal{K}(m,n) \cap \mathcal{K}(m', n'))^2 \left(\frac{1}{P_Z(\mathcal{K}(m,n))} + \frac{1}{P_Z(\mathcal{K}(m', n'))} \right).$$

As for every $a, b > 0$ $\frac{1}{a} + \frac{1}{b} \geq \frac{2}{\sqrt{ab}}$ (with equality iff $a = b$), we obtain

$$P_{S,Y_1'}(m', m) + P_{S,Y_1'}(m, m')$$

$$\geq 2 \sum_{n,n'} P_Z(\mathcal{K}(m,n) \cap \mathcal{K}(m',n')) \frac{P_Z(\mathcal{K}(m,n) \cap \mathcal{K}(m',n'))}{\sqrt{P_Z(\mathcal{K}(m,n)) P_Z(\mathcal{K}(m',n'))}}.$$

Note that $\{1,\ldots,K\} = \bigcup_{n,n'} \mathcal{K}(m,n) \cap \mathcal{K}(m',n')$ and the sets are disjoint. Therefore $\sum_{n,n'} P_Z(\mathcal{K}(m,n) \cap \mathcal{K}(m',n')) = 1$ and we can exploit the $\cap$-convexity of ln and get

$$\begin{aligned}
&\ln \left(P_{S,Y_1'}(m',m) + P_{S,Y_1'}(m,m')\right) \\
&\geq \ln 2 + \sum_{n,n'} P_Z(\mathcal{K}(m,n) \cap \mathcal{K}(m',n')) \ln \frac{P_Z(\mathcal{K}(m,n) \cap \mathcal{K}(m',n'))}{\sqrt{P_Z(\mathcal{K}(m,n)) P_Z(\mathcal{K}(m',n'))}} \\
&= \ln 2 + \frac{1}{2} \sum_{n,n'} P_Z(\mathcal{K}(m,n) \cap \mathcal{K}(m',n')) \ln P_Z(\mathcal{K}(m,n) \cap \mathcal{K}(m',n')) \\
&\quad \underbrace{+ \frac{1}{2} \sum_{n,n'} P_Z(\mathcal{K}(m,n) \cap \mathcal{K}(m',n')) \ln \frac{P_Z(\mathcal{K}(m,n) \cap \mathcal{K}(m',n'))}{P_Z(\mathcal{K}(m,n)) P_Z(\mathcal{K}(m',n'))}}_{\star} \\
&\geq \ln 2 + \frac{1}{2} \sum_{n,n'} P_Z(\mathcal{K}(m,n) \cap \mathcal{K}(m',n')) \ln P_Z(\mathcal{K}(m,n) \cap \mathcal{K}(m',n')),
\end{aligned}$$

where we used in the last step that the term $\star$ is greater than or equal to 0, which follows from the inequality $\ln x \geq 1 - \frac{1}{x}$ (it can also be seen directly by the observation that the sum is up to a positive factor an I-divergence, which is always nonnegative).

Multiplying both sides of the inequality with $\log e$ and applying the grouping axiom of the entropy function yields the desired result.

$$\begin{aligned}
&\log \left(P_{S,Y_1'}(m',m) + P_{S,Y_1'}(m,m')\right) \\
&\geq \log 2 + \frac{1}{2} \sum_{z} P_Z(z) \log P_Z(z) = 1 - \frac{1}{2} H(Z).
\end{aligned}$$

□

Theorem 35 (Gilbert, Mac Williams, Sloane-Bassalygo) *If the distribution P_X satisfies $P_X(m) \leq \frac{1}{2}$ for all $m \in \mathcal{M}$, then*

$$P_S \geq 2^{-\frac{H(Z)}{2}} \geq \frac{1}{\sqrt{K}}.$$

Proof

$$P_S = \sum_{m\in\mathcal{M}} P_X(m)\,P_S(m) \geq \sum_{m\in\mathcal{M}} P_X(m) \max_{m'\neq m} P_{S,Y_1'}(m', m). \tag{2.3.11}$$

Let $q \triangleq \min_{m\in\mathcal{M}} \max_{m'\neq m} P_{S,Y_1'}(m', m)$.

If $q \geq 2^{-\frac{H(Z)}{2}}$, then we are done and, as we did not use any restriction on P_X in this case, the theorem is valid for any distribution P_X. So let us assume that $q < 2^{-\frac{H(Z)}{2}}$.

Let $m_0 \in \mathcal{M}$ be a message such that

$$q = \max_{m'\neq m_0} P_{S,Y_1'}(m', m_0)$$

and let $m \in \mathcal{M}$ be any message with $m \neq m_0$. Then from the definition of m_0 and Lemma 5 it follows that

$$q + \max_{m'\neq m} P_{S,Y_1'}(m', m) \geq P_{S,Y_1'}(m, m_0) + P_{S,Y_1'}(m_0, m) \tag{2.3.12}$$

$$\geq 2^{1-\frac{H(Z)}{2}}. \tag{2.3.13}$$

Hence, for all $m \in \mathcal{M}$ with $m \neq m_0$ we have

$$\max_{m'\neq m} P_{S,Y_1'}(m', m) \geq 2^{1-\frac{H(Z)}{2}} - q.$$

Together with (2.3.11) this implies

$$\begin{aligned} P_S &\geq P_X(m_0)\,q + (1 - P_X(m_0))(2^{1-\frac{H(Z)}{2}} - q) \\ &= (1 - P_X(m_0))2^{1-\frac{H(Z)}{2}} - q\underbrace{(1 - 2P_X(m_0))}_{\geq 0} \\ &\geq (1 - P_X(m_0))2^{1-\frac{H(Z)}{2}} - 2^{-\frac{H(Z)}{2}}(1 - 2P_X(m_0)) \\ &= 2^{1-\frac{H(Z)}{2}} - 2^{-\frac{H(Z)}{2}} = 2^{-\frac{H(Z)}{2}}. \end{aligned} \tag{2.3.14}$$

□

Impossibility of a Generalization

In this section we show that the constant $\frac{1}{2}$ in the assumptions of Theorem 35 is best possible, i.e., a generalization of the theorem in the form that the condition "$P_X(m) \leq \frac{1}{2}$ for all m" is weakened to "$P_X(m) \leq c$ for all m" where c is a constant $> \frac{1}{2}$ is not possible.

We need the following auxiliary result.

Lemma 6

$$\lim_{a\to\infty}\left(1+a-\sqrt{a^2+a}\right)=\frac{1}{2}$$

Proof

$$1+a-\sqrt{a^2+a}-\frac{1}{2}=\sqrt{a^2+a+\frac{1}{4}}-\sqrt{a^2+a}\geq 0$$

and on the other hand the $\cap$-convexity of the square-root function implies

$$\sqrt{a^2+a+\frac{1}{4}}-\sqrt{a^2+a}\leq\frac{1}{4}\frac{1}{2\sqrt{a^2+a}}.$$

□

Now let $a\in\mathbb{N}$. We define an authentication code with two messages, $\mathcal{M}\triangleq\{1,2\}$, and $K\triangleq a^2+a$ keys, which are chosen according to the uniform distribution.

The enciphering is defined by specifying the bundles $\mathcal{K}(m,n)$ in the following way:

$$\mathcal{K}(1,n)\triangleq\{(n-1)(a+1)+1,\ldots,n(a+1)\}$$

for all $n\in\{1,\ldots,a\}$ and

$$\mathcal{K}(2,n)\triangleq\{n,n+(a+1),n+2(a+1),\ldots,n+(a-1)(a+1)\}$$

for all $n\in\{1,\ldots,a+1\}$.

For the first message we have a bundles of cardinality $a+1$ and
for the second message we have $a+1$ bundles of cardinality a. Note that

$$|\mathcal{K}(1,n)\cap\mathcal{K}(2,n')|=|\{(n-1)(a+1)+n'\}|=1$$

for all $n\in\{1,\ldots,a\}$ and $n'\in\{1,\ldots,a+1\}$. Therefore we can easily calculate P_S. According to (2.3.9) we obtain

$$P_S(1)=\sum_{n=1}^{a}\frac{1}{K}=\frac{a}{a^2+a}=\frac{1}{a+1}$$

$$\text{and } P_S(2)=\sum_{n=1}^{a+1}\frac{1}{K}=\frac{a+1}{a^2+a}=\frac{1}{a}.$$

Let $c\triangleq P_X(1)$, then

$$P_S=c\frac{1}{a+1}+(1-c)\frac{1}{a}$$

and we have $P_S < \frac{1}{\sqrt{K}} = \frac{1}{\sqrt{a^2+a}}$, if $c\frac{1}{a+1} + (1-c)\frac{1}{a} < \frac{1}{\sqrt{a^2+a}}$ or equivalently

$$c > 1 + a - \sqrt{a^2 + a}.$$

Hence, with Lemma 6 we get that if $P_X(1) > \frac{1}{2}$, then for large enough a, we obtain $P_S < \frac{1}{\sqrt{K}}$.

Conditions for Equality

Now we concentrate on the case where P_Z is the uniform distribution. For this case necessary and sufficient conditions for the equality $P_S = \frac{1}{\sqrt{K}}$ were given in [12]. As there the bound was proved for equiprobable messages and the conditions were derived from that proof, we have to give a new proof which is based on our derivation on the bound on P_S. Therefore we will make use of two lemmas stated in [2].

Definition 27 For any message $m \in \mathcal{M}$ we denote by $\mathcal{N}(m) = \{n : (m, n) = c_z(m)$ for some $z \in \{1, \ldots, K\}\}$ the set of possible authenticators attached to message m.

Lemma 7 *For given P_Z and any two messages $m, m' \in \mathcal{M},\ m \neq m'$*

$$P_{S,Y_1'}(m', m) \geq \frac{1}{|\mathcal{N}(m')|}.$$

Proof From the $\cup$-convexity of $x \mapsto x^2$ it follows that for any finite index set $\mathcal{I}$

$$\sum_{i\in\mathcal{I}} z_i^2 \geq \frac{1}{|\mathcal{I}|}\left(\sum_{i\in\mathcal{I}} z_i\right)^2, \tag{2.3.15}$$

with equality exactly if all z_i are equal. Applying this to (2.3.10) we obtain

$$\begin{aligned}
P_{S,Y_1'}(m', m) &= \sum_{n\in\mathcal{N}(m)} \sum_{n'\in\mathcal{N}(m')} \frac{P_Z(\mathcal{K}(m,n) \cap \mathcal{K}(m',n'))^2}{P_Z(\mathcal{K}(m,n))} \\
&\geq \sum_{n\in\mathcal{N}(m)} \frac{1}{P_Z(\mathcal{K}(m,n))} \frac{1}{|\mathcal{N}(m')|} \left(\sum_{n'\in\mathcal{N}(m')} P_Z(\mathcal{K}(m,n) \cap \mathcal{K}(m',n')) \right)^2 \\
&= \sum_{n\in\mathcal{N}(m)} \frac{1}{|\mathcal{N}(m')|} P_Z(\mathcal{K}(m,n)) \\
&= \frac{1}{|\mathcal{N}(m')|}.
\end{aligned}$$

□

Lemma 8 *If P_Z is the uniform distribution then for any two messages $m, m' \in \mathcal{M}$, with $m \neq m'$*

$$P_{S,Y_1'}(m',m) \geq \frac{|\mathcal{N}(m)|}{K}.$$

Proof

$$\begin{aligned} P_{S,Y_1'}(m',m) &= \sum_{n,n'} \frac{|\mathcal{K}(m,n) \cap \mathcal{K}(m',n')|^2}{K\,|\mathcal{K}(m,n)|} \\ &\geq \sum_{n}\sum_{n'} \frac{|\mathcal{K}(m,n) \cap \mathcal{K}(m',n')|}{K\,|\mathcal{K}(m,n)|} = \sum_{n\in\mathcal{N}(m)} \frac{1}{K} \\ &= \frac{|\mathcal{N}(m)|}{K}, \end{aligned}$$

with equality exactly if $|\mathcal{K}(m,n) \cap \mathcal{K}(m',n')| \leq 1$ for all n, n'. □

Now we can derive necessary and sufficient conditions that an authentication code achieves $P_S = \frac{1}{\sqrt{K}}$. These conditions are as follows:

1. $|\mathcal{N}(m)| = \sqrt{K}$ for all $m \in \mathcal{M}$.
2. $|\mathcal{K}(m,n) \cap \mathcal{K}(m',n')| = 1$ for all $m \neq m', n \in \mathcal{N}(m), n' \in \mathcal{N}(m')$.
3. $|\mathcal{K}(m,n)| = \sqrt{K}$ for all $m \in \mathcal{M},\ n \in \mathcal{N}(m)$.

Note that condition 1 and 2 imply condition 3 and therefore one could as well eliminate 3. from this list ($|\mathcal{K}(m,n)| = \sum_{n'\in\mathcal{N}(m')} |\mathcal{K}(m,n) \cap \mathcal{K}(m',n')| = \sum_{n'\in\mathcal{N}(m')} 1 = |\mathcal{N}(m')| = \sqrt{K}$).

Theorem 36 *Let P_Z be the uniform distribution. If conditions 1. and 2. are satisfied, then $P_S = \frac{1}{\sqrt{K}}$ and on the other hand if $P_S = \frac{1}{\sqrt{K}}$ and the assumption of Theorem 35 holds, then conditions 1. and 2. are satisfied.*

Proof First of all we show that condition 1. and 2. are sufficient. From (2.3.9) it follows that for every message $m \in \mathcal{M}$

$$\begin{aligned} P_S(m) &= \sum_{n} \sum_{\mathcal{K}(m,n)\cap\mathcal{K}(m^*,n^*)} \frac{1}{K} \\ &= |\mathcal{N}(m)|\,\frac{1}{K} = \frac{1}{\sqrt{K}}. \end{aligned}$$

Therefore also $P_S = \frac{1}{\sqrt{K}}$.

Now we show the necessity. Assume that $P_S = \frac{1}{\sqrt{K}}$.

Case 1: In the proof of Theorem 35 we have $q \geq \frac{1}{\sqrt{K}}$.

Then it follows

$$\max_{m' \neq m} P_{S,Y_1'}(m', m) = \frac{1}{\sqrt{K}} \qquad \text{for all } m \in \mathcal{M}.$$

Hence, for any $m' \neq m$ Lemma 7 implies

$$\frac{1}{\sqrt{K}} \geq P_{S,Y_1'}(m', m) \geq \frac{1}{|\mathcal{N}(m')|}.$$

Therefore $|\mathcal{N}(m)| \geq \frac{1}{\sqrt{K}}$ for all $m \in \mathcal{M}$ and Lemma 8 implies

$$\frac{1}{\sqrt{K}} = \max_{m' \neq m} P_{S,Y_1'}(m', m) \geq \frac{|\mathcal{N}(m)|}{K} \text{ for all } m \in \mathcal{M}.$$

Hence, we also have $|\mathcal{N}(m)| \leq \frac{1}{\sqrt{K}}$ for all $m \in \mathcal{M}$ and therefore $|\mathcal{N}(m)| = \frac{1}{\sqrt{K}}$ for all $m \in \mathcal{M}$. Furthermore Lemmas 7 and 8 hold with equality for every $m, m', m \neq m'$. Thus, the corresponding conditions for equality imply $|\mathcal{K}(m, n) \cap \mathcal{K}(m', n')| = 1$ for all $m \neq m', n \in \mathcal{N}(m), n' \in \mathcal{N}(m')$, which shows that conditions 1. and 2. are satisfied.

Case 2: $q < \frac{1}{\sqrt{K}}$.

Then in the proof of Theorem 35 for every $m \neq m_0$, (2.3.14) implies that equality holds in (2.3.12) and (2.3.13), i.e.,

$$\max_{m' \neq m_0} P_{S,Y_1'}(m', m_0) + \max_{m' \neq m} P_{S,Y_1'}(m', m)$$

$$= P_{S,Y_1'}(m, m_0) + P_{S,Y_1'}(m_0, m) = \frac{2}{\sqrt{K}}.$$

Then Lemma 7 implies $\frac{1}{\sqrt{K}} > q = P_{S,Y_1'}(m, m_0) \geq \frac{1}{|\mathcal{N}(m)|}$ or

$$|\mathcal{N}(m)| > \sqrt{K}$$

and Lemma 8 implies $\frac{1}{\sqrt{K}} > P_{S,Y_1'}(m, m_0) \geq \frac{|\mathcal{N}(m_0)|}{K}$ or

$$|\mathcal{N}(m_0)| < \sqrt{K}.$$

Together we have

$$|\mathcal{N}(m_0)| < |\mathcal{N}(m)|. \tag{2.3.16}$$

But note that for m and m_0 Lemma 5 holds with equality. For instance, the first inequality in the proof of this lemma must hold with equality and this means:

$$\text{If } \mathcal{K}(m_0, n) \cap \mathcal{K}(m, n') \neq \emptyset, \text{ then } |\mathcal{K}(m_0, n)| = |\mathcal{K}(m, n')|,$$

for all $m, n' \in \mathcal{N}(m), n \in \mathcal{N}(m_0)$.

As this is a contradiction to (2.3.16), we see that if $P_S = \frac{1}{\sqrt{K}}$, then $q \geq \frac{1}{\sqrt{K}}$ and therefore conditions 1. and 2. are necessarily satisfied. □

A Construction

We will come now to a construction which is taken from [12]. We will define an authentication code which achieves $P_S = \frac{1}{\sqrt{K}}$ (for certain values of K) and possesses the maximal possible number of messages under that constraint.

In order to see what is the maximal number of messages M, assume that we are given an authentication code with $P_S = \frac{1}{\sqrt{K}}$. Then we know that conditions 1. and 2. (and therefore also 3.) are satisfied. Now we list all unordered pairs of key-indices which are together in some bundle $\mathcal{K}(m, n)$, where $m \in \mathcal{M}, n \in \mathcal{N}(m)$. As we have M messages, $\sqrt{K}$ bundles for each message and $\sqrt{K}$ elements in each bundle, we get with this procedure $M\sqrt{K}\binom{\sqrt{K}}{2}$ pairs. Condition 2. implies that all these pairs are different and therefore their number must be less or equal the total number of unordered pairs of key-indices. This shows that

$$M\sqrt{K}\binom{\sqrt{K}}{2} \leq \binom{K}{2} \text{ or equivalently } M \leq \sqrt{K} + 1. \tag{2.3.17}$$

Our construction applies for the case that K is an even prime power. So, let us assume that $K = p^{2k}$ where p is prime and $k \in \mathbb{N}$. We make use of the projective plane constructed from $GF(q)$, where $q = p^k$. This has

- $q^2 + q + 1$ points
- $q^2 + q + 1$ lines
- $q + 1$ points on each line
- $q + 1$ lines through each point.

Recall that for every projective plane two different lines intersect in exactly one point and two different points uniquely determine a line, on which both points lie.

We select arbitrarily a line to play a special role. According to [12] we call this line the equator. The points on the equator represent the messages. All other points in the projective plane represent the keys ($K = q^2 + q + 1 - (q + 1) = q^2 = p^{2k}$). Then a message and a key uniquely determine a line through their representations in the projective plane. Therefore this line will stand for the cryptogram to which the message is encrypted using the key. From now on we will make no difference anymore between message, key, cryptogram and their representation in the projective plane.

This authentication system provides no secrecy as a cryptogram and the equator intersect in exactly one point, which is therefore the encrypted message.

In order to see if $P_S = \frac{1}{\sqrt{K}}$, we have to check if conditions 1. and 2. are satisfied:

1. As through the point m we have $q+1$ lines of which one is the equator, it follows
$$|\mathcal{N}(m)| = q + 1 - 1 = q = \sqrt{K}.$$
2. Let $m \neq m'$, $n \in \mathcal{N}(m)$, $n' \in \mathcal{N}(m')$. The lines (m, n) and (m', n') are different (if not m and m' would lie on this line and therefore (m, n) and (m', n') would be the equator, which is impossible). Hence, there is exactly one intersection point of the lines (m, n) and (m', n') (which again cannot lie on the equator because $m \neq m'$) and we obtain
$$|\mathcal{K}(m, n) \cap \mathcal{K}(m', n')| = 1.$$

Therefore the authentication code satisfies conditions 1. and 2. and we have $P_S = \frac{1}{\sqrt{K}}$. Note that equality holds in (2.3.17),
$$M = q + 1 = \sqrt{K} + 1.$$

2.3.3 Estimates on the Number of Messages Given the Success Probability of the Opponent

In this section we ask how many messages can be included in an authentication code under some constraints on the success probabilities of the opponent. We saw in the last section that a first result for this sort of question was already given in [12]. In [3] Bassalygo and Burnashev considered the case of authentication codes without secrecy. These results will be presented in Section "The Number of Messages for Authentication Codes Without Secrecy Given the Probability of Deception". Recently they gave in [4] an approach for the problem under a slightly modified constraint by connecting it to the problem of identification and the problem of the maximal cardinality of pairwise separated measures in the L_1-metric. This approach includes also cases of authentication codes without secrecy. We present the results relevant for the authentication problem in the Section on "Pairwise Separated Measures".

The Number of Messages for Authentication Codes Without Secrecy Given the Probability of Deception

Definition 28 Let $P_S^{max} \triangleq \max_{m \in \mathcal{M}} P_S(m)$ denote the maximal probability of successful substitution.

Burnashev and Bassalygo [3] require for the authentication codes under consideration to have the property that P_S^{max} does not exceed some given (usually small) constant $p \geq 0$ and ask for the maximal number of messages under this constraint. This requirement can be justified because an authentication code with $P_S^{max} \leq p$ has the property $P_D \leq p$ as well. Clearly, if $P_S^{max} \leq p$, then also $P_S \leq p$ but this holds for P_I as well, which is shown in the next theorem.

Theorem 37 *For any authentication code without secrecy*

$$P_S^{max} \geq P_I. \tag{2.3.18}$$

Proof Let $m_0 \in \mathcal{M}$ and $n_0 \in \mathcal{N}(m_0)$ such that (m_0, n_0) is an optimal choice for the impersonation attack, i.e.,

$$P_I = \Pr((m_0, n_0) \text{ valid}) = P_Z(\mathcal{K}(m_0, n_0)).$$

Now the idea is to bound for any $m \neq m_0$ the value of $P_S(m)$ below by choosing the strategy to substitute always (m_0, n_0). Let $m \in \mathcal{M},\ m \neq m_0$. Then with (2.3.9) it follows

$$P_S(m) \geq \sum_{n \in \mathcal{N}(m)} P_Z(\mathcal{K}(m, n) \cap \mathcal{K}(m_0, n_0)).$$

Therefore, as $\{\mathcal{K}(m, n) : n \in \mathcal{N}(m)\}$ is a partition of $\{1, \ldots, K\}$, we obtain

$$P_S(m) \geq P_Z(\mathcal{K}(m_0, n_0)) = P_I.$$

Hence, the statement follows from $P_S^{max} \geq P_S(m) \geq P_I$. □

Remark 13 We have seen in Example 3 that there are authentication codes (with secrecy) for which the statement (2.3.18) does not hold.

Corollary 2 *If for an authentication code without secrecy there exist $m_0, m_1 \in \mathcal{M},\ m_0 \neq m_1$ and $n_0 \in \mathcal{N}(m_0),\ n_1 \in \mathcal{N}(m_1)$ such that $P_I = P_Z(\mathcal{K}(m_0, n_0)) = P_Z(\mathcal{K}(m_1, n_1))$, i.e., if the optimal choice for an impersonation attack is not unique with respect to messages, then*

$$P_S \geq P_I.$$

Proof In this case it follows directly from the proof of Theorem 37 that for any $m \in \mathcal{M}$ we have that $P_S(m) \geq P_I$ and therefore $P_S = \sum_{m \in \mathcal{M}} P_X(m) P_S(m) \geq P_I$. □

Clearly P_S^{max} depends on the number of messages M, the definition of the K keys in $\mathcal{C}$ and the distribution P_Z, i.e., $P_S^{max} = P_S^{max}(M, \mathcal{C}, P_Z)$. If the parameters M, K and P_Z are given, then sender and receiver try to minimize P_S^{max} by using the K keys in the best possible way. Therefore it is natural to introduce the minimal achievable probability $p(M, K, P_Z)$ of successful substitution as

$$p(M, K, P_Z) \triangleq \min_{\mathcal{C}} P_S^{max}(M, \mathcal{C}, P_Z).$$

Now the question is how large can M be if K and P_Z are given and we require that $p(M, K, P_Z)$ does not exceed a given value p. The maximal M with this property will be denoted as

$$M(K, P_Z, p).$$

In other words if $M \leq M(K, P_Z, p)$, then there exists $\mathcal{C} = \{c_1, \ldots, c_K\}$ with $P_S^{max}(M, \mathcal{C}, P_Z) \leq p$.

If P_Z is the uniform distribution, $M(K, P_Z, p)$ will be denoted as

$$M_e(K, p).$$

As $P_S^{max} \geq \frac{1}{\sqrt{K}}$, we have to analyze the cases where $Kp^2 \geq 1$. We saw in Section "A Construction" that $M_e(K, \frac{1}{\sqrt{K}}) = \sqrt{K} + 1$, if K is an even prime power. Burnashev and Bassalygo studied in [3] the asymptotic behaviour of $M(K, P_Z, p)$ for large values of $K p^2$ and gave the following results.

Theorem 38 *For* $0 < p \leq \frac{1}{2}$ *the following inequality holds*

$$\log M_e(K, p) \geq \frac{K p^2}{8} + 2 \log p - 6.2\,.$$

Theorem 39 *For* $0 < p < 1$ *the following inequality holds*

$$\log M(K, P_Z, p) \leq 64\, K p^2 \log \frac{2}{p} + 2 \log K.$$

Derivation of the Lower Bound The lower bound will be proved by a construction. The idea is the following. For given $\mathcal{C}$, every message $m \in \mathcal{M}$ induces a partition of the set $\{1, \ldots, K\}$ into sets $\mathcal{K}(m, n)$, $n \in \mathcal{N}(m)$. If we have equiprobable keys, (2.3.9) implies that a "rather good" authentication code (with small P_S^{max}) must have the property that all the intersections of partition elements of the different partitions are sufficiently small.

$\mathcal{C}$ is completely determined by specifying partitions of $\{1, \ldots, K\}$ for each message. We do this by dividing the set $\{1, \ldots, K\}$ for every message $m \in \mathcal{M}$ into sets of cardinality a (the parameter a will be chosen later and we assume for the moment that $\frac{K}{a}$ is an integer). With this property each of our partitions has $\frac{K}{a}$ elements and we want to form the partitions additionally in such a way that the following condition is satisfied.

Any two elements of any two different partitions have no more than ap_0 *common elements.*

Here $0 < p_0 < p \leq \frac{1}{2}$ is a parameter, which will be chosen later. We will refer to these properties by saying that a collection of partitions satisfies the intersection property.

After adjusting the parameters we will have to show that our construction leads to an authentication code with the desired property $P_S^{max} \leq p$ but first of all, in order to get a bound on M we ask how many partitions of the described form we can find. Let $N(K, a)$ denote the number of all possible partitions of the set $\{1, \ldots, K\}$ into sets with a elements. Clearly, we have

$$N(K,a)=\frac{\binom{K}{a}\binom{K-a}{a}\cdots\binom{a}{a}}{\left(\frac{K}{a}\right)!}. \qquad (2.3.19)$$

If M is the maximal number of partitions with the intersection property, then

$$N(K,a)\le M\frac{K}{a}N(K-a,a)\sum_{i=i_0}^{a}\binom{a}{i}\binom{K-a}{a-i}, \qquad (2.3.20)$$

where i_0 is the smallest integer strictly greater than ap_0.

The validity of this inequality can be seen as follows. Take a maximal collection of partitions with the intersection property. The maximality implies that if we take any of the $N(K,a)$ partitions we find an element in it and an element of one of the partitions of our maximal collection such that they have more than ap_0 common elements. Therefore we can get any of the $N(K,a)$ partitions by a transformation of a partition of the maximal collection in the following way. First we choose one of the M partitions from which we choose one of its $\frac{K}{a}$ partition elements. From the a elements of this set we keep i in it ($i\ge i_0$) and exchange the remaining $a-i$ with some of the $K-a$ other elements. Then the other partition elements are formed from the $K-a$ remaining elements.

The right-hand side of the inequality (2.3.20) counts the number of such transformations.

From (2.3.20) we get a lower bound on M, which is

$$M\ge\frac{aN(K,a)}{KN(K-a,a)\sum_{i=i_0}^{a}\binom{a}{i}\binom{K-a}{a-i}}=\frac{a^2\binom{K}{a}}{K^2\sum_{i=i_0}^{a}\binom{a}{i}\binom{K-a}{a-i}}.$$

Now we use the following inequality which can easily be verified[2]

$$\frac{\binom{K-a}{a-i}}{\binom{K}{a}}\le\left(\frac{K-a}{K}\right)^{a-i}\left(\frac{a}{K-a}\right)^{i}$$

and we obtain

$$M\ge\frac{a^2}{K^2\sum_{i=i_0}^{a}\binom{a}{i}\left(\frac{K-a}{K}\right)^{a-i}\left(\frac{a}{K-a}\right)^{i}}=\frac{a^{2-a}(K-a)^a}{K^2\sum_{i=i_0}^{a}\binom{a}{i}\left(\frac{(K-a)^2}{aK}\right)^{a-i}}$$

2

$$\frac{\binom{K-a}{a-i}}{\binom{K}{a}}=\underbrace{\frac{K-a}{K}\cdots\frac{K-2a+i+1}{K-a+i+1}}_{a-i\text{ factors}}\underbrace{\frac{a}{K-a+i}\cdots\frac{a-i+1}{K-a+1}}_{i\text{ factors}}\le\left(\frac{K-a}{K}\right)^{a-i}\left(\frac{a}{K-a}\right)^{i}.$$

$$= \frac{a^{2-a}(K-a)^a}{K^2 \sum_{j=0}^{a-i_0} \binom{a}{j} \left(\frac{aK}{(K-a)^2}\right)^{-j}}.$$

Further we use the inequality

$$\sum_{j=0}^{b} \binom{a}{j} z^{-j} \leq z^{-b} \exp\left(a\, h\left(\frac{b}{a}\right)\right) \text{ for } 0 < z \leq \frac{a-b}{b},\ b \leq a,$$

which holds because for any $0 < x \leq 1$

$$\sum_{j=0}^{b} \binom{a}{j} z^{-j} \leq \frac{1}{x^b} \sum_{j=0}^{b} \binom{a}{j} \left(\frac{x}{z}\right)^j \leq \frac{1}{x^b} \sum_{j=0}^{a} \binom{a}{j} \left(\frac{x}{z}\right)^j = \frac{1}{x^b} \left(1 + \frac{x}{z}\right)^a$$

and with the substitution $x = \frac{bz}{a-b}$ we get

$$= \frac{1}{z^b} \left(\frac{a-b}{b}\right)^b \left(\frac{a}{a-b}\right)^a = \frac{1}{z^b} \exp(a\, h(\frac{b}{a})).$$

In our case the condition $z \leq \frac{a-b}{b}$ turns out to be

$$\frac{aK}{(K-a)^2} \leq \frac{i_0}{a-i_0}, \tag{2.3.21}$$

which we have to check after our choice of the parameters. If it holds we can bound M by

$$\begin{aligned} M &\geq \frac{a^{2-a} \left(\frac{(K-a)^2}{aK}\right)^{i_0 - a} (K-a)^a}{K^2 \exp\left(a\, h\left(\frac{a-i_0}{a}\right)\right)} \\ &\geq \frac{a^{2-a} \left(\frac{(K-a)^2}{aK}\right)^{ap_0 - a} (K-a)^a}{K^2 \exp\left(a\, h(p_0)\right)} \\ &= \frac{a^2}{K^2} \exp\left(a \underbrace{(1-2p_0)}_{\geq 0} \log \frac{K}{K-a} + ap_0 \log \frac{K}{a} - a\, h(p_0)\right) \\ &\geq \frac{a^2}{K^2} \exp\left(ap_0 \log \frac{Kp_0}{ae}\right), \end{aligned} \tag{2.3.22}$$

Now we pass to the general case, where K is not necessarily a multiple of a.

Let

$$p_0 \triangleq \frac{pe^2}{1+e^2} \quad \text{and} \quad a \triangleq \left\lceil \frac{pK}{1+e^2} \right\rceil.$$

Let $K_0 \leq K$ be the largest integer divisible by a, i.e., $K_0 \triangleq \lfloor \frac{K}{a} \rfloor a$. Now we define $\mathcal{C}$ by choosing the partitions as follows. We select an arbitrary subset of $\{1, \dots, K\}$ with $K - K_0$ elements to form a partition element for every partition. From the remaining K_0 elements of $\{1, \dots, K\}$ we form a collection of partitions such that the intersection property holds.

First of all we show that the resulting authentication code possesses the desired property $P_S^{max} \leq p$. Let $m \in \mathcal{M}$. Then

$$P_S(m) = \sum_n \frac{1}{K} \left| \mathcal{K}(m, n) \cap \mathcal{K}(m^\star, n^\star) \right| \leq \frac{K - K_0}{K} + \frac{K_0}{a} \cdot \frac{ap_0}{K}.$$

By definition of K_0 we have $K - K_0 \leq a - 1$ and therefore

$$P_S(m) \leq \frac{a-1}{K} + \frac{K_0}{K} p_0 \leq \frac{a-1}{K} + p_0 \leq \frac{pK}{K(1+e^2)} + \frac{pe^2}{1+e^2} = p.$$

In order to apply our estimate for M we have to check if (2.3.21) is satisfied. W.l.o.g. we may assume that $pK > 70$ (see (2.3.25)). Then $a = \lceil \frac{p_0 K}{e^2} \rceil \leq \frac{p_0 K}{e}$ and $\frac{K_0}{K} = \lfloor \frac{K}{a} \rfloor \frac{a}{K} \geq 1 - \frac{a}{K} \geq 1 - \frac{p_0}{e}$. Therefore

$$\frac{aK_0}{(K_0 - a)^2} \leq \frac{\frac{p_0}{e}}{(\frac{K_0}{K} - \frac{p_0}{e})^2} \leq \frac{\frac{p_0}{e}}{(1 - 2\frac{p_0}{e})^2} \leq \frac{p_0}{1 - p_0} \leq \frac{i_0}{a - i_0},$$

where we used that $p_0 \leq \frac{1}{2}$ and $ap_0 \leq i_0$. As (2.3.21) holds the number $M_e(K, p)$ must satisfy the last inequality for M, which is (2.3.22), with K replaced by K_0, i.e.,

$$\begin{aligned} M_e(K, p) &\geq \frac{a^2}{K_0^2} \exp\left(ap_0 \log \frac{K_0 p_0}{ae} \right) \\ &\geq \frac{p^2}{(1+e^2)^2} \exp\left(\frac{pK}{1+e^2} \cdot \frac{pe^2}{1+e^2} \log \frac{K_0 pe}{a(1+e^2)} \right) \\ &= \frac{p^2}{(1+e^2)^2} \exp\left(\frac{Kp^2 e^2}{(1+e^2)^2} \log \frac{1}{z} \right), \end{aligned} \tag{2.3.23}$$

with $z \triangleq \frac{a(1+e^2)}{K_0 pe}$.

The value z satisfies the following inequalities

$$z = \frac{\left\lceil \frac{pK}{1+e^2} \right\rceil (1+e^2)}{K_0 pe} \geq \frac{K}{K_0 e} \geq \frac{1}{e}$$

and

$$z \leq \frac{\left(\frac{pK}{1+e^2} + 1\right)(1+e^2)}{K_0 pe} = \frac{1}{e} \cdot \frac{1 + \frac{1+e^2}{pK}}{1 - \frac{K-K_0}{K}}$$
$$\leq \frac{1}{e} \cdot \frac{1 + \frac{1+e^2}{pK}}{1 - \frac{a-1}{K}} \leq \frac{1}{e} \cdot \frac{1 + \frac{1+e^2}{pK}}{1 - \frac{p}{1+e^2}}.$$

Combining these two inequalities, yields

$$\frac{1}{e} \leq z \leq \frac{1}{e} \cdot \frac{1 + \frac{1+e^2}{pK}}{1 - \frac{p}{1+e^2}} \tag{2.3.24}$$

and as $\log \frac{1}{z}$ is monotonically decreasing in z, it attains its minimal value at the right-hand side of (2.3.24). Substituting this into (2.3.23) we get

$$M_e(K,p) \geq \frac{p^2}{(1+e^2)^2} \exp\left(\frac{Kp^2e^2}{(1+e^2)^2} (\log e) \left(1 + \ln \frac{1 - \frac{p}{1+e^2}}{1 + \frac{1+e^2}{pK}} \right)\right).$$

Taking the logarithm on both sides of the inequality we get that if $pK > 70$ and $p \leq \frac{1}{2}$

$$\log M_e(K,p) \geq 2\log p - \underbrace{2\log(1+e^2)}_{\approx 6.14} + Kp^2 \underbrace{\frac{e^2 \log e}{(1+e^2)^2} \left(1 + \ln \frac{1 - \frac{\frac{1}{2}}{1+e^2}}{1 + \frac{1+e^2}{70}} \right)}_{\approx 0.12502}$$
$$\geq \frac{Kp^2}{8} + 2\log p - 6.2.$$

If $pK \leq 70$, the statement is trivial because in this case

$$\exp(\frac{Kp^2}{8} - 6.2)p^2 \leq \exp(\frac{70}{16} - 6.2) \leq 0.3. \tag{2.3.25}$$

Hence, the proof of Theorem 38 is complete.

Derivation of the Upper Bound First of all we will derive an upper bound for $M_e(K,p)$ and then generalize this bound to arbitrary key distributions P_Z.

Let us assume we are given an authentication code with $P_S^{max} \leq p$. As we have seen before every message $m \in \mathcal{M}$ induces a partition of $\{1, \dots, K\}$ into sets $\mathcal{K}(m, n)$. Every such partition element must have a cardinality less than pK. Assume on the contrary that $|\mathcal{K}(m, n)| > pK$ for some $m \in \mathcal{M}, n \in \mathcal{N}(m)$. Then for any $m' \in \mathcal{M}$ with $m' \neq m$ we would have $P_S(m') \geq \sum_{n'} \frac{1}{K} |\mathcal{K}(m', n') \cap \mathcal{K}(m, n)| = \frac{1}{K} |\mathcal{K}(m, n)| > p$, which is a contradiction.

Moreover, there exists for every message $m \in \mathcal{M}$ a certain element $\mathcal{A}_m$ of the corresponding partition with the property that the cardinality of the intersection of $\mathcal{A}_m$ with any element of any other partition does not exceed $p|\mathcal{A}_m|$. This follows from the next lemma.

Lemma 9 *If for an authentication code without secrecy* $P_S^{max} \leq p$*, then for every* $m \in \mathcal{M}$ *there exists* $n \in \mathcal{N}(m)$ *with the property*

$$P_Z(\mathcal{K}(m, n) \cap \mathcal{K}(m', n')) \leq p P_Z(\mathcal{K}(m, n)) \quad \text{for any } m' \neq m, n' \in \mathcal{N}(m').$$

Proof Assume on the contrary that for $m \in \mathcal{M}$ there exists no such $\mathcal{K}(m, n)$. This means that for every $n \in \mathcal{N}(m)$ there exists $m' \in \mathcal{M} \quad m' \neq m$ and $n' \in \mathcal{N}(m')$ such that $P_Z(\mathcal{K}(m, n) \cap \mathcal{K}(m', n')) > p P_Z(\mathcal{K}(m, n))$. Therefore we get by substituting (m, n) with (m', n') the desired contradiction $P_S(m) > \sum_n p P_Z(\mathcal{K}(m, n)) = p$. □

From the obtained set $\{\mathcal{A}_m : m \in \mathcal{M}\}$ we can take out a maximal subset $\{\mathcal{A}_{m_1}, \dots, \mathcal{A}_{m_N}\}$ such that all the $\mathcal{A}_{m_i}$ have the same cardinality. We denote this cardinality by w. Then this subset has the following properties:

1. $|\mathcal{A}_{m_i}| = w \leq \lfloor pK \rfloor$ for all $i = 1, \dots, N$.
2. $|\mathcal{A}_{m_i} \cap \mathcal{A}_{m_j}| \leq \lfloor pw \rfloor$ for all $i, j = 1, \dots, N \quad i \neq j$.
3. $N \geq \frac{M}{pK}$.

Properties 1 and 2 are clear by construction of the set $\{\mathcal{A}_{m_1}, \dots, \mathcal{A}_{m_N}\}$. Property 3 follows from the fact that all the sets $\mathcal{A}_m$ have cardinalities less than pK and the number of sets $\mathcal{A}_m$ with some same cardinality is less than N. Therefore $N \cdot pK \geq M$.

We can also give an upper bound for N, which is well known in coding theory and combinatorics (see the remark below) but we will give its derivation here. Let $l \triangleq \lfloor pw \rfloor$ and let $t > l$. Then property 2 implies that all possible subsets of the sets $\mathcal{A}_{m_1}, \dots, \mathcal{A}_{m_N}$, which have t elements, are different. Therefore the total number of subsets obtained in this way is less than the total number of t-elementary subsets of $\{1, \dots, K\}$, i.e., $N \cdot \binom{w}{t} \leq \binom{K}{t}$, or

$$N \leq \frac{\binom{K}{t}}{\binom{w}{t}} \qquad \text{for all } t > l.$$

As the right hand side attains its minimal value for $t = l + 1$ we obtain

$$N \leq \frac{K(K-1)\cdots(K-l)}{w(w-1)\cdots(w-l)}. \tag{2.3.26}$$

Remark 14 If we consider the characteristic vectors of the sets $\mathcal{A}_{m_i}$, then we obtain a constant weight code with weight w and Hamming distance between the codewords at least $2(w-l)$. The upper bound in (2.3.26) is nothing else than the Johnson bound (see [17], pp. 527) for the cardinality of such a code.

If we combine the two estimates for N ((2.3.26) and property 3.) we get an upper bound for M. As we do not know the concrete value of w we maximize over w.

$$\begin{aligned}
M &\le pK \max_{1\le w\le pK} \frac{K(K-1)\cdots(K-l)}{w(w-1)\cdots(w-l)} \\
&\le pK \max_{1\le w\le pK} \frac{K}{w}\cdot\left(\frac{K-l}{w-l}\right)^{l} \\
&\le pK^2 \max_{1\le w\le pK} \left(\frac{K-pw}{w-pw}\right)^{pw} \\
&= pK^2 \exp\left(p \max_{1\le w\le pK} w \log\frac{K-pw}{w-pw}\right).
\end{aligned}$$

The maximized function is $\cap$-convex in w and the first derivative is positive at $w = pK$ provided that $p \le 0.42$. Hence, in this case the function attains its maximum at $w = pK$. By substituting this into the last term we obtain the following Proposition.

Proposition 3 *If $p \le 0.42$, then the following inequality holds*

$$M_e(K,p) \le pK^2 \exp\left(Kp^2 \log\frac{1+p}{p}\right).$$

Now we would like to transform this result to the case of an arbitrary key distribution P_Z.

Definition 29 If P_Z is the uniform distribution then let

$$p_e(M,K) \triangleq p(M,K,P_Z)$$

and let $p(M,K)$ denote the minimal achievable probability of successful substitution for K keys and M messages, i.e.,

$$p(M,K) \triangleq \min_{P_Z} p(M,K,P_Z).$$

Lemma 10 *Let $\mathcal{K} \subset \{1,\dots,K\}$ with $|\mathcal{K}| = N$. Then the following statements hold.*

(a)

$$p(M,K,P_Z) \ge P_Z(\mathcal{K})\, p(M,N).$$

(b) If P_Z satisfies also the condition $P_Z(z) \geq \beta$ for all $z \in \mathcal{K}$, then

$$p(M, K, P_Z) \geq \beta N\, p_e(M, N).$$

Proof We start with (a). Recall that

$$p(M, K, P_Z) = \min_{\mathcal{C}} P_S^{max}(M, \mathcal{C}, P_Z). \tag{2.3.27}$$

Let $\mathcal{C}$ be a minimizer in (2.3.27). Then for all $m \in \mathcal{M}$ it follows

$$p(M, K, P_Z) \geq P_S(m) \geq \sum_n P_Z(\mathcal{K}(m, n) \cap \mathcal{K}(m', n'))$$

where $m' = m'(m, n) \neq m$ and $n' = n'(m, n)$ are chosen according to some not necessarily optimal decision rule. Hence,

$$p(M, K, P_Z) \geq P_Z(\mathcal{K}) \sum_n \frac{P_Z(\mathcal{K}(m, n) \cap \mathcal{K}(m', n') \cap \mathcal{K})}{P_Z(\mathcal{K})}$$

Let $\mathcal{C}' \subset \mathcal{C}$ be the subset of keys with index in $\mathcal{K}$. If we take for $m'(m, n)$ and $n'(m, n)$ the opponents's optimal decision rule for the authentication code, where the keys are chosen from $\mathcal{C}'$ according to the distribution $\frac{P_Z(\cdot)}{P_Z(\mathcal{K})}$, then we can conclude from the last inequality and the definition of $p(M, N)$ that

$$\begin{aligned} p(M, K, P_Z) &\geq P_Z(\mathcal{K})\, P_S^{max}(M, \mathcal{C}', \frac{P_Z(\cdot)}{P_Z(\mathcal{K})}) \\ &\geq P_Z(\mathcal{K})\, p(M, N, \frac{P_Z(\cdot)}{P_Z(\mathcal{K})}) \geq P_Z(\mathcal{K})\, p(M, N), \end{aligned}$$

which completes the proof of (a).

Now we prove (b). Let $\mathcal{C}$ be a minimizer in (2.3.27) again. Then

$$\begin{aligned} p(M, K, P_Z) = P_S^{max}(M, \mathcal{C}, P_Z) &\geq \max_{m \in \mathcal{M}} \sum_n P_Z(\mathcal{K}(m, n) \cap \mathcal{K}(m', n')) \\ &\geq \max_{m \in \mathcal{M}} \sum_n P_Z(\mathcal{K}(m, n) \cap \mathcal{K}(m', n') \cap \mathcal{K}) \\ &\geq \beta N \max_{m \in \mathcal{M}} \sum_n \frac{|\mathcal{K}(m, n) \cap \mathcal{K}(m', n') \cap \mathcal{K}|}{N} \end{aligned}$$

and if $m'(m,n)$ and $n'(m,n)$ are chosen such that the last expression is maximized, then we obtain

$$p(M,K,P_Z) \geq \beta N p_e(M,N). \qquad \square$$

In order to prove Theorem 39 we will derive a sequence of upper bounds for $M(K,P_Z,p)$ and in the limit we get the bound of the theorem.

Let us start with the following result.

Proposition 4 *The following statements hold.*

(a) *If $M \geq 2^K$ then $p(M,K) = 1$.*
(b) *If $0 < p < 1$ then for arbitrary P_Z*

$$\lfloor \log M(K,P_Z,p) \rfloor \leq Kp.$$

Proof Let $M \geq 2^K$ and suppose $P_S(m) \leq p$ for all $m \in \mathcal{M}$ and some $p \leq 1$. In order to prove (a) we have to show that $p = 1$.

We know from Lemma 9 that for every $m \in \mathcal{M}$ there exists an element $\mathcal{A}_m$ of the corresponding partition with $P_Z(\mathcal{A}_m \cap \mathcal{K}(m',n')) \leq p P_Z(\mathcal{A}_m)$ for any $m' \neq m$ and $n' \in \mathcal{N}(m')$.

In particular we have

$$P_Z(\mathcal{A}_m \cap \mathcal{A}_{m'}) \leq p P_Z(\mathcal{A}_m) \qquad \text{for all } m' \neq m.$$

As there are $2^K - 1$ nonempty subsets of $\{1,\dots,K\}$ and as $M > 2^K - 1$ we can find $m' \neq m$ with $\mathcal{A}_m = \mathcal{A}_{m'}$. If $p < 1$, then it follows

$$P_Z(\mathcal{A}_m) = P_Z(\mathcal{A}_m \cap \mathcal{A}_{m'}) \leq p P_Z(\mathcal{A}_m) < P_Z(\mathcal{A}_m)$$

which is a contradiction and therefore necessarily $p = 1$.

In order to prove (b) let $\mathcal{K} \subset \{1,\dots,K\}$ be the subset with the $\lfloor \log M \rfloor$ most probable key-indices. Then we apply part (a) of Lemma 10 and get

$$p(M,K,P_Z) \geq P_Z(\mathcal{K}) p(M, \lfloor \log M \rfloor).$$

By the choice of $\mathcal{K}$ it follows that $P_Z(\mathcal{K}) \geq \frac{\lfloor \log M \rfloor}{K}$ and we have already proved in (a) that $p(M, \lfloor \log M \rfloor) = 1$. Therefore

$$\lfloor \log M \rfloor \leq K p(M,K,P_Z). \qquad \square$$

In the sequel we only have to consider the case $p < \frac{1}{4}$ because for $p \geq \frac{1}{4}$ the bound in Proposition 4 (b) is stronger than the bound of Theorem 39 (for $p \geq \frac{1}{4}$ it holds that $64Kp^2 \log \frac{2}{p} + 2\log K \geq 64Kp^2 \geq 16Kp \geq Kp$).

We assume the keys to be enumerated such that

$$P_Z(1) \geq \cdots \geq P_Z(K).$$

Then necessarily $P_Z(1) < \frac{1}{4}$ because otherwise $P_S^{max} \geq \frac{1}{4}$ and therefore $p \geq \frac{1}{4}$.

Let $\mathcal{K} \subset \{1, \ldots, K\}$ be the maximal subset consisting of the first N key-indices such that $P_Z(\mathcal{K}) \leq \frac{1}{2}$.

Then clearly $P_Z(\mathcal{K}) > \frac{1}{4}$ and $P_Z(z) \geq \frac{1}{2K}$ for all $z \in \mathcal{K}$ (assume on the contrary that $P_Z(N) < \frac{1}{2K}$ then $P_Z(z) < \frac{1}{2K}$ for all $z \geq N$ and therefore $P_Z(\mathcal{K}) > 1 - (K - N)\frac{1}{2K} > \frac{1}{2}$, which is a contradiction).

We now apply Lemma 10 (a) and get $p \geq p(M, K, P_Z) \geq P_Z(\mathcal{K})p(M, N) \geq \frac{p(M,N)}{4}$ and therefore

$$p(M, N) \leq 4p < 1. \tag{2.3.28}$$

From part (b) of Lemma 10 we get $p \geq \frac{1}{2K} N\, p_e(M, N)$ or

$$p_e(M, N) \leq \frac{2Kp}{N}. \tag{2.3.29}$$

Combining (2.3.28) and Proposition 4 (b) we see that M must satisfy the inequality

$$\log M \leq 4pN + 1 \leq 4pN \log \frac{2}{p} + 2 \log K. \tag{2.3.30}$$

Combining (2.3.29) and the bound of Proposition 3 we get

$$\begin{aligned}\log M &\leq \log M_e(N, p_e(M, N)) \\ &\leq \log M_e(N, \frac{2Kp}{N}) \leq \frac{4K^2p^2}{N} \log \frac{2}{p} + 2 \log K,\end{aligned} \tag{2.3.31}$$

where we have to assume that $\frac{2Kp}{N} \leq 0.42$ in order to apply Proposition 3 but otherwise $4pN \leq 64Kp^2$ and therefore the bound in (2.3.30) would be sharper then the bound of Theorem 39.

Combining (2.3.30) and (2.3.31) yields

$$\begin{aligned}\log M &\leq 4p \min\{N, \frac{K^2p}{N}\} \log \frac{2}{p} + 2 \log K \\ &\leq 4Kp^{\frac{3}{2}} \log \frac{2}{p} + 2 \log K,\end{aligned} \tag{2.3.32}$$

where the last inequality can be verified as follows: if $N \leq \frac{K^2p}{N}$, then $N \leq Kp^{\frac{1}{2}}$ and if $\frac{K^2p}{N} < N$, then $\frac{K}{N} < \frac{1}{\sqrt{p}}$ and therefore $\frac{K^2p}{N} \leq Kp^{\frac{1}{2}}$.

So we have obtained from the bound Kp the stronger bound (for sufficiently small p) $4Kp^{\frac{3}{2}} \log \frac{2}{p} + 2 \log K$. We now repeat the procedure using instead of the bound Kp the new bound, i.e., we combine the inequalities

$$\log M \leq 4N(4p)^{\frac{3}{2}} \log \frac{2}{p} + 2 \log K$$

and

$$\log M \leq 4\frac{K^2 p^2}{N} \log \frac{2}{p} + 2 \log K \leq 8\frac{K^2 p^2}{N} \log \frac{2}{p} + 2 \log K$$

to

$$\begin{aligned} \log M &\leq (4p)^{\frac{3}{2}} 4 \min\{N, \frac{K^2 p^{2-\frac{3}{2}}}{N}\} \log \frac{2}{p} + 2 \log K \\ &\leq 16Kp^{\frac{7}{4}} \log \frac{2}{p} + 2 \log K. \end{aligned}$$

Generally, if after the nth step we have the inequality

$$\log M \leq C_n K p^{\alpha_n} \log \frac{2}{p} + 2 \log K$$

then in the $(n+1)$th step we obtain the same type of inequality with coefficients C_{n+1} and α_{n+1} that satisfy

$$C_{n+1}^2 = 64 C_n \quad \text{and} \quad \alpha_{n+1} = 1 + \frac{\alpha_n}{2}.$$

(Note that in the $(n+1)$th step the inequality $\log M \leq 4\frac{K^2p^2}{N} \log \frac{2}{p} + 2 \log K$ has to be weakened to $\log M \leq \beta_{n+1} \frac{K^2 p^2}{N} \log \frac{2}{p} + 2 \log K$ with $\beta_{n+1} \triangleq \frac{64}{4^{\alpha_n}} \geq 4$ to adjust the min term in the right way.)

As $\lim_{n \to \infty} \alpha_n = 2$ and $\lim_{n \to \infty} C_n = 64$ we obtain that $M(K, P_Z, p)$ must satisfy the inequality

$$\log M(K, P_Z, p) \leq 64\, K\, p^2 \log \frac{2}{p} + 2 \log K,$$

which completes the proof of Theorem 39.

Remark 15 1. For small p the principal difference of the upper and the lower bound consists of an additional factor of order $\ln \frac{1}{p}$. Burnashev and Bassalygo [3, 4] say that they do not know which of the bounds can be improved.

2. The estimates on the number $M(K, P_Z, p)$ should certainly depend on the distribution P_Z. Burnashev and Bassalygo [3] conjectured that this dependence is as follows

$$C_1 + C_2 p^2 \exp(H(Z)) < \log M(K, P_Z, p) < C_3 + C_4 p^2 \log \frac{1}{p} \exp(H(Z))$$

where $C_1, \ldots, C_4$ are constants.

Pairwise Separated Measures

Now we will return to the general case and no longer restrict ourselves to the class of authentication codes without secrecy. If we consider the problem of the last section, then the lower bound we gave there remains valid as we have only enlarged our possibilities to build authentication codes. The problem how the secrecy provided by an authentication code attaches the answer to the question of the maximal number of messages given the probability of deception has not been treated rigorously. In this section the constraint on the success probability of the opponent, which has to be fulfilled by the authentication codes, is sharpened compared to the last section. This will allow us to use as an upper bound for the maximal number of messages the maximal number of pairwise separated measures.

Definition 30 Let $\mathcal{K} \triangleq \{1, \ldots, K\}$ and $\mu_1, \ldots, \mu_M$ be probability measures on $\mathcal{K}$. Further let p be a constant with $0 \leq p \leq 1$. The L_1-norm of a function $\mu : \mathcal{K} \rightarrow \mathbb{R}$ is

$$||\mu|| \triangleq \sum_{z \in \mathcal{K}} |\mu(z)|.$$

The set $\{\mu_i \ : \ i = 1, \ldots, M\}$ is called p-pairwise separated if

$$||\mu_i - \mu_j|| \geq 2(1 - p)$$

for any $i, j = 1, \ldots, M \quad i \neq j$.

When working with the L_1 distance of probability measures the following identity is useful.

Lemma 11 *For two probability measures μ and ν on $\mathcal{K}$*

$$||\mu - \nu|| = 2\left(1 - \sum_{z \in \mathcal{K}} \min\{\mu(z), \nu(z)\}\right).$$

Proof

$$\begin{aligned} ||\mu - \nu|| &= \sum_{z:\mu(z)\geq\nu(z)} (\mu(z) - \nu(z)) + \sum_{z:\nu(z)>\mu(z)} (\nu(z) - \mu(z)) \\ &= \sum_{z:\mu(z)\geq\nu(z)} \mu(z) + \sum_{z:\nu(z)>\mu(z)} \nu(z) - \sum_{z\in\mathcal{K}} \min\{\mu(z), \nu(z)\} \\ &= \sum_{z\in\mathcal{K}} \mu(z) + \sum_{z\in\mathcal{K}} \nu(z) - 2\sum_{z\in\mathcal{K}} \min\{\mu(z), \nu(z)\} = 2 - 2\sum_{z\in\mathcal{K}} \min\{\mu(z), \nu(z)\}. \quad \square \end{aligned}$$

Definition 31 For a given constant $0 \leq p < 1$ we denote by $M_{sep}(K, p)$ the maximal cardinality of a set of p-pairwise separated probability measures on $\mathcal{K}$.

In [6] the following inequality for the value $M_{sep}(K, p)$ was proved

$$M_{sep}(K, p) \leq \left(\frac{2}{1-p}\right)^{K-1}. \tag{2.3.33}$$

The main analytical result in [4] consists of an improvement of this bound for small p, which makes it valuable for the problem of the maximal number of messages in an authentication code.

Theorem 40 (Burnashev and Bassalygo) *For any* $0 < p < 1$ *the following inequality holds*

$$M_{sep}(K, p) \leq K + \frac{1}{p^2} + \frac{1}{2p^2} \exp\left(\frac{p^2 K}{(1-\sqrt{p})^3} \log \frac{2e}{p^2}\right).$$

In order to prove Theorem 40 we need the following Lemma.

Lemma 12 *Let* $\{\mu_1, \ldots, \mu_M\}$ *be a set of* δ*-pairwise separated probability measures on* $\mathcal{K}$ *and let* $\mathcal{K}_i \triangleq \{z \in \mathcal{K} : \mu_i(z) > 0\}$ *be the support of* μ_i *for any* $i = 1, \ldots, M$. *Then the following statements hold.*

(a) If $\max\{\mu_i(z) : z \in \mathcal{K}, i = 1, \ldots, M\} \leq \mu$, *then*

$$M \leq \frac{(1-\delta)\mu K}{1-\delta\mu K}, \qquad \textit{provided that } 1 - \delta\mu K > 0.$$

(b) If $\mu_i(z) \geq \mu \geq \frac{\delta}{K}$ *for all* $z \in \mathcal{K}_i$ *and all* $i = 1, \ldots, M$, *then*

$$M \leq \frac{(1-\delta)\mu K}{2\delta} \exp\left(\frac{\delta}{\mu} \log \frac{2e\mu K}{\delta(1-\delta)}\right).$$

Proof We start with (a). As $\{\mu_1, \ldots, \mu_M\}$ is δ-pairwise separated it follows that

$$\sum_{i=1}^{M} \sum_{j=1}^{M} ||\mu_i - \mu_j|| \geq 2(1-\delta)M(M-1). \tag{2.3.34}$$

Now we bound this sum from above using the identity of Lemma 11 and the inequality $\min\{\mu_i(z), \mu_j(z)\} \geq \frac{\mu_i(z)\mu_j(z)}{\mu}$, which holds by the assumption made in (a).

$$\sum_{i=1}^{M}\sum_{j=1}^{M}||\mu_i - \mu_j|| \leq 2\left(M^2 - \frac{1}{\mu}\sum_{i=1}^{M}\sum_{j=1}^{M}\sum_{z\in\mathcal{K}}\mu_i(z)\mu_j(z)\right)$$
$$= 2\left(M^2 - \frac{1}{\mu}\sum_{z\in\mathcal{K}}\left(\sum_{i=1}^{M}\mu_i(z)\right)^2\right)$$
$$\leq 2\left(M^2 - \frac{1}{\mu K}\left(\sum_{z\in\mathcal{K}}\sum_{i=1}^{M}\mu_i(z)\right)^2\right) = 2M^2\left(1 - \frac{1}{\mu K}\right), \tag{2.3.35}$$

where we applied (2.3.15) to get the last inequality. Combining (2.3.34) and (2.3.35) leads to

$$\left(\frac{1-\delta\mu K}{\mu K}\right)M \leq (1-\delta)$$

and this proves (a).

Now we prove (b). As $\{\mu_1, \ldots, \mu_M\}$ is δ-pairwise separated and the assumption made in (b) implies that $\mu \leq \min\{\mu_i(z), \mu_j(z)\}$ for all $z \in \mathcal{K}_i \cap \mathcal{K}_j$ it follows that for $i \neq j$

$$|\mathcal{K}_i \cap \mathcal{K}_j| \cdot \mu \leq \sum_{z\in\mathcal{K}} \min\{\mu_i(z), \mu_j(z)\} \leq \delta.$$

Therefore $|\mathcal{K}_i \cap \mathcal{K}_j| \leq \left\lfloor \frac{\delta}{\mu} \right\rfloor$ for $i \neq j$. Let $T \triangleq \left\lfloor \frac{\delta}{\mu} \right\rfloor$. This implies that the number of measures μ_i with $|\mathcal{K}_i| > T$ does not exceed $\binom{K}{T+1}$ (otherwise there would be two measures μ_i and μ_j ($i \neq j$) with $|\mathcal{K}_i \cap \mathcal{K}_j| \geq T+1$) and clearly the number of measures μ_i with $|\mathcal{K}_i| \leq T$ does not exceed $\binom{K}{T}M_{sep}(T, \delta)$. Therefore

$$M \leq \binom{K}{T+1} + \binom{K}{T}M_{sep}(T,\delta) \leq \frac{K}{T}\binom{K}{T}M_{sep}(T,\delta).$$

Using the bound given in (2.3.33) for the value $M_{sep}(T, \delta)$ and the inequality $\binom{n}{k} \leq \left(\frac{ne}{k}\right)^k$, which can be verified using Stirling's formula,[3] we obtain

$$M \leq \frac{K}{T}\left(\frac{Ke}{T}\right)^T\left(\frac{2}{1-\delta}\right)^{T-1} = \frac{(1-\delta)K}{2T}\exp\left(T\log\left(\frac{2Ke}{(1-\delta)T}\right)\right)$$

[3] $\binom{n}{k} \leq \left(\frac{n}{k}\right)^k (1+\frac{k}{n-k})^{n-k} e^{\frac{1}{12n}-\frac{1}{12k+1}-\frac{1}{12(n-k)+1}+\frac{1}{2}\ln(\frac{n}{2\pi k(n-k)})}$
$\leq \left(\frac{ne}{k}\right)^k e^{\frac{1}{2n}-\frac{1}{6n+1}+\frac{1}{2}\ln(\frac{n}{2\pi(n-1)})} \leq \left(\frac{ne}{k}\right)^k.$

$$\leq \frac{(1-\delta)\mu K}{2\delta} \exp\left(\frac{\delta}{\mu} \log\left(\frac{2e\mu K}{\delta(1-\delta)}\right)\right). \qquad \square$$

Proof of the theorem. Let $\{\mu_1, \dots, \mu_M\}$ be a set of p-pairwise separated probability measures on $\mathcal{K}$. It contains not more than K measures μ_i with $\max_z \mu_i(z) > p$ because otherwise there would be some $i \neq j$ and a z with $\min\{\mu_i(z), \mu_j(z)\} > p$, which implies $||\mu_i - \mu_j|| < 2(1-p)$. Therefore below we assume that all the measures μ_i satisfy $\max_z \mu_i(z) \leq p$ and derive for that case an upper bound to which we have to add K in the end.

Fix now parameters μ and ϵ such that $0 < p < \epsilon < 1$ and $0 < \mu$ (the parameters will be chosen later) and let $\mathcal{K}_i(\mu) \triangleq \{z \in \mathcal{K} : \mu_i(z) \geq \mu\}$. First we upper bound the number M_1 of measures μ_i with $\mu_i(\mathcal{K}_i^c(\mu)) \geq 1 - \epsilon$. We may assume that these measures are $\mu_1, \dots, \mu_{M_1}$ and introduce on their basis new probability measures ν_i with supports $\mathcal{K}_i^c(\mu)$ in the following way.

$$\nu_i(z) \triangleq \frac{\mu_i(z)}{\mu_i(\mathcal{K}_i^c(\mu))} \qquad \text{for all } z \in \mathcal{K}_i^c(\mu) \text{ and } i = 1, \dots, M_1.$$

For these measures we obtain the following relation.

$$||\nu_i - \nu_j|| \geq 2\left(1 - \sum_{z \in \mathcal{K}_i^c(\mu) \cap \mathcal{K}_j^c(\mu)} \frac{\min\{\mu_i(z), \mu_j(z)\}}{1-\epsilon}\right) \geq 2\left(1 - \frac{p}{1-\epsilon}\right)$$

for all $i, j = 1 \dots, M_1$, $i \neq j$. Furthermore

$$\max_{z \in \mathcal{K}} \nu_i(z) < \frac{\mu}{\mu_i(\mathcal{K}_i^c(\mu))} \leq \frac{\mu}{1-\epsilon} \qquad \text{for all } i = 1, \dots, M_1.$$

Thus we can apply Lemma 12 (a) to bound M_1.

$$M_1 \leq \frac{(1 - \frac{p}{1-\epsilon})\frac{\mu}{1-\epsilon}K}{1 - \frac{p}{1-\epsilon}\frac{\mu}{1-\epsilon}K} = \frac{(1-\epsilon-p)\mu K}{(1-\epsilon)^2 - p\mu K} \leq \frac{\mu K}{(1-\epsilon)^2 - p\mu K}, \tag{2.3.36}$$

provided that

$$(1-\epsilon)^2 - p\mu K > 0. \tag{2.3.37}$$

Now we consider the remaining $M_2 = M - M_1$ measures μ_i with $\mu_i(\mathcal{K}_i(\mu)) \geq \epsilon$. As all the values $\mu_i(z)$ do not exceed p there exists in every set $\mathcal{K}_i(\mu)$ a subset $\mathcal{K}_i'(\mu)$ such that $\epsilon \leq \mu_i(\mathcal{K}_i(\mu)) \leq \epsilon + p$. We introduce new probability measures σ_i with supports $\mathcal{K}_i'(\mu)$ in the following way.

$$\sigma_i(z) \triangleq \frac{\mu_i(z)}{\mu_i(\mathcal{K}_i'(\mu))} \qquad \text{for all } z \in \mathcal{K}_i'(\mu) \text{ and } i = M_1 + 1, \dots, M.$$

For these measure we obtain the following relation.

$$||\sigma_i - \sigma_j|| \geq 2\left(1 - \sum_{z \in \mathcal{K}_i'(\mu)\cap\mathcal{K}_j'(\mu)} \frac{\min\{\mu_i(z), \mu_j(z)\}}{\epsilon}\right) \geq 2\left(1 - \frac{p}{\epsilon}\right)$$

for all $i, j = M_1 + 1, \ldots, M \quad i \neq j$. Furthermore

$$\sigma_i(z) \geq \frac{\mu}{\epsilon + p} \qquad \text{for all } z \in \mathcal{K}_i'(\mu) \text{ and } i = M_1 + 1, \ldots, M.$$

Thus we can apply Lemma 12 (b) to bound M_2.

$$\begin{aligned} M_2 &\leq \frac{(1-\frac{p}{\epsilon})\frac{\mu}{\epsilon+p}K}{2\frac{p}{\epsilon}} \exp\left(\frac{\frac{p}{\epsilon}}{\frac{\mu}{\epsilon+p}} \log \frac{2e\frac{\mu}{\epsilon+p}K}{\frac{p}{\epsilon}(1-\frac{p}{\epsilon})}\right) \\ &\leq \frac{\mu K}{2p} \exp\left(\frac{p(p+\epsilon)}{\mu\epsilon} \log \frac{2e\mu K\epsilon}{p(\epsilon - \frac{p^2}{\epsilon})}\right) \\ &\leq \frac{\mu K}{2p} \exp\left(\frac{p(p+\epsilon)}{\mu\epsilon} \log \frac{2e\mu K\epsilon}{p(\epsilon - p)}\right), \end{aligned} \tag{2.3.38}$$

provided that the assumption made in Lemma 12 (b) holds, which is in this case

$$\frac{\mu}{\epsilon + p} \geq \frac{p}{\epsilon K}. \tag{2.3.39}$$

We choose the parameters ϵ and μ as follows

$$\epsilon \triangleq \sqrt{p} \text{ and } \mu \triangleq \frac{(1-\sqrt{p})^3(1+\sqrt{p})}{pK}.$$

Then clearly $0 < p < \epsilon < 1$ and $0 < \mu$.

Furthermore we have to check for this choice of parameters that (2.3.37) and (2.3.39) hold. (2.3.37) holds as

$$(1-\epsilon)^2 - p\mu K = (1-\sqrt{p})^2 - (1-\sqrt{p})^3(1+\sqrt{p}) = (1-\sqrt{p})^2 p > 0$$

and (2.3.39) holds, provided that $p \leq \frac{1}{4}$ because then

$$\frac{\mu}{\epsilon + p} = \frac{(1-p)(1-\sqrt{p})^2}{pK(p+\sqrt{p})} \geq \frac{(1-\frac{1}{4})(1-\sqrt{\frac{1}{4}})^2}{\frac{1}{4}K(\frac{1}{4}+\sqrt{\frac{1}{4}})} \geq \frac{1}{K} \geq \frac{\sqrt{p}}{K} = \frac{p}{\epsilon K}.$$

Hence, if $p \leq \frac{1}{4}$ we get from (2.3.36) and (2.3.38) that

$$\begin{aligned} M &\leq K + M_1 + M_2 \\ &\leq K + \frac{1-p}{p^2} + \frac{(1-\sqrt{p})^3(1+\sqrt{p})}{2p^2} \exp\left(\frac{Kp^2}{(1-\sqrt{p})^3} \log \frac{2e(1-\sqrt{p})(1-p)}{p^2}\right) \\ &\leq K + \frac{1}{p^2} + \frac{1}{2p^2} \exp\left(\frac{Kp^2}{(1-\sqrt{p})^3} \log \frac{2e}{p^2}\right) \end{aligned}$$

If $\frac{1}{4} < p < 1$, then the last bound is weaker than (2.3.33), as we have the factor $\frac{p^2}{(1-\sqrt{p})^3}$ in the exponent. This completes the proof of Theorem 40. □

Now we will require that the authentication codes satisfy the condition

$$P_S^\star \triangleq \max_{y \in \mathcal{M}'} P_S(y) \leq p \tag{2.3.40}$$

for some given constant $p > 0$, i.e., (recall Definition 25 and (2.3.3)) that for any cryptogram $y \in \mathcal{M}'$ the probability of a successful substitution with any cryptogram $y' \in \mathcal{M}', y' \neq y$, does not exceed p.

In the case of an authentication code without secrecy we have $P_S^{max} \leq P_S^\star$. Therefore the requirement made in (2.3.40) is stronger than $P_S^{max} \leq p$ and we have $P_D \leq p$ if (2.3.40) holds. However the deficiency of this approach is that, in general (for authentication codes with some degree of secrecy), we cannot assure $P_D \leq p$ if (2.3.40) holds, which can be seen in Example 3 again.

Definition 32 For any $0 < p < 1$ let $M^\star(K, p)$ denote the maximal number of messages in an authentication code with K keys such that $P_S^\star \leq p$.

The next lemma enables us to use as an upper bound for $M^\star(K, p)$ upper bounds for the maximal cardinality of a set of pairwise separated probability measures.

Lemma 13 *Let $0 < p < 1$. If $P_S^\star \leq p$ for an authentication code, then the set $\{P_{Z|Y}(\cdot|y) : y \in \mathcal{M}'\}$ of probability measures on the set $\{1, \ldots, K\}$ is p-pairwise separated.*

Proof Let $y, y' \in \mathcal{M}'$, $y \neq y'$. According to Definition 25 the support of $P_{Z|Y}(\cdot|y')$ is $\mathcal{K}(y')$. As $P_S(y) \leq p$ it follows from (2.3.3) that

$$P_{Z|Y}(\mathcal{K}(y')|y) \leq p \tag{2.3.41}$$

Using Lemma 11 and (2.3.41) we obtain

$$\begin{aligned}||P_{Z|Y}(\cdot\,|y) - P_{Z|Y}(\cdot\,|y')|| &= 2\left(1 - \sum_{z=1}^{K} \min\{P_{Z|Y}(z|y), P_{Z|Y}(z|y')\}\right)\\ &= 2\left(1 - \sum_{z\in\mathcal{K}(y')} \min\{P_{Z|Y}(z|y), P_{Z|Y}(z|y')\}\right)\\ &\geq 2\left(1 - P_{Z|Y}(\mathcal{K}(y')|y)\right) \geq 2(1-p).\end{aligned}$$ □

With this notion the next theorem is immediate.

Theorem 41 *For any* $0 < p < 1$ *the following inequality holds*

$$M^{\star}(K,p) \leq K + \frac{1}{p^2} + \frac{1}{2p^2}\exp\left(\frac{p^2K}{(1-\sqrt{p})^3}\log\frac{2e}{p^2}\right).$$

Proof The statement follows directly from the previous Lemma, the bound on the cardinality of a set of pairwise separated measures given in Theorem 40 and the fact that for any authentication code $M \leq |\mathcal{M}'|$. □

Remark 16 1. We exploited the fact that an authentication code induces a probability distribution P_{ZY} on the set $\{1,\ldots,K\} \times \mathcal{M}'$ such that the measure of the support of $P_{Z|Y}(\cdot\,|y')$ under $P_{Z|Y}(\cdot\,|y)$ is less than p for any $y' \neq y$. For the moment let us denote such a configuration as a $(|\mathcal{M}'|, K, p)$-configuration. Burnashev and Bassalygo [4] looked abstractly on such configurations, i.e., where not necessarily the probability distribution is induced by some cipher and a message source, and denoted as $M_{aut,1}(K,p)$ the maximal M such that there exists a (M, K, p)-configuration. Furthermore they denoted as $M_{aut,2}(K,p)$ the maximal number of messages in a generalized authentication code (where keys and messages are not necessarily generated independently) such that $P_S^{\star} \leq p$. Clearly, $M_{aut,1}(K,p) \leq M_{aut,2}(K,p)$, because we can define for an optimal (M, K, p)-configuration the encryption by $c_z(m) = m$ for all $z = 1,\ldots,K$. On the other hand we saw already that an authentication code with $P_S^{\star} \leq p$ induces a $(|\mathcal{M}'|, K, p)$-configuration (this is also true if messages and keys are no longer chosen independently). As for any authentication code we have $M \leq |\mathcal{M}'|$ it follows $M_{aut,2}(K,p) \leq M_{aut,1}(K,p)$. Therefore the values $M_{aut,1}(K,p)$ and $M_{aut,2}(K,p)$ coincide.

2. In [4] the value $M_{aut,1}(K,p)$ was bounded by $M^{sep}(K,2p)$ but it is also possible to bound it directly by $M^{sep}(K,p)$ similarly to the derivation of Lemma 13 and Theorem 41. This gives a better result as $M^{sep}(K,p) \leq M^{sep}(K,2p)$.

2.3.4 Authentication as an Hypothesis Testing Problem

In this paragraph we present an elegant approach by Maurer [19] to give information-theoretic lower bounds on the success probabilities of the opponent in a generalized model. The key point is the interpretation of the receiver's decision whether the received cryptogram is authentic or not as a decision for one of two hypotheses.

Generalizations

We generalize the model in the following ways.

- The sender wants to inform the receiver about a sequence of messages produced by a source at some time instances. We denote by $X_1, X_2, \dots, X_i, \dots$ the random variables for those messages.
- Each message is encrypted separately to some cryptogram. We denote by $Y_1, Y_2, \dots, Y_i, \dots$ the corresponding random variables. The cryptogram sent at time i depends on the secret key, the message produced at time i and possibly also on the previous messages. Therefore in this context a key c_z can be described as a mapping $c_z : \bigcup_{i=1}^{\infty} \mathcal{M}^i \longrightarrow \mathcal{M}'$ such that $y_i = c_z(m_1, \dots, m_i)$.
- We assume that the receiver is synchronized, i.e., he knows the message number i. In order to enable the receiver to decrypt correctly we have to assume that the message m_i produced at time i is uniquely determined by the previous messages $m_1, \dots, m_{i-1}$ and cryptograms $y_1, \dots, y_i$ and the secret key. Therefore, by induction, m_i is uniquely determined by $m_1, \dots, m_{i-1}, y_i$ and the secret key (also by $y_1, \dots, y_i$ and the secret key itself). In other words we require that for all $i \in \mathbb{N}$ and all $m_1, \dots, m_i, m_i' \in \mathcal{M}$ with $m_i \neq m_i'$ we have $c_z(m_1, \dots, m_i) \neq c_z(m_1, \dots, m_{i-1}, m_i')$ for all $z \in \{1, \dots, K\}$.
- The opponent can choose between impersonation and substitution. In an *impersonation attack at time i* he waits until he has seen the first $i-1$ cryptograms $y_1, \dots, y_{i-1}$, which he lets pass unchanged to the receiver and then sends a fraudulent cryptogram y_i'. We denote by Y_i' the corresponding random variable. In a *substitution attack at time i* the opponent lets pass the first $i-1$ cryptograms $y_1, \dots, y_{i-1}$, intercepts y_i and replaces it by a different cryptogram y_i'.
- Up to now the receiver has accepted a cryptogram as authentic if and only if it is consistent with the secret key. Now we will allow, at least for purposes of calculation, the receiver to reject a valid cryptogram with some probability. This generalization is important because it establishes the link to the standard hypothesis testing scenario.

We will also refine our notion when the opponent is considered to be successful in an impersonation attack and substitution attack, respectively. Suppose the receiver accepted the fraudulent cryptogram y_i' as a valid cryptogram. Then he decodes $y_1, \dots, y_{i-1}, y_i'$ to some message m_i'. We distinguish now three cases. The opponent is considered to be successful when

(a) the receiver accepts the fraudulent cryptogram y_i' as a valid cryptogram (this is the case we considered so far).

(b) the receiver accepts the fraudulent cryptogram y_i' as a valid cryptogram and the message m_i' is known to the opponent. In other words the opponent is only considered to be successful if he also guesses the message m_i' correctly.
(c) the receiver accepts the fraudulent cryptogram y_i' as a valid cryptogram and the message m_i' was chosen by the opponent before. Of course this type of attack depends on the particular value m_i'.

Note that in an authentication code without secrecy case (a) and (b) coincide as the cryptograms uniquely determine the message and therefore the opponent will always guess correctly.

Definition 33 We distinguish the three described cases by denoting the corresponding attacks as impersonation attack and substitution attack of type (a), (b) and (c), respectively. We denote the success probabilities for the opponent using an optimal strategy for an attack of the type (a), (b) and (c) by

$$P_{I,i}^a,\ P_{I,i}^b \text{ and } P_{I,i,m_i'}^c,$$

for an impersonation attack at time i, respectively, and by

$$P_{S,i}^a,\ P_{S,i}^b \text{ and } P_{S,i,m_i'}^c,$$

for a substitution attack at time i, respectively.

For a particular observed sequence $y_1, \dots, y_{i-1}$ of cryptograms and, in case of a substitution attack also for a fixed intercepted cryptogram y_i, we denote the corresponding success probabilities by

$$P_{I,i}^a(y_1, \dots, y_{i-1}),\ P_{I,i}^b(y_1, \dots, y_{i-1}) \text{ and } P_{I,i}^c(y_1, \dots, y_{i-1}),$$

respectively, for an impersonation attack at time i and by

$$P_{S,i}^a(y_1, \dots, y_i),\ P_{S,i}^b(y_1, \dots, y_i) \text{ and } P_{S,i,m_i'}^c(y_1, \dots, y_i),$$

respectively, for a substitution attack at time i.

With this notion, for instance, $P_{I,i}^a$ is the expected value of $P_{I,i}^a(y_1, \dots, y_{i-1})$, i.e.,

$$P_{I,i}^a = \sum_{(y_1,\dots,y_{i-1})} P_{Y_1\dots Y_{i-1}}(y_1, \dots, y_{i-1}) P_{I,i}^a(y_1, \dots, y_{i-1}).$$

Some Results on Hypothesis Testing

We collect some results of the theory of hypothesis testing. Suppose we have to decide which of two hypotheses, H_0 or H_1, is true and we know from some random experiment the outcome of a random variable U with values in some set $\mathcal{U}$. The distribution of U depends on which of the two hypotheses is true. Under H_0 let U

be distributed according to P and under H_1 let U be distributed according to Q. A decision rule assigns to each possible value $u \in \mathcal{U}$ one of the two hypotheses. Therefore a decision rule may be viewed as a partition of $\mathcal{U}$ into two sets $\mathcal{U}_0$ and $\mathcal{U}_1$ such that we vote for H_0 if $U \in \mathcal{U}_0$ and vote for H_1 otherwise. There are two types of possible errors that may occur when making a decision. Accepting hypothesis H_1 when actually H_0 is true is called an error of the first kind and we will typically denote the probability of this event by α. Accepting hypothesis H_0 when actually H_1 is true is called an error of the second kind and we will typically denote the probability of this event by β. The optimal decision rule is given by the Neyman–Pearson Theorem which states that, for a given maximal tolerable probability β of an error of the second kind, α can be minimized by assuming hypothesis H_0 if and only if

$$\log \frac{P(u)}{Q(u)} \geq T \tag{2.3.42}$$

for some threshold T (see for instance [5]).

Note that only the existence of T, but not its specific value is given by the theorem. The term on the left-hand side of (2.3.42) is called the log-likelihood ratio. The expected value of the log-likelihood ratio with respect to P is the I-divergence

$$D(P||Q) = \sum_{u \in \mathcal{U}} P(u) \log \frac{P(u)}{Q(u)},$$

which is nonnegative and equal to zero exactly if the two distributions P and Q are identical.

The I-divergence and the error probabilities in an hypothesis test of the described form are related at follows.

Lemma 14 *The probabilities α and β of an error of the first and second kind, respectively, satisfy*

$$d(\alpha, \beta) \leq D(P||Q),$$

where $d(\alpha, \beta) \triangleq \alpha \log \frac{\alpha}{1-\beta} + (1-\alpha) \log \frac{1-\alpha}{\beta}$.

In particular, for $\alpha = 0$ we have

$$\beta \geq 2^{-D(P||Q)}.$$

Proof Let $\{\mathcal{U}_0, \mathcal{U}_1\}$ be the partition of $\mathcal{U}$ induced by the used decision rule. Then

$$\alpha = \sum_{u \in \mathcal{U}_1} P(u) \quad \text{and} \quad \beta = \sum_{u \in \mathcal{U}_0} Q(u).$$

Therefore

$$d(\alpha,\beta)=\left(\sum_{u\in\mathcal{U}_1}P(u)\right)\log\frac{\sum_{u\in\mathcal{U}_1}P(u)}{\sum_{u\in\mathcal{U}_1}Q(u)}+\left(\sum_{u\in\mathcal{U}_0}P(u)\right)\log\frac{\sum_{u\in\mathcal{U}_0}P(u)}{\sum_{u\in\mathcal{U}_0}Q(u)}$$

$$\leq\sum_{u\in\mathcal{U}_1}P(u)\log\frac{P(u)}{Q(u)}+\sum_{u\in\mathcal{U}_0}P(u)\log\frac{P(u)}{Q(u)}=D(P||Q),$$

where we applied the log-sum inequality. □

Later we will deal with the case where the random variable U is given as a random couple $U=(S,T)$, the distribution P will be the actual joint distribution P_{ST} and the distribution Q will be the product of the marginal distributions $P_S P_T$. In that case the I-divergence $D(P||Q)$ turns out to be the mutual information $I(S\wedge T)$.

$$D(P||Q)=\sum_{s,t}P_{ST}(s,t)\log\frac{P_{ST}(s,t)}{P_S(s)P_T(t)}$$

$$=H(S)+H(T)-H(S,T)=H(S)-H(S|T)=I(S\wedge T).$$

Suppose now that the distributions P and Q depend on the value v of an additional random variable V with values in $\mathcal{V}$, which is known to the testing person, i.e., we have a collection of pairs (P_v,Q_v) of conditional distributions each pair occurring with probability $P_V(v)$. The decision rule may depend on the value v of V and for each $v\in\mathcal{V}$ we denote by $\alpha(v)$ and $\beta(v)$ the probabilities of an error of the first and second kind, respectively, given that $V=v$.

Lemma 15 *The average probabilities of an error of the first and second kind given by*

$$\alpha\triangleq\sum_{v\in\mathcal{V}}P_V(v)\alpha(v)\quad and\quad\beta\triangleq\sum_{v\in\mathcal{V}}P_V(v)\beta(v),$$

respectively, satisfy

$$d(\alpha,\beta)\leq\sum_{v\in\mathcal{V}}P_V(v)D(P_v||Q_v).$$

Proof As the function d is $\cup$-convex we can apply Jensen's inequality and get

$$d(\alpha,\beta)\leq\sum_{v\in\mathcal{V}}P_V(v)d(\alpha(v),\beta(v)).$$

Lemma 14 implies that for every $v\in\mathcal{V}$

$$d(\alpha(v),\beta(v))\leq D(P_v||Q_v)$$

and this completes the proof. □

We may go another step further. Lemma 15 holds of course also for distributions conditioned on the event that a further random variable W takes on a particular value w known to the testing person, i.e., for pairs $(P_{v,w}, Q_{v,w})$ of distributions. We denote by $\alpha(v, w)$ and $\beta(v, w)$ the two error probabilities. The following corollary follows directly from Lemma 15.

Corollary 3 *The average probabilities (over $\mathcal{V}$) of an error of the first and second kind given by*

$$\alpha(w) \triangleq \sum_{v \in \mathcal{V}} P_V(v)\alpha(v, w) \quad \textit{and} \quad \beta(w) \triangleq \sum_{v \in \mathcal{V}} P_V(v)\beta(v, w),$$

respectively, satisfy

$$d(\alpha(w), \beta(w)) \leq \sum_{v \in \mathcal{V}} P_V(v) D(P_{v,w}||Q_{v,w}).$$

Let us look again at the special case where $U = (S, T)$ and the distributions $P_v = P_{ST|V}(\cdot|v)$ and $Q_v = P_{S|V}(\cdot|v)P_{T|V}(\cdot|v)$ depend on the value of the random variable V. Then the expression on the right-hand side in the statement of Lemma 15 becomes

$$\sum_{v \in \mathcal{V}} P_V(v) D(P_v||Q_v) = \sum_{v \in \mathcal{V}} P_V(v)\, I(S \wedge T|V = v) = I(S \wedge T|V).$$

Similarly if $P_{v,w} = P_{ST|VW}(\cdot|v, w)$ and $Q_{v,w} = P_{S|VW}(\cdot|v, w)P_{T|VW}(\cdot|v, w)$ then the right-hand side in Corollary 3 becomes

$$\begin{aligned}\sum_{v \in \mathcal{V}} P_V(v) D(P_{v,w}||Q_{v,w}) &= \sum_{v \in \mathcal{V}} P_V(v)\, I(S \wedge T|V = v, W = w) \\ &= I(S \wedge T|V, W = w).\end{aligned}$$

The Receivers Hypothesis Testing Problems

Let us now describe how we can make these methods applicable to the authentication problem.

Basically the receiver is faced with the following two hypotheses:

H_0—the received cryptogram is authentic.

H_1—the received cryptogram has been inserted by the opponent.

The two probabilities α and β of an error of the first and second kind, respectively, become:

α—probability of rejecting a valid cryptogram.

β—probability of accepting a fraudulent cryptogram.

Note that the behavior of the receiver considered so far implies $\alpha = 0$.

Let us consider an impersonation attack of the type (a) at time i. The receiver and the opponent have seen the first $i-1$ cryptograms $Y_1 = y_1, \dots, Y_{i-1} = y_{i-1}$. Let us denote by $\bar{Y}_i$ the random variable for the ith cryptogram (under H_0 we have $\bar{Y}_i = Y_i$ and under H_1 we have $\bar{Y}_i = Y_i'$). The receiver knows the secret key, i.e., he knows the value of Z. Given the value of the random couple $(\bar{Y}_i, Z)$ the receiver has to decide which of the two hypotheses is true. If H_0 is true then $(\bar{Y}_i, Z)$ is distributed according to

$$P_{Y_i Z|Y_1 \dots Y_{i-1}}(\cdot\,|y_1, \dots, y_{i-1}). \tag{2.3.43}$$

The opponent chooses the fraudulent cryptogram y_i' depending on $y_1, \dots, y_{i-1}$ but without further knowledge about the value of Z. Therefore, if H_1 is true, then $(\bar{Y}_i, Z)$ is distributed according to

$$P_{Y_i'|Y_1 \dots Y_{i-1}}(\cdot\,|y_1, \dots, y_{i-1}) P_{Z|Y_1 \dots Y_{i-1}}(\cdot\,|y_1, \dots, y_{i-1}). \tag{2.3.44}$$

One possible but generally not optimal impersonation strategy for the opponent would be to select y_i' according to the actual distribution of Y_i given $Y_1 = y_1, \dots, Y_{i-1} = y_{i-1}$, i.e., he chooses

$$P_{Y_i'|Y_1 \dots Y_{i-1}}(\cdot\,|y_1, \dots, y_{i-1}) = P_{Y_i|Y_1 \dots Y_{i-1}}(\cdot\,|y_1, \dots, y_{i-1}). \tag{2.3.45}$$

Now we can derive the following theorem.

Theorem 42 *For every authentication system*

$$P_{I,i}^a(y_1, \dots, y_{i-1}) \geq 2^{-I(Y_i \wedge Z|Y_1 = y_1, \dots, Y_{i-1} = y_{i-1})}$$

and

$$P_{I,i}^a \geq 2^{-I(Y_i \wedge Z|Y_1, \dots, Y_{i-1})}. \tag{2.3.46}$$

Proof Let $Y_1 = y_1, \dots, Y_{i-1} = y_{i-1}$ be given. Suppose the opponent chooses his impersonation strategy according to (2.3.45). Let us denote by $P_{I,Y'}(y_1, \dots, y_{i-1})$ his success probability when following this strategy and by $P_{I,Y'}$ the corresponding average success probability. Suppose the receiver selects some decision rule giving him $\alpha(y_1, \dots, y_{i-1})$ as the probability of rejecting a valid cryptogram and $\beta(y_1, \dots, y_{i-1})$ as the probability of accepting a fraudulent cryptogram.

Then Lemma 14 implies

$$d(\alpha(y_1, \dots, y_{i-1}), \beta(y_1, \dots, y_{i-1})) \leq I(Y_i \wedge Z|Y_1 = y_1, \dots, Y_{i-1} = y_{i-1}).$$

Denoting by α and β the corresponding average error probability we get from Lemma 15

$$d(\alpha, \beta) \leq I(Y_i \wedge Z|Y_1, \dots, Y_{i-1}).$$

Selecting the decision rule for the receiver as before which means that he accepts the cryptogram exactly if it is consistent with the secret key and the previous $i-1$ cryptograms we get $\alpha(y_1, \dots, y_{i-1}) = 0$ and $\beta(y_1, \dots, y_{i-1}) = P_{I,Y'}(y_1, \dots, y_{i-1})$. This implies

$$P_{I,Y'}(y_1, \dots, y_{i-1}) \geq 2^{-I(Y_i \wedge Z | Y_1 = y_1, \dots, Y_{i-1} = y_{i-1})}$$

and

$$P_{I,Y'} \geq 2^{-I(Y_i \wedge Z | Y_1, \dots, Y_{i-1})}.$$

Therefore we obtain from

$$P^a_{I,i}(y_1, \dots, y_{i-1}) \geq P_{I,Y'}(y_1, \dots, y_{i-1}) \text{ and } P^a_{I,i} \geq P_{I,Y'}$$

the desired result. □

Remark 17 Note that in the case when $i = 1$, (2.3.46) is Simmons' bound of Theorem 32.

Let us analyze an impersonation attack of type (b) at time i, i.e., the opponent is only considered to be successful if he also guesses the message to which the receiver decodes the fraudulent cryptogram to correctly. Now a strategy for the opponent consists of a distribution $P_{X'_i Y'_i | Y_1, \dots, Y_{i-1}}(\cdot | y_1, \dots, y_{i-1})$ where the value of Y'_i is the fraudulent cryptogram and the value of X'_i is the message the opponent guesses. Consider now the 'fictive' hypothesis testing scenario, where in addition to values of the random variables $\bar{Y}_i$ and Z the receiver also gets a value of $\bar{X}_i$, which is under hypothesis H_0 equal to X_i and under H_1 equal to X'_i. This means that if H_0 is true than the receiver is told the correct message and if H_1 is true the receiver is told the message the opponent guesses. One possible but generally not optimal impersonation strategy for the opponent would be to select the pair (m'_i, y'_i) according to the actual distribution of (X_i, Y_i) given $Y_1 = y_1, \dots, Y_{i-1} = y_{i-1}$, i.e., he chooses

$$P_{X'_i Y'_i | Y_1 \dots Y_{i-1}}(\cdot | y_1, \dots, y_{i-1}) = P_{X_i Y_i | Y_1 \dots Y_{i-1}}(\cdot | y_1, \dots, y_{i-1}). \tag{2.3.47}$$

Then it follows that if H_0 is true then $(\bar{X}_i, \bar{Y}_i, Z)$ is distributed according to

$$P_{X_i Y_i Z | Y_1 \dots Y_{i-1}}(\cdot | y_1, \dots, y_{i-1})$$

and if H_1 is true then $(\bar{X}_i, \bar{Y}_i, Z)$ is distributed according to

$$P_{X_i Y_i | Y_1 \dots Y_{i-1}}(\cdot | y_1, \dots, y_{i-1}) P_{Z | Y_1 \dots Y_{i-1}}(\cdot | y_1, \dots, y_{i-1}).$$

Now we can derive the following theorem.

Theorem 43 *For every authentication system*

$$P^b_{I,i}(y_1, \dots, y_{i-1}) \geq 2^{-I(X_i Y_i \wedge Z | Y_1 = y_1, \dots, Y_{i-1} = y_{i-1})}$$

and

$$P^b_{I,i} \geq 2^{-I(X_iY_i \wedge Z|Y_1,\ldots,Y_{i-1})}.$$

Proof Let $Y_1 = y_1, \ldots, Y_{i-1} = y_{i-1}$ be given. Suppose the opponent chooses his impersonation strategy according to (2.3.47). Let us denote by $P_{I,Y'}(y_1, \ldots, y_{i-1})$ his success probability when following this strategy and by $P_{I,Y'}$ the corresponding average success probability. Suppose the receiver selects some decision rule giving him $\alpha(y_1, \ldots, y_{i-1})$ as the probability of an error of the first kind and $\beta(y_1, \ldots, y_{i-1})$ as the probability of an error of the second kind in the above described hypothesis testing scenario. Then Lemmas 14 and 15 imply

$$d(\alpha(y_1, \ldots, y_{i-1}), \beta(y_1, \ldots, y_{i-1})) \leq I(X_iY_i \wedge Z|Y_1 = y_1, \ldots, Y_{i-1} = y_{i-1})$$

and

$$d(\alpha, \beta) \leq I(X_iY_i \wedge Z|Y_1, \ldots, Y_{i-1})$$

for the average error probabilities α and β.

Now suppose the receiver selects the decision rule in such a way that he votes for H_0 exactly if the value of $\bar{Y}_i$ is a valid cryptogram under the secret key and he would decode it to the message given by $\bar{X}_i$.

Then we get $\alpha(y_1, \ldots, y_{i-1}) = \alpha = 0,\ \beta(y_1, \ldots, y_{i-1}) = P_{I,Y'}(y_1, \ldots, y_{i-1})$ and $\beta = P_{I,Y'}$. As $P^b_{I,i}(y_1, \ldots, y_{i-1}) \geq P_{I,Y'}(y_1, \ldots, y_{i-1})$ and $P^b_{I,i} \geq P_{I,Y'}$, we obtain the desired result. □

Let us analyze an impersonation attack of type (c), when the opponent is only considered to be successful if the receiver accepts the fraudulent cryptogram and decodes it to some message, which was chosen by the opponent. Let this message be $m'_i \in \mathcal{M}$. We consider the following 'fictive' hypothesis testing scenario. Suppose $Y_1 = y_1, \ldots, Y_{i-1} = y_{i-1}$ are given and the message source produces at time i the message m'_i, i.e., $X_i = m'_i$. Let us assume the receiver knows this. As in case (a) the receiver now sees some value of the random couple $(\bar{Y}_i, Z)$ and has to decide if the cryptogram he got is authentic or not. Again we may consider a generally not optimal impersonation strategy for the opponent given by

$$P_{Y'_i|Y_1\ldots Y_{i-1}}(\cdot\,|y_1, \ldots, y_{i-1}) = P_{Y_i|Y_1\ldots Y_{i-1}X_i}(\cdot\,|y_1, \ldots, y_{i-1}, m'_i). \tag{2.3.48}$$

If H_0 is true than $(\bar{Y}_i, Z)$ is distributed according to

$$P_{Y_iZ|Y_1\ldots Y_{i-1}X_i}(\cdot\,|y_1, \ldots, y_{i-1}, m'_i)$$

and if H_1 is true then $(\bar{Y}_i, Z)$ is distributed according to

$$P_{Y_i|Y_1\ldots Y_{i-1}X_i}(\cdot\,|y_1, \ldots, y_{i-1}, m'_i)P_{Z|Y_1\ldots Y_{i-1}}(\cdot\,|y_1, \ldots, y_{i-1}),$$

which is (as Z and X_i are independent) the same as

$$P_{Y_i|Y_1\dots Y_{i-1}X_i}(\cdot\,|y_1,\dots,y_{i-1},m_i')P_{Z|Y_1\dots Y_{i-1}X_i}(\cdot\,|y_1,\dots,y_{i-1},m_i')$$

With this the following conclusion is no more difficult.

Theorem 44 *For every authentication system*

$$P^c_{I,i}(y_1,\dots,y_{i-1}) \geq 2^{-I(Y_i\wedge Z|Y_1=y_1,\dots,Y_{i-1}=y_{i-1},X_i=m_i')}$$

and

$$P^c_{I,i} \geq 2^{-I(Y_i\wedge Z|Y_1,\dots,Y_{i-1},X_i=m_i')}.$$

Proof We proceed analogously to the proofs of the Theorems 42 and 43 using instead of Lemma 15 the Corollary 3 for the above described hypothesis test. Then the desired result is obtained for the receiver's decision rule to accept H_0 exactly if the observed cryptogram is valid under the secret key and would be decoded to m_i'. □

For the substitution attacks of the three described forms (a), (b) and (c), respectively, we can derive a lower bound on the success probability simply by giving a lower bound on the opponent's probability to guess the correct value of Z because, when guessing the secret key correctly, the opponent can launch any of the described attacks.

Let S be a random variable with values in some finite set $\mathcal{S}$. The probability to guess a value of S correctly knowing only P_S is $\max_{s\in\mathcal{S}} P_S(s)$. As the entropy of S is the expected value of $-\log P_S(S)$ we obtain

$$-\log\left(\max_{s\in\mathcal{S}} P_S(s)\right) = \min_{s\in\mathcal{S}}\Big(-\log P_S(s)\Big) \leq H(S)$$

and therefore

$$\max_{s\in\mathcal{S}} P_S(s) \geq 2^{-H(S)}.$$

Knowing in addition the value of a further random variable T we get by applying Jensen's inequality that the (average) probability of guessing S correctly is bounded by

$$\sum_t P_T(t)2^{-H(S|T=t)} \geq 2^{-H(S|T)}.$$

This applies to our situation in the following way.

Theorem 45 *For every authentication system*

$$P^a_{S,i}(y_1,\dots,y_i) \geq 2^{-H(Z|Y_1=y_1,\dots,Y_i=y_i)}$$

and

$$P^a_{S,i} \geq 2^{-H(Z|Y_1,\dots,Y_i)}.$$

These bounds also hold for the types (b) and (c) of substitution attacks.

Proof In a substitution attack at time i the opponent knows a sequence of values of $Y_1, \dots, Y_i$ and therefore the result follows from the previously made remarks. □

We can combine the bounds derived for impersonation attacks and substitution attacks in the following way.

Theorem 46 *For every authentication system*

$$\max\{P^a_{I,1}, \dots, P^a_{I,n}, P^a_{S,n}\} \geq 2^{-\frac{H(Z)}{n+1}} \quad \textit{for all} \quad n \in \mathbb{N}.$$

Proof Recall that

$$\begin{aligned}
&\sum_{i=1}^{n} I(Y_i \wedge Z|Y_1 \dots Y_{i-1}) \\
&\quad = \Big(H(Z) - H(Z|Y_1)\Big) + \Big(H(Z|Y_1) - H(Z|Y_1Y_2)\Big) + \cdots \\
&\qquad \cdots + \Big(H(Z|Y_1 \dots Y_{n-1}) - H(Z|Y_1 \dots Y_n)\Big) \\
&\quad = H(Z) - H(Z|Y_1 \dots Y_n) = I(Y_1 \dots Y_n \wedge Z).
\end{aligned}$$

(Sometimes this is called "Chain Rule of Mutual Information".)

Applying the bound of Theorem 42 for $P^a_{I,i}$ and the bound of Theorem 45 for $P^a_{S,n}$ we obtain that

$$-\sum_{i=1}^{n} \log P^a_{I,i} - \log P^a_{S,n} \leq \sum_{i=1}^{n} I(Y_i \wedge Z|Y_1 \dots Y_{i-1}) + H(Z|Y_1 \dots Y_n) = H(Z)$$

and therefore

$$\begin{aligned}
&-\log\left(\max\{P^a_{I,1}, \dots, P^a_{I,n}, P^a_{S,n}\}\right) \\
&\leq -\log\left(\frac{1}{n+1}\left(\sum_{i=1}^{n} P^a_{I,1} + P^a_{S,n}\right)\right) \leq \frac{H(Z)}{n+1},
\end{aligned}$$

where we used the fact that $-\log$ is a monotonically decreasing and ∪-convex function. □

Remark 18 The last result can be interpreted as follows. If an authentication system is used to authenticate n messages the opponent can choose the type of attack that gives him the highest success probability. For a cipher of a given size (measured in terms of the entropy $H(Z)$) Theorem 46 states that the achievable authenticity for n messages corresponds at most to the difficulty of guessing the secret key of a cipher whose size is $n+1$ times smaller than the size of the actual cipher.

2.4 Secret-Key Cryptology

The information-theoretic approach to secret-key cryptology was introduced by Shannon [24] as already mentioned. The problems of these "classical" secrecy systems were further discussed in papers by Ahlswede [1] and Hellman [13]. In this section we concentrate on some new results and approaches of Shtarkov [25] concerning the following problems.

1. Evaluation or estimation of $H(X|Y)$ for a given cipher $(\mathcal{C}, Q)$ and different distributions P_X. This is meaningful for incomplete information on the distribution P_X and/or different constraints on the choice of the cipher.
2. Determination of the optimal (or close to optimal) cipher, if the number of keys and the message distribution is given.

Furthermore the model is extended with a source coder and a randomizer.

2.4.1 Preliminaries

Conditions for Perfectness and Upper Bounds for Secrecy

We start with the derivation of some general upper bounds for the secrecy measured by the opponent's average uncertainty about the message after observing the cryptogram. These are combined in the next theorem.

Theorem 47 *For every secrecy system*

$$\begin{aligned} H(X|Y) &\leq \min\{H(X), H(Z|Y)\} \leq \min\{H(X), H(Z)\} \\ &\leq \min\{H(X), \log K\} \leq \log K. \end{aligned} \tag{2.4.1}$$

Proof The statement immediately follows if we can show $H(X|Y) \leq H(Z|Y)$. Recall that cryptogram and key determine the message, i.e., $H(X|Y, Z) = 0$ and therefore

$$H(X|Y) \leq H(X, Z|Y) = H(X|Y, Z) + H(Z|Y) = H(Z|Y). \qquad \square$$

Keeping this in mind we can derive necessary conditions for the perfectness of a cipher.

Theorem 48 *If a secrecy system is perfect, then*

$$H(Z) \geq H(X).$$

Proof Recall that a secrecy system is said to be perfect, if the random variables for the message and the cryptogram are independent, i.e., $H(X) = H(X|Y)$. Combining this with (2.4.1) yields the desired result. □

Theorem 49 *If a secrecy system is perfect, then*

$$K \geq M.$$

Proof Recall that we have assumed all messages and keys to occur with probability strictly greater than 0. Therefore the fact that X and Y are independent implies for any $y \in \mathcal{M}$

$$P_{X|Y}(m|y) = P_X(m) \text{ for all } m \in \mathcal{M}.$$

Hence, for every $m \in \mathcal{M}$ there exists at least one key $z \in \{1, \ldots, K\}$ such that $m = c_z^{-1}(y)$. As the keys are injective this implies $K \geq |\mathcal{M}|$. □

These are quite pessimistic results, which tell us that perfect secrecy requires that the uncertainty about the key must be at least as big as the uncertainty about the message and that the secrecy system must contain more keys than messages.

Example 7 We show that it is possible to guarantee perfect secrecy with $K = M$ keys. Let

$$c_z(m) \triangleq (m + z) \bmod M \quad \text{for all } m, z \in \{1, \ldots, M\}$$

and let the keys be equiprobable, i.e., $P_Z(z) \triangleq \frac{1}{M}$ for all $z \in \{1, \ldots, M\}$.

This cipher has the property that for every message $m \in \mathcal{M}$ and every cryptogram $y \in \mathcal{M}$ there exists exactly one key c_z with $c_z(m) = y$ and therefore we immediately get that $P_{X|Y}(m|y) = P_X(m)$. Hence, $H(X|Y) = H(X)$, which means that the secrecy system is perfect. Moreover it is perfect independent of the kind of distribution P_X and one can speak therefore of a robustly perfect cipher. Note that if $K = M$, then every regular and canonical cipher (what will be defined in the next section) has the here described properties.

The idea to use of $K = M$ keys in such a way that a message and a cryptogram is consistent with exactly one key was first developed by G.S. Vernam in 1926 ([18], pp. 7). He enciphered messages given as binary strings by adding binary strings of the same length componentwise modulo 2, that is, in the Vernam cipher each single message bit is enciphered with a new randomly chosen key bit. As the key bits are used only one time those systems are called *One-Time Systems* (or *One-Time Pads* in some contexts). They are only used for transmission of highly confidential information because of the large number of keys.

Regular and Canonical Ciphers

Usually we will restrict ourselves to ciphers where the keys are equiprobable.

Definition 34 A cipher $(\mathcal{C}, Q)$ is canonical, if Q is the uniform distribution.

From now on we will *always* assume ciphers to be canonical. This restriction is usually done [1, 13, 24, 25] and it does not seem to be severe but this has not been proved.

Definition 35 A cipher $(\mathcal{C}, Q)$ is regular, if $|\{c_z^{-1}(y) : z \in \{1, \dots, K\}\}| = K$ for any cryptogram $y \in \mathcal{M}$.

Now suppose we are given a number $S \in \mathbb{N}$ and two partitions $\mathcal{X} = \{\mathcal{X}_i : i = 1, \dots, S\}$ and $\mathcal{Y} = \{\mathcal{Y}_i : i = 1, \dots, S\}$ of the set $\mathcal{M}$.

Definition 36 A cipher $(\mathcal{C}, Q)$ is locally regular (with respect to $(\mathcal{X}, \mathcal{Y})$) if:

1. $|\mathcal{X}_i| = |\mathcal{Y}_i|$ for all $i \in \{1, \dots, S\}$.
2. $c_z(\mathcal{X}_i) \subset \mathcal{Y}_i$ for all $z \in \{1, \dots, K\},\ i \in \{1, \dots, S\}$.
3. $(\mathcal{C}, Q)$ is a regular cipher.

Remark 19 By definition every locally regular cipher is regular and every regular cipher is locally regular at least with respect to the trivial partitions $(\mathcal{X}, \mathcal{Y})$ which consist only of the set $\mathcal{M}$.

Using "random ciphers" Shannon [24] gave the following lower bound on $H(X|Y)$ (under the additional AEP hypothesis on the message source).

$$H(X|Y) \geq \log K + H(X) - \log M.$$

With our notion of regular ciphers, we get this bound for every regular cipher and without any assumption on the message source $(\mathcal{M}, P_X)$, just by observing that $H(Y|X) = \log K$ in those situations and therefore

$$\begin{aligned} H(X|Y) &= H(X, Y) - H(Y) \\ &= H(Y|X) + H(X) - H(Y) = \log K + H(X) - H(Y). \\ &\geq \log K + H(X) - \log M \end{aligned} \tag{2.4.2}$$

If $H(X) = \log M$, i.e., if the source is compressed, then the bound is tight but for general X it is rather poor. In the Sect. 2.4.3 we give a better bound by evaluating $H(X|Y)$ for a certain cipher. Ahlswede considers in [1] the class of message sources $(\mathcal{M}, P_X)$ with $H(X) \geq H_0$ for some constant $0 \leq H_0 \leq \log M$. Then (2.4.2) obviously implies for any such source

$$H(X|Y) \geq \log K + H_0 - \log M. \tag{2.4.3}$$

This bound reflects a robustified model, where one drops the assumption that sender and receiver know the message statistics. The opponent is still granted to know it exactly but sender and receiver only have to know a lower bound on the entropy of the source. In [1] it was also shown that the bound (2.4.3) is essentially best possible for this class of sources.

2.4.2 The Lower Bound for Locally Regular Ciphers

We now derive the fundamental result of Shtarkov in [25], where he gives a lower bound on $H(X|Y)$ for any locally regular cipher, which uses as information about the message statistics the relation of the greatest to the smallest probability of the messages in each of the sets $\mathcal{X}_i$ (recall Definition 36). Essential for the derivation of this bound is the Schur-concavity of the entropy function.

Lemma 16 *Let P and Q be two probability distributions on $\{1, \dots, K\}$ with $P(1) \geq \cdots \geq P(K)$ and $Q(1) \geq \cdots \geq Q(K)$. Furthermore let $P(1) = Q(1)$ and $P(K) = Q(K)$. If P has the property that all the probabilities $P(i)$ are equal to $P(1)$ or $P(K)$, i.e., if there exists an $n \in \{1, \dots, K-1\}$ with $P(1) = \cdots = P(n)$ and $P(n+1) = \cdots = P(K)$, then*

$$H(P) \leq H(Q).$$

Proof The statement follows from the Schur-concavity of the entropy function, if we can show that P Schur-dominates Q, i.e., if

$$\sum_{i=1}^{j} P(i) \geq \sum_{i=1}^{j} Q(i) \qquad \text{for all } j \in \{1, \dots, K\}.$$

Let $j \in \{1, \dots, K\}$.

Case 1: $j \leq n$, then

$$\sum_{i=1}^{j} P(i) = jP(1) = jQ(1) \geq \sum_{i=1}^{j} Q(i).$$

Case 2: $j > n$, then

$$\sum_{i=1}^{j} P(i) = 1 - \sum_{i=j+1}^{K} P(i) = 1 - (K-j)\,P(K) = 1 - (K-j)\,Q(K)$$

$$\geq 1 - \sum_{i=j+1}^{K} Q(i) = \sum_{i=1}^{j} Q(i).\square$$

□

Theorem 50 (Shtarkov) *Let $(\mathcal{C}, Q)$ be a locally regular cipher with respect to $(\mathcal{X}, \mathcal{Y})$. Let*

$$\rho_i \triangleq \frac{\max\limits_{m \in \mathcal{X}_i} P_X(m)}{\min\limits_{m \in \mathcal{X}_i} P_X(m)} \quad \text{for all} \quad i = 1, \dots, S. \tag{2.4.4}$$

Then

$$H(X|Y) \geq \log K - (\log e) \sum_{i=1}^{S} P_i\, \delta(\rho_i), \tag{2.4.5}$$

where

$$P_i \triangleq \sum_{m \in \mathcal{X}_i} P_X(m)$$

and

$$\delta : [1, \infty[\to \mathbb{R}$$

$$\delta(1) \triangleq 0$$

$$\delta(\rho) \triangleq \begin{cases} \ln(\rho - 1) - \ln \ln \rho - 1 + \frac{\ln \rho}{\rho - 1}, & 1 < \rho \leq T \\ \ln\left(\frac{\rho K}{\rho + K - 1}\right) - \frac{K-1}{\rho + K - 1} \ln \rho, & \rho > T \end{cases} \tag{2.4.6}$$

and $T = T(K)$ is the greatest solution of the equation

$$(T \ \ln T - T + 1)\, K = (T - 1)^2 \tag{2.4.7}$$

Proof $\delta'(\rho) = \frac{(\rho \ln \rho + 1 - \rho)(\rho - 1 - \ln(\rho))}{\rho \ln \rho (\rho - 1)^2} \geq 0, \quad \text{if } 1 < \rho < T(K).$

$\delta'(\rho) = \frac{(K-1) \ln \rho}{(K + \rho - 1)^2} \geq 0, \quad \text{if } T(K) < \rho.$

Hence, as the function δ is continuous we see that it is monotonically increasing. From the local regularity of the cipher follows that

$$\begin{aligned} H(X|Y) &= \sum_{y \in \mathcal{M}} P_Y(y)\, H(X|Y = y) = \sum_{i=1}^{S} \sum_{y \in \mathcal{Y}_i} P_Y(y) H(X|Y = y) \\ &\geq \sum_{i=1}^{S} P_i \min_{y \in \mathcal{Y}_i} H(X|Y = y). \end{aligned}$$

Thus we are done if we can show that for any $i \in \{1, \dots, S\}$

$$H(X|Y = y) \geq \log K - (\log e) \delta(\rho_i) \qquad \text{for all } y \in \mathcal{Y}_i. \tag{2.4.8}$$

because this implies (2.4.5). So let $i \in \{1, \dots, S\}$ and $y \in \mathcal{Y}_i$.

Case 1: $\rho_i = 1$.

In this case all messages in $\mathcal{X}_i$ are equiprobable and therefore for any $m \in \mathcal{X}_i$ $P_{X|Y}(m|y) = \frac{1}{K}$ provided that $P_{X|Y}(m|y) > 0$.

This implies $H(X|Y = y) = \log K$ and as $\delta(\rho_i)$ is defined to be 0 in this case the estimate (2.4.8) holds.

Case 2: $\rho_i > 1$.

Let

$$\rho_i(y) \triangleq \frac{\max\limits_{m \in \mathcal{X}_i} P_{X|Y}(m|y)}{\min\limits_{m \in \mathcal{X}_i} P_{X|Y}(m|y)},$$

where the minimum is taken only over terms strictly greater than 0.

If for $m, m' \in \mathcal{X}_i$ $P_{X|Y}(m|y) > 0$ and $P_{X|Y}(m'|y) > 0$, then the local regularity of the cipher implies that $\frac{P_{X|Y}(m|y)}{P_{X|Y}(m'|y)} = \frac{P_X(m)\frac{1}{K}}{P_X(m')\frac{1}{K}} = \frac{P_X(m)}{P_X(m')}$. If all these conditional probabilities would be greater than 0, then we would have $\rho_i(y) = \rho_i$, but if $|\mathcal{X}_i| > K$ then some of the conditional probabilities are equal to 0 and therefore we get $\rho_i(y) \leq \rho_i$, in general. If we take into account that δ is monotonically increasing then we see that it suffices to show (2.4.8) with ρ_i replaced with $\rho_i(y)$. In order to get this lower estimate we ask for what probability distribution $P_{X|Y}(\cdot|y)$ the entropy $H(X|Y = y)$ is minimal if ρ_i is given.

Let c_i denote the smallest probability of such a distribution (then $\rho_i\, c_i$ is the largest) then we know from Lemma 16 a lower bound on the entropy given by the entropy of the distribution with n_i values equal to $\rho_i\, c_i$ and $K - n_i$ values equal to c_i, which is

$$-\, n_i\, \rho_i\, c_i \log \rho_i\, c_i - (K - n_i) c_i \log c_i, \tag{2.4.9}$$

where n_i is determined by the equation $n_i\, \rho_i\, c_i + (K - n_i)\, c_i = 1$ and therefore

$$n_i = \frac{1 - K\, c_i}{c_i\, (\rho_i - 1)}. \tag{2.4.10}$$

If we substitute (2.4.10) into (2.4.9), we can minimize over c_i. The first and second derivative of (2.4.9) with respect to c_i are

$$\frac{1}{\ln 2}\left(\frac{\rho_i\, K}{\rho_i - 1} \ln \rho_i - \frac{1}{c_i}\right)$$

and

$$\frac{1}{c_i^2 \ln 2} > 0.$$

In this way we obtain that (2.4.9) is minimal for $c_i^\star$ and $n_i^\star$, where

$$c_i^\star = \frac{\rho_i - 1}{K\, \rho_i \ln \rho_i} \text{ and } n_i^\star = K \frac{\rho_i \ln \rho_i - \rho_i + 1}{(\rho_i - 1)^2}.$$

If we substitute these values in (2.4.9), then we get as a lower bound for $H(X|Y=y)$ the bound in (2.4.8), where δ is defined by the first expression in (2.4.6).

Now notice that we have $n_i \geq 1$ as an additional restriction. So if $n_i^\star < 1$, which is the case if $\rho_i > T(K)$, then we get a sharper lower bound by taking $n_i^\star = 1$ and correspondingly $c_i^\star = \frac{1}{\rho_i + K - 1}$. Substituting these terms into (2.4.9) we obtain again the bound (2.4.8) now with δ defined in the second expression of (2.4.6). □

Corollary 4 *Let*

$$\rho \triangleq \frac{\max\limits_{m \in \mathcal{M}} P_X(m)}{\min\limits_{m \in \mathcal{M}} P_X(m)}. \tag{2.4.11}$$

With the assumptions of Theorem 50 it follows

$$\begin{aligned} H(X|Y) &\geq \log K - (\log e)\, \delta(\max_{1 \leq i \leq S} \rho_i) \\ &\geq \log K - (\log e)\, \delta(\rho). \end{aligned} \tag{2.4.12}$$

Proof The bounds follow from (2.4.5), $\rho_i \leq \rho$ for all $i \in \{1, \ldots, S\}$ and the fact that the function δ is monotonically increasing. □

Remark 20 1. Equation (2.4.7) has always the solution $T = 1$. For $K \geq 3$ there exists exactly one other solution greater 1.

2. The lower bound on $H(X|Y)$ is always nontrivial, in the sense that the term in (2.4.12) is always nonnegative because we have seen that it is a value of the entropy function.

2.4.3 A Simple Cipher

Suppose that the probabilities $P_X(m)$ are ordered in such a way that

$$P_X(1) \geq \cdots \geq P_X(M). \tag{2.4.13}$$

Furthermore let $K \leq M$. We consider now the problem of constructing a good cipher if the distribution P_X and the number of keys K is given. A natural approach to the solution of this problem was given by Ahlswede [1], who defined a locally regular cipher with respect to $(\mathcal{X}, \mathcal{Y})$ with

$$\mathcal{X} \triangleq \{\mathcal{X}_i : i = 1, \ldots, S\} \text{ and } \mathcal{Y} \triangleq \{\mathcal{Y}_i : i = 1, \ldots, S\},$$

where $S \triangleq \left\lfloor \frac{M}{K} \right\rfloor$,

$$\mathcal{X}_i \triangleq \mathcal{Y}_i \triangleq \{(i-1)K + 1, \ldots, iK\} \qquad i = 1, \ldots, S-1$$

$$\text{and } \mathcal{X}_S \triangleq \mathcal{Y}_S \triangleq \{(S-1)K+1, \dots, M\}.$$

Let $(\mathcal{C}, Q)$ be any locally regular cipher with respect to $(\mathcal{X}, \mathcal{Y})$.

It is clear that this choice of the cipher provides the minimal or close to the minimal values of the ρ_i and therefore yields the maximal or close to maximal estimate of $H(X|Y)$ in (2.4.5). Recall that for an regular cipher $H(X|Y) = \log K + H(X) - H(Y)$. Therefore the optimal choice of the cipher is it to minimize $H(Y)$. P_Y is a "smoothed" version of P_X. For the construction above almost equiprobable messages are put together in the sets $\mathcal{X}_i$ and the resulting P_Y is the corresponding "step approximation" of P_X. Hence, it is clear that the above choice of the partitions tries to minimize the action of the smoothing and therefore should be the best or close to the best one.

But before analyzing $H(X|Y)$ for this cipher let us take a look at the other secrecy criterion introduced in Section "Measurements for Secrecy" of Sect. 2.2. We proved in Theorem 1 that the cryptanalyst's error probability λ satisfies

$$\frac{K-1}{K}(1 - P_X(1)) \leq \lambda \leq (1 - P_X(1)). \tag{2.4.14}$$

It was shown in [1] that if M is a multiple of K and $P_X(m) \leq \frac{1}{K}$ for all $m \in \mathcal{M}$, then for the described cipher

$$H(X|Y) \geq \log K - 1. \tag{2.4.15}$$

Using Lemma 4 and (2.4.14) we can prove that this holds also if M is not a multiple of K.

Theorem 51 *For the cipher $(\mathcal{C}, Q)$ described above*

$$H(X|Y) \geq \log K - \log\Big((K-1)P_X(1) + 1\Big). \tag{2.4.16}$$

Proof From Lemma 4 it follows that $H(X|Y) \geq -\log \lambda_c = -\log(1-\lambda)$ and with (2.4.14) we obtain

$$H(X|Y) \geq -\log\left(1 - \frac{K-1}{K}(1 - P_X(1))\right) = \log K - \log\Big((K-1)P_X(1) + 1\Big).$$

□

Corollary 5 *If $P_X(1) \leq \frac{1}{K}$ then for the cipher $(\mathcal{C}, Q)$*

$$H(X|Y) \geq \log K - 1.$$

Proof If $P_X(1) \leq \frac{1}{K}$, then we get from (2.4.16)

$$H(X|Y) \geq \log K - \log\left(2 - \frac{1}{K}\right) \geq \log K - 1. \qquad \square$$

Shtarkov [25] derives the following lower bound for this cipher.

Theorem 52 *If M is a multiple of K then for the cipher* $(\mathcal{C}, Q)$ *described above*

$$H(X|Y) \geq \log K - \frac{K}{2}(\log e)\Big(P_X(1) - P_X(M)\Big). \tag{2.4.17}$$

Proof Let $m \in \mathcal{X}_i$ and $y \in \mathcal{Y}_i$ for some $i \in \{1, \dots, S\}$. By construction of the cipher it follows that

$$P_{X|Y}(m|y) = \frac{P_{X,Y}(m,y)}{P_Y(y)} = \frac{\frac{1}{K}P_X(m)}{\sum_{m \in \mathcal{X}_i} P_X(m)\frac{1}{K}} = \frac{P_X(m)}{P_i},$$

with $P_i \triangleq \sum_{m \in \mathcal{X}_i} P_X(m)$. Note that for $m \in \mathcal{X}_i$ $P_{X|Y}(m|y)$ is independent of $y \in \mathcal{Y}_i$. Hence, we know from Lemma 16 that for given $y \in \mathcal{Y}_i$ that $H(X|Y = y)$ is minimal if P_X is concentrated on two values in $\mathcal{X}_i$. In order to get a lower bound on $H(X|Y)$ we may therefore assume that for all $i \in \{1, \dots, S\}$ there exist numbers $n_i \in \{1, \dots, K - 1\}$ with the property

$$\alpha_i \triangleq P_X(K(i-1)+1) = \cdots = P_X(K(i-1)+n_i)$$

and

$$\beta_i \triangleq P_X(K(i-1)+n_i+1) = \cdots = P_X(K\, i).$$

Then (2.4.13) implies that

$$\alpha_1 \geq \beta_1 \geq \alpha_2 \geq \beta_2 \geq \cdots \geq \alpha_S \geq \beta_S$$

and $P_i = n_i\, \alpha_i + (K - n_i)\beta_i$. With these preliminaries we calculate now $H(X|Y)$.

$$\begin{aligned}
H(X|Y) &= \sum_{i=1}^{S} \sum_{y \in \mathcal{Y}_i} P_Y(y) H(X|Y = y) \\
&= -\sum_{i=1}^{S} P_i \sum_{m \in \mathcal{X}_i} \frac{P_X(m)}{P_i} \log \frac{P_X(m)}{P_i}
\end{aligned}$$

$$= -\sum_{i=1}^{S} n_i\,\alpha_i \log\frac{\alpha_i}{P_i} + (K-n_i)\beta_i \log\frac{\beta_i}{P_i}$$

$$= \sum_{i=1}^{S} -P_i \log\frac{\alpha_i+\beta_i}{P_i} - n_i\,\alpha_i \log\frac{\alpha_i}{\alpha_i+\beta_i} - (K-n_i)\beta_i \log\frac{\beta_i}{\alpha_i+\beta_i}$$

$$= \sum_{i=1}^{S} -P_i \log\frac{\alpha_i+\beta_i}{2P_i} - n_i\,\alpha_i \log\frac{2\,\alpha_i}{\alpha_i+\beta_i} - (K-n_i)\beta_i \log\frac{2\,\beta_i}{\alpha_i+\beta_i}$$

$$= \log K + \sum_{i=1}^{S} -P_i \log\frac{K(\alpha_i+\beta_i)}{2P_i} - n_i\,\alpha_i \log\frac{2\,\alpha_i}{\alpha_i+\beta_i} - (K-n_i)\beta_i \log\frac{2\,\beta_i}{\alpha_i+\beta_i}.$$

Now we use the inequality $-\ln x \geq 1-x$ and obtain

$$H(X|Y) \geq \log K - (\log e)\frac{K}{2}\sum_{i=1}^{S}\frac{(\alpha_i-\beta_i)^2}{\alpha_i+\beta_i}.$$

Recall that $\alpha_i \geq \beta_i \geq \alpha_{i+1} \geq 0$ and therefore

$$H(X|Y) \geq \log K - (\log e)\frac{K}{2}\left(\sum_{i=1}^{S-1}(\alpha_i-\alpha_{i+1})\frac{\alpha_i-\beta_i}{\alpha_i+\beta_i} + (\alpha_S-\beta_S)\frac{\alpha_S-\beta_S}{\alpha_S+\beta_S}\right)$$

$$\geq \log K - (\log e)\frac{K}{2}\left(\sum_{i=1}^{S-1}(\alpha_i-\alpha_{i+1}) + \alpha_S - \beta_S\right)$$

$$= \log K - (\log e)\frac{K}{2}(\alpha_1-\beta_S) = \log K - (\log e)\frac{K}{2}\Big(P_X(1)-P_X(M)\Big). \qquad \square$$

Remark 21 1. If $P_X(m) \leq \frac{1}{K}$ for all $m \in \mathcal{M}$, then (2.4.15) is improved to

$$H(X|Y) \geq \log K - \frac{1}{2}(\log e)K\frac{1}{K} = \log K - \frac{1}{2}\log e \approx \log K - 0.72\,.$$

2. If P_X is the uniform distribution, then it follows $H(X|Y) \geq \log K$ and therefore $H(X|Y) = \log K$.
3. The bound in (2.4.5) is ≥ 0 exactly if $P_X(1) - P_X(M) \leq \frac{2\ln K}{K}$. Therefore it may happen that this bound is weaker than the bound of Theorem 50.
4. In order to construct the described cipher it is not necessary that sender and receiver know the message distribution P_X exactly. They (only) have to know

the information about the ordering of messages according to probability which is needed to form the partitions $\mathcal{X}$ and $\mathcal{Y}$.

2.4.4 Data Compression

We would like to analyze the effects of data compression in a cryptographic system. In all our previous considerations a message was an element of some set of other messages which occur with some probabilities. We have not been interested in the description of the messages. In a lot of applications the messages are given as a sequence of letters over a finite alphabet and we will assume that these sequences are produced by a source. This allows to install a source coder before using a cipher. The idea behind this is to remove the redundancy that helps a cryptanalyst.

Before we proceed we need some definitions to formalize the described scenario.

Preliminaries

In the sequel let $\mathcal{A} \triangleq \{0, \ldots, a-1\}$ for some $a \in \mathbb{N}$ with $a \geq 2$.

Definition 37 We call the set $\mathcal{A}$ an alphabet. An element of $\mathcal{A}$ is referred to as a letter and an element of $\mathcal{A}^n$ is called a word (of length n over $\mathcal{A}$). We denote the set of all words (over $\mathcal{A}$) by

$$\mathcal{A}^\star \triangleq \bigcup_{n=0}^{\infty} \mathcal{A}^n.$$

For a word $u \in \mathcal{A}^\star$ we denote by $l(u)$ its length.

Remark 22 Note that also the word with length 0 belongs to $\mathcal{A}^\star$. This is called the empty word.

We define the concatenation of two words and the prefix property.

Definition 38 Let $u = (u_1, \ldots, u_n),\ v = (v_1, \ldots, v_m) \in \mathcal{A}^\star$ be two words. We denote by

$$uv \triangleq (u_1, \ldots, u_n, v_1, \ldots, v_m)$$

their concatenation.

We say that u is a prefix of v if their exists a $w \in \mathcal{A}^\star$ such that $uw = v$ and we write in this case $u \preccurlyeq v$. We say that a set of words $\mathcal{W} \subset \mathcal{A}^\star$ has the prefix property (or shortly is a prefix set) if no element of $\mathcal{W}$ is prefix of another element, i.e., $u \preccurlyeq v$ for two elements $u, v \in \mathcal{W}$ necessarily implies $u = v$.

Remark 23 A well known fact is that a prefix set $\mathcal{W}$ satisfies the Kraft inequality, which is

$$\sum_{u \in \mathcal{W}} a^{-l(u)} \le 1.$$

(See, for instance, [5], pp. 41.)

We would like to describe the output of a source which is a sequence of letters as a sequence of elements of a prefix set. Therefore the next definition is important.

Definition 39 We call a set $\mathcal{W} \subset \mathcal{A}^\star$ complete if for all $v \in \mathcal{A}^\star$ there exists a $u \in \mathcal{W}$ with $u \preccurlyeq v$ or $v \preccurlyeq u$.

This implies that given a complete set $\mathcal{W}$ we can find for any word $v \in \mathcal{A}^\star$ words $u_1, \dots, u_n \in \mathcal{W}$ such that

$$v \preccurlyeq u_1 \dots u_n \text{ and } u_1 \dots u_{n-1} \preccurlyeq v. \tag{2.4.18}$$

If $\mathcal{W}$ is in addition a prefix set than this decomposition of v is unique except, maybe, for the last word u_n.

Remark 24 A complete prefix set has $i(a-1)+1$ elements (for some $i \in \mathbb{N}_0$) and a prefix set is complete exactly if we have equality in the Kraft inequality ([5], pp. 41).

Definition 40 For some finite set $\mathcal{V}$ we call a mapping $\phi : \mathcal{V} \to \mathcal{A}^\star$ a code. The words $\phi(v)$, $v \in \mathcal{V}$, are called codewords.

A code ϕ is said to be uniquely decodable if every word in $\mathcal{A}^\star$ has at most one representation as a sequence of code words, i.e., if the mapping

$$\Phi : \bigcup_{n=1}^{\infty} \mathcal{V}^n \to \mathcal{A}^\star \text{ defined by } \Phi(v_1, \dots, v_n) \triangleq \phi(v_1)\phi(v_2)\dots\phi(v_n)$$

is injective.

A code is called a prefix code if the set of codewords is a prefix set.

Remark 25 Every prefix code is uniquely decodable. The opposite is not true but if a uniquely decodable code is given, then it is always possible to find a prefix code with the same codeword lengths (see for instance [5], pp. 51).

Definition 41 A (discrete) source over the alphabet $\mathcal{A}$ is a sequence $(U_n)_{n=1}^{\infty}$ of random variables with values in $\mathcal{A}$.

A source is called stationary if $P_{U_1 \dots U_n}(u_1, \dots, u_n) = P_{U_m \dots U_{n+m-1}}(u_1, \dots, u_n)$ for all $n, m \in \mathbb{N}$, i.e., if the joint distribution of $(U_m, \dots, U_{n+m-1})$ does not depend on m (for all $n \in \mathbb{N}$).

Remark 26 A special case of a stationary source is the so called discrete memoryless source where the random variables are independent and identically distributed.

Definition 42 If for a given source $\lim_{n\to\infty} \frac{1}{n} H(U_1 \dots U_n)$ exists then this limit is called the entropy rate of the source.

Remark 27 For a stationary source $\frac{1}{n} H(U_1 \dots U_n)$ is nonincreasing in n and therefore the entropy rate always exist (see, for instance, [8], pp. 65).

The Extension of the Model with a Source Coder

Let $\mathcal{A} \triangleq \{0, \dots, a-1\}$ and $\mathcal{B} \triangleq \{0, \dots, b-1\}$, where $a, b \in \mathbb{N}$ with $a, b \geq 2$, be two alphabets. Suppose now that the messages to be securely transmitted consist of sequences over the alphabet $\mathcal{A}$, which are generated by a source $(U_n)_{n=0}^{\infty}$. The transmission of this source output to the receiver is implemented in three steps.

1. **Source Coding**
   The output of the source is encoded in the following way. Let $\mathcal{V} \subset \mathcal{A}^\star$ be a complete prefix set. According to [25] the elements of the set $\mathcal{V}$ are referred to as 'segments'. With these segments the output of the source is decomposed, i.e., any word $u \in \mathcal{A}^\star$ is split into a sequence of segments from $\mathcal{V}$.

$$\underbrace{u_1, u_2, \dots, u_{l(v_1)}}_{v_1 \in \mathcal{V}}, \underbrace{u_{l(v_1)+1}, \dots, u_{l(v_1)+l(v_2)}}_{v_2 \in \mathcal{V}}, \dots$$

   Then using a uniquely decodable code $\phi : \mathcal{V} \to \mathcal{B}^\star$ every segment $v \in \mathcal{V}$ is replaced by its codeword $\phi(v)$ over $\mathcal{B}$.
   Thus the source coding allows to transform the sequence of letters from $\mathcal{A}$ into a sequence of letters from $\mathcal{B}$ ruled by a modified probability law.
2. **Encryption**
   The sequence of letters from $\mathcal{B}$ is encrypted in the following way. We take a set $\mathcal{M} \subset \mathcal{B}^\star$ such that we can decompose every possible sequence of letters over $\mathcal{B}$ generated by the encoding procedure and the source into elements from $\mathcal{M}$ (of course it always suffices to choose a complete set $\mathcal{M}$, usually, the set of words over $\mathcal{B}$ with a fixed length n is taken for $\mathcal{M}$, i.e., $\mathcal{M} = \mathcal{B}^n$). Then the elements of $\mathcal{M}$ are encrypted with a cipher $(\mathcal{C}, Q)$ in the usual way. This means the encoded sequence of letters of $\mathcal{B}$ is decomposed into a sequence of elements from $\mathcal{M}$ and each of this elements is encrypted with a secret key $c_z \in \mathcal{C}$ known to the sender and the receiver. Again we will refer to the elements of $\mathcal{M}$ as messages although it has to be remembered that these are only encoded versions of the original messages.
3. **Decryption**
   The receiver can reconstruct the original source output as $c_z : \mathcal{M} \to \mathcal{M}$ is bijective and ϕ is uniquely decodable.

Remark 28 We make Kerckhoffs' assumption (see Section "The Opponent's Knowledge" in Sect. 2.2) that the only thing the opponent does not know about the described secrecy system is which of the keys is used by sender and receiver. In particular this means that the opponent knows the method how the source is encoded by means of the set of segments $\mathcal{V}$ and the code ϕ.

The described secrecy system is shown schematically in Fig. 2.4. We would like to define a random variable X with values in $\mathcal{M}$ whose distribution is induced by the source and the coding procedure and for the cryptograms a random variable Y with values in $\mathcal{M}$ whose distribution is induced as usual by $\mathcal{C}$ and the distributions of X and Z. (Note that in some cases the distribution of X may not be well defined because the probability that message $m \in \mathcal{M}$ occurs may be dependent upon the point of occurrence of m in the sequence of letters from $\mathcal{B}$ produced by the source and the coding method. Later we will be in a context where this problem does not occur.) Then in [25] the security of such a secrecy system is measured by

$$H(X|Y).$$

In the sequel we restrict ourselves to stationary sources. We say that the source coding is absent if $\mathcal{A} = \mathcal{B}$ and $\mathcal{V} = \mathcal{M} = \mathcal{A}^n$ for some $n \in \mathbb{N}$. If the source coding is absent and the number of keys K satisfies

$$\log K \geq c\,H(X) = c\,H(U_1 \dots U_n), \tag{2.4.19}$$

for some constant $c > 1$ then from Remark 27 it follows that

$$\log K - H(X|Y) \geq \log K - H(X) \geq (c-1)H(X) \geq (c-1)\,n\,H_\infty, \tag{2.4.20}$$

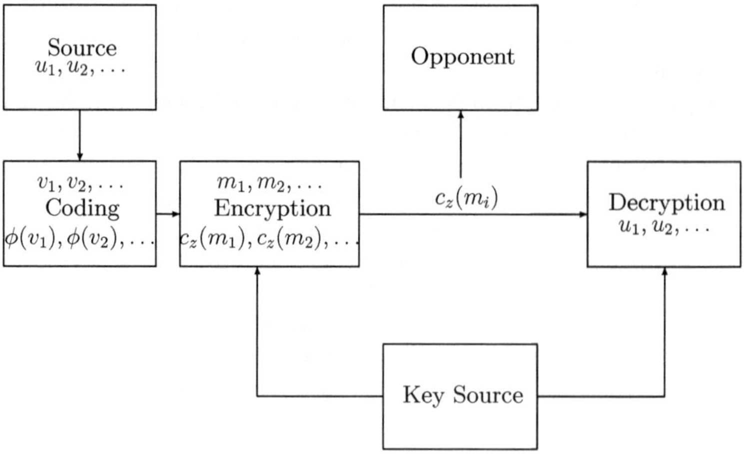

Fig. 2.4 A secrecy system with a source coder

where $H_\infty \triangleq \lim_{n\to\infty} \frac{1}{n} H(U_1 \dots U_n)$ is the entropy rate of the stationary source. It follows from (2.4.20) that if n tends to infinity the difference between $\log K$ and $H(X|Y)$ tends to infinity. It has to be remembered that "n tends to infinity" means according to (2.4.19) that the number of keys K and the number of messages $M \triangleq |\mathcal{M}| \,(= a^n)$ grow in such a way that

$$K \geq \exp(cnH_\infty) = M^{\frac{cH_\infty}{\log a}}.$$

We will see in the next section that the source coding allows to bound the difference $\log K - H(X|Y)$ above by a constant, which is independent of n. Therefore the source coding seems to be reasonable at least for numbers of keys K satisfying (2.4.19) and also the other cases require a special analysis.

If we use a cipher, which is locally regular with respect to $(\mathcal{X}, \mathcal{Y})$, then, in order to get a large value of $H(X|Y)$, we should use a source coding procedure such that the resulting distribution P_X is as uniform as possible within each of the sets $\mathcal{X}_i$, but quite different for different $\mathcal{X}_i$. This criterion has not been treated so far and Shtarkov [25] says that *in general the redundancy cannot characterize the efficiency of the source coding for the information protection.*

In the way we introduced the source coding the segments $v \in \mathcal{V}$ may have different lengths and also the codewords $\phi(v)$ may have different lengths. Then we speak of a variable-to-variable length coding. Beside the above mentioned problem that the distribution of X may not be well defined also the analysis of the value $H(X|Y)$ encounter some difficulties in this case because a given message $m \in \mathcal{M}$ may begin with a suffix of different codewords of ϕ or end with the prefix of different codewords of ϕ.

These problems do not arise if we consider the variable-to-fixed length coding procedure of the next section.

Variable-to-Fixed Length Coding

We now use codes ϕ such that all the codewords $\phi(v)$ have the same length. If we take $n \in \mathbb{N}$ for the length, then ϕ has the property that

$$\phi(\mathcal{V}) \subset \mathcal{B}^n.$$

We take

$$\mathcal{M} \triangleq \phi(\mathcal{V}).$$

Then $M = |\mathcal{V}|$ and the distribution of X is given by $P_X(m) = P_{U_1 \dots U_{l(v)}}(v)$ for $m \in \mathcal{M}$ and $v \in \mathcal{V}$ with $\phi(v) = m$.

A minimization of the average description length of the source output in the context of variable-to-fixed length coding means, as the length of the codewords is given, that one has to maximize the average length of the segments (in contrast to the minimization of the average codeword length in fixed-to-variable length coding). The solution to this problem under the constraints that the number of segments $|\mathcal{V}|$ is given and that the set of segments has to be complete is known as Tunstall's method of coding which is a recursively defined procedure (of course the number of segments must be of the in Remark 24 described form because otherwise one cannot find a complete prefix set with this cardinality).

Tunstall's Method of Coding Define complete prefix sets $\mathcal{V}_i \subset \mathcal{A}^\star$ in the following way. Let

$$\mathcal{V}_1 \triangleq \mathcal{A}, \tag{2.4.21}$$

i.e., we take for $\mathcal{V}_1$ the set of all one letter words. If $\mathcal{V}_i$ ($i \in \mathbb{N}$) is already defined then let

$$\mathcal{V}_{i+1} \triangleq \mathcal{V}_i \backslash \{v_i^\star\} \cup \{v_i^\star u : u \in \mathcal{A}\}, \tag{2.4.22}$$

where $v_i^\star \in \mathcal{V}_i$ is chosen such that $P_{U_1 \dots U_{l(v_i^\star)}}(v_i^\star) = \max_{v \in \mathcal{V}} P_{U_1 \dots U_{l(v)}}(v)$ (if the choice of $v_i^\star$ is not unique we take any such element). Thus $\mathcal{V}_{i+1}$ is constructed by appending to the most probable element in $\mathcal{V}_i$ one letter in all possible ways.

Clearly, by construction $\mathcal{V}_i$ is a complete prefix set with $|\mathcal{V}_i| = i(a-1)+1$. The associated variable-to-fixed length code is a mapping $\phi_i : \mathcal{V}_i \to \mathcal{B}^n$, which is injective and $n \triangleq \lceil \log_b(i(a-1)+1) \rceil$ is the minimal possible codeword length.

The proof for the optimality of Tunstall's method of coding can be found in ([30], see also [11], pp. 418). For our purposes we need only the following property of the sets $\mathcal{V}_i$. Let V_i be a random variable with values in $\mathcal{V}_i$ and distribution $P_{V_i}(v) \triangleq P_{U_1 \dots U_{l(v)}}(v)$ for any $v \in \mathcal{V}_i$.

Lemma 17 *Let $(U_n)_{n=0}^\infty$ be a discrete memoryless source and let $\mathcal{V}_i$ be constructed according to (2.4.21) and (2.4.22) for some $i \in \mathbb{N}$. Then*

$$\frac{\max_{v \in \mathcal{V}_i} P_{V_i}(v)}{\min_{v \in \mathcal{V}_i} P_{V_i}(v)} \leq \frac{1}{\min_{u \in \mathcal{A}} P_{U_1}(u)}, \tag{2.4.23}$$

where the minima are taken only over terms greater than zero.

Proof Clearly the statement holds for $i = 1$ because

$$\frac{\max_{u \in \mathcal{A}} P_{U_1}(u)}{\min_{u \in \mathcal{A}} P_{U_1}(u)} \leq \frac{1}{\min_{u \in \mathcal{A}} P_{U_1}(u)}.$$

Suppose now that the lemma is proved for $i \in \mathbb{N}$. From (2.4.22) follows that

$$\max_{v \in \mathcal{V}_{i+1}} P_{V_{i+1}}(v) \leq \max_{v \in \mathcal{V}_i} P_{V_i}(v).$$

This implies that if $\min_{v \in \mathcal{V}_{i+1}} P_{V_{i+1}}(v) = \min_{v \in \mathcal{V}_i} P_{V_i}(v)$ the statement holds also for $i+1$. Therefore we may assume that there exists an $u \in \mathcal{A}$ such that $P_{V_{i+1}}(v_i^\star u) = \min_{v \in \mathcal{V}_{i+1}} P_{V_{i+1}}(v)$. But then it follows

$$\frac{\max\limits_{v\in\mathcal{V}_{i+1}} P_{V_{i+1}}(v)}{\min\limits_{v\in\mathcal{V}_{i+1}} P_{V_{i+1}}(v)} \le \frac{P_{V_i}(v_i^\star)}{P_{V_{i+1}}(v_i^\star u)} = \frac{1}{P_{U_1}(u)} \le \frac{1}{\min\limits_{u\in\mathcal{A}} P_{U_1}(u)}. \qquad \square$$

Remark 29 It is easy to generalize Lemma 17 (and therefore also the next theorem) to Markovian sources. In these cases the minimum on the right-hand side of (2.4.23) has to be taken over the transition probabilities ([11], pp. 423).

Theorem 53 *Let* $(U_n)_{n=0}^{\infty}$ *be a discrete memoryless source. Let* $\mathcal{V}_i$ *and* ϕ_i *be given by Tunstall's method of coding. Then for any regular cipher* $(\mathcal{C}, \mathcal{Q})$

$$\log K - H(X|Y) \le (\log e)\,\delta\Big(\frac{1}{\min\limits_{u\in\mathcal{A}} P_{U_1}(u)}\Big),$$

where δ *is the function defined in (2.4.6).*

Proof The statement follows by combining Corollary 4 and Lemma 17. Lemma 17 implies that in (2.4.11) we get

$$\rho \le \frac{1}{\min\limits_{u\in\mathcal{A}} P_{U_1}(u)}$$

and therefore, as the function δ is monotonically increasing, the estimate in (2.4.12) implies that

$$H(X|Y) \ge \log K - (\log e)\,\delta\Big(\frac{1}{\min\limits_{u\in\mathcal{A}} P_{U_1}(u)}\Big). \qquad \square$$

Note that we have bounded the difference $\log K - H(X|Y)$ by a constant, which does not depend on M and K for any regular cipher.

Next we consider a simple example, which is taken from [25]. Suppose we are given a binary memoryless source, i.e., $\mathcal{A} \triangleq \{0, 1\}$ and the random variables U_i are independent and identically distributed. Let $P_{U_i}(0) \triangleq \frac{59}{64}$ and $P_{U_i}(1) \triangleq \frac{5}{64}$ for all $i \in \mathbb{N}$. We take 64 segments and messages, respectively, i.e., $|\mathcal{V}| \triangleq M \triangleq 64$ and as we take also a binary coding alphabet $\mathcal{B} \triangleq \{0, 1\}$ the lengths of the codewords is 6 and $\mathcal{M} = \mathcal{A}^6$. We consider two possible choices of the set of segments $\mathcal{V}$.

(a) *Absence of Source Coding*
Let $\mathcal{V} \triangleq \mathcal{A}^6$.
(b) *Optimal Variable-to-Fixed Length Coding for the given Source*
Let $\mathcal{V} \triangleq \mathcal{V}_{63}$, i.e., $\mathcal{V}$ is constructed by Tunstall's method for the given source. Then $\mathcal{V}$ contains the following segments:

$$0^i10^j1 \quad \text{and} \quad 0^i10^{6-i}, \quad \text{for } i = 0, 1, 2 \quad j = 0, 1, \dots, 5 - i,$$

$$0^i10^j1 \quad \text{and} \quad 0^i10^{7-i}, \quad \text{for } i = 3, 4, 5, 6 \quad j = 0, 1, \dots, 6 - i,$$

$$0^i 1, \quad \text{for } i = 7, 8, \dots, 37$$

$$\text{and } 0^{38},$$

where we denote by $u^i \triangleq \underbrace{(u, \dots, u)}_{i\text{–times}}$ the word of length i with letters all equal to u $(u \in \mathcal{A})$.

For these two choices of the segments we take the cipher of Sect. 2.4.3 with K= 2, 4, 8, ..., 64 keys. The calculated values of $H(X|Y)$ are presented in Table 2.1. The values in row (c) will be treated in Sect. 2.4.5. Now we can take a look at the performance of the bounds we derived in Theorems 50 and 52. Let us first look at the case (a) when the source coding is absent. The values that the bound in (2.4.5) returns and the deviation from the actual value of $H(X|Y)$ are shown in the Table 2.2. The estimates are good for $K < 8$ because then $\rho_1 = 11.8$ and many of the values ρ_i are equal to 1 since in the blocks of length K often occur words with the same number of zeros. The bound in (2.4.17) degenerates in case (a), as $P_X(0^6) - P_X(1^6) = 0.614$ is very large.

For the case (b) we consider the simpler bound in (2.4.12) and the bound in (2.4.17). The values of these bounds and the deviation to $H(X|Y)$ are shown in Table 2.3. Already the simpler bound in (2.4.12) returns values that are approximately not more than 1 bit away from $H(X|Y)$. The bound in (2.4.17) becomes worse with increasing K but as the difference of the probabilities of the most probable segments $0^i 10^{6-i}$ $(i = 0, 1, 2)$ and the most unlikely segment $0^i 10^{6-i} 1$ $(i = 3, 4, 5, 6)$ is only 0.044 it beats the bound (2.4.12) for all K up to 32.

Table 2.1 Calculated values of $H(X|Y)/H(V|Y)$

$\log K$	1	2	3	4	5	6
(a)	0.563	1.217	1.901	2.137	2.334	2.373
(b)	0.999	1.997	2.987	3.961	4.802	5.407
(c)	0.156	0.254	0.340	0.389	0.393	0.396

Table 2.2 Performance of the bound in (2.4.5) for (a)

$\log K$	1	2	3	4	5	6
Bound in (2.4.5)	0.563	1.105	0.563	0.913	0.225	1.842
Difference to $H(X|Y)$	≈ 0	0.112	1.338	1.224	2.109	0.532

Table 2.3 Performance of the bounds in (2.4.12) and (2.4.17) for (b)

$\log K$	1	2	3	4	5	6
Bound in (2.4.12)	0.375	1.375	1.921	2.921	3.921	4.921
Difference to $H(X\|Y)$	0.624	0.322	1.066	1.04	0.881	0.486
Bound in (2.4.17)	0.936	1.872	2.745	3.49	3.98	3.959
Difference to $H(X\|Y)$	0.063	0.125	0.242	0.471	0.822	1.448

2.4.5 Randomization

An old cryptographic method is the usage of randomized ciphers known as multiple-substitution ciphers or homophonic ciphers. The idea is the substitution of highly probable words by randomly chosen representatives. For instance in a typical English text the letter e appears with the highest frequency. If the letters e are randomly substituted by different symbols all representing the e, then the new text over this larger alphabet may have a more balanced frequency distribution of letters and therefore an enciphering of this modified text can increase the secrecy.

We will extend our model of Sect. 2.4.4 in the following way. Let V be a random variable for the occurrence of the segments, i.e., V has values in $\mathcal{V}$ and the distribution is given by $P_V(v) \triangleq P_{U_1 \dots U_{l(v)}}(v)$ for all $v \in \mathcal{V}$. We assume that with each occurrence of a segment $v \in \mathcal{V}$ the sender gets to know the value of an additional random variable R with values in some finite set $\mathcal{R}$. In general R and V are not independent. We make the encoding dependent upon the value of R, i.e., we replace the code $\phi : \mathcal{V} \to \mathcal{B}^\star$ by a code $\phi : \mathcal{V} \times \mathcal{R} \to \mathcal{B}^\star$ such that the decoding of a sequence over $\mathcal{B}$ is unique with respect to v. The rest of the model is as treated before. The receiver knowing the secret key can reconstruct the output of the source.

The introduction of the randomization results of course in an enlargement of the codeword lengths (if we take them all equal as before) compared to an absence of the randomization. Therefore we are dealing with to different approaches to increase the secrecy. The first is the elimination of redundancy by means of an effective source coding and the second is the randomization, which can be regarded as a special form of source coding increasing the description length and the redundancy. These approaches seem to be contradictory in principle. However, sometimes this contradiction can be eliminated.

We restrict ourselves again to a variable-to-fixed length encoding. This means we assume

$$\phi(\mathcal{V} \times \mathcal{R}) \subset \mathcal{B}^n \qquad \text{for some } n \in \mathbb{N}$$

and we define

$$\mathcal{M} \triangleq \phi(\mathcal{V} \times \mathcal{R}).$$

Furthermore let

$$\mathcal{M}(v) \triangleq \{m \in \mathcal{M} : m = \phi(v, r),\ r \in \mathcal{R}\} \subset \mathcal{M} \quad \text{for any } v \in \mathcal{V}$$

be the set of all possible messages if the segment v occurs. The decoding is unique with respect to v if the sets $\mathcal{M}(v),\ v \in \mathcal{V}$, are disjoint. Then it follows for the number of messages that

$$M = |\mathcal{M}| = \sum_{v \in \mathcal{V}} |\mathcal{M}(v)| \geq |\mathcal{V}|.$$

Shtarkov [25] notes that in this context the above mentioned contradiction can be eliminated rather simply. The secrecy of such a cryptosystem is related to the value $H(V|Y)$ rather than to the value of $H(X|Y)$ because a message $m \in \mathcal{M}$ is only an auxiliary description for some segment $v \in \mathcal{V}$ and therefore for a part of the original output sequence of the source. Without randomization, i.e., if we consider the secrecy system with the variable-to-fixed length coding scheme of the last section we have

$$H(X|Y) = H(V|Y),$$

but with the introduction of the randomization these values become different and we are interested in the behaviour of $H(V|Y)$. We would like to investigate, if the randomization allows to increase $H(V|Y)$. The inequality $H(V|Y) \leq H(V)$ gives an obvious upper bound and we know from Example 7 that this bound can be achieved *without* randomization if we are allowed to use $K = |\mathcal{V}|$ keys. With randomization the analogous bounds to (2.4.1) hold which is shown by

$$H(V|Y) \leq H(VZ|Y) = H(Z|VY) + \underbrace{H(V|ZY)}_{=0}$$

$$= H(Z|VY) \leq H(Z) \leq \log K.$$

This shows that also with randomization a necessary condition for $H(V) = H(V|Y)$ is that $H(Z) \geq H(V)$.

Under what conditions the randomization allows that the value of $H(V|Y)$ reaches the upper bound $\log K$ is treated in the next theorem.

Theorem 54 *If*

$$K \max_{v \in \mathcal{V}} |\mathcal{M}(v)| \leq M, \tag{2.4.24}$$

then there exists a regular cipher $(\mathcal{C}, Q)$ *with* K *keys such that*

$$H(V|Y) = H(X|Y).$$

If condition (2.4.24) does not hold then for any cipher $(\mathcal{C}, Q)$ *with* K *keys*

$$H(V|Y) < H(X|Y) \leq \log K.$$

Proof From the grouping axiom of the entropy function it follows that

$$H(X|Y = y) = H(V|Y = y) + \sum_{v \in \mathcal{V}} P_{V|Y}(v|y) H(P_v),$$

where P_v is the distribution on $\mathcal{M}(v)$ given by $P_v(m) \triangleq \frac{P_{X|Y}(m|y)}{\sum_{m' \in \mathcal{M}(v)} P_{X|Y}(m'|y)}$. Therefore in general we have $H(V|Y = y) \leq H(X|Y = y)$ with equality exactly if for every $v \in \mathcal{V}$ and $y \in \mathcal{M}$ with $P_{V|Y}(v|y) > 0$, there exist only one $m \in \mathcal{M}(v)$ with $P_{X|Y}(m|y) > 0$.

Now let us enumerate the segments, the messages and the cryptograms

$$v_1, \ldots, v_{|\mathcal{V}|} \in \mathcal{V} \quad m_0, \ldots, m_{M-1} \in \mathcal{M} \quad y_0, \ldots, y_{M-1} \in \mathcal{M}.$$

The enumeration of segments and cryptograms is arbitrary. The messages should be enumerated such that the first messages are those of the set $\mathcal{M}(v_1)$, the next are in $\mathcal{M}(v_2)$ and so on. More precisely the following condition has to be satisfied.

$$\mathcal{M}(v_i) = \{m_j \in \mathcal{M} : \eta(i-1) \leq j < \eta(i)\} \quad \text{for all } i = 1, \ldots, |\mathcal{V}|,$$

where

$$\eta(i) \triangleq \sum_{l=1}^{i} |\mathcal{M}(v_l)|$$

for all $i = 0, \ldots, |\mathcal{V}|$ (with the convention that $\sum_{l=1}^{0} \cdots = 0$).

Let $(\mathcal{C}, Q)$ be any regular cipher with K keys such that a message m_j is mapped to the K different cryptograms y_n with

$$n \in \{(K(j-1)+0) \bmod M, (K(j-1)+1) \bmod M, \ldots, (Kj-1) \bmod M\}.$$

Thus for every $v \in \mathcal{V}$ the messages $m \in \mathcal{M}(v)$ are mapped onto $|\mathcal{M}(v)|$ consecutive (modulo M) cryptograms. Therefore (2.4.24) implies that for every $y \in \mathcal{M}$ the set $\{c_z^{-1}(y) \in \mathcal{M} : z = 1, \ldots, K\}$ contains at most one message of every set $\mathcal{M}(v)$. Therefore $H(V|Y) = H(X|Y)$ and the first statement is proved.

On the other hand if (2.4.24) does not hold then for the segment $v \in \mathcal{V}$ with maximal $|\mathcal{M}(v)|$ there exists for any cipher with K keys a cryptogram $y \in \mathcal{M}$ such that the set $\{c_z^{-1}(y) \in \mathcal{M} \ z = 1, \dots, K\}$ contains at least two different messages belonging both to $\mathcal{M}(v)$. Therefore we have for this cryptogram

$$H(V|Y = y) < H(X|Y = y)$$

and this proves the second statement. □

If $P_V(v) < \frac{1}{M}$, then it follows for $m \in \mathcal{M}(v)$ that $P_X(m) \leq P_V(v) < \frac{1}{M}$. Therefore only if the minimal nonnegative probability $P_V(v)$ of a segment is not less than $\frac{1}{M}$ it may be possible to get a uniformly distributed random variable X on $\mathcal{M}$. In this case, when M is large enough such that $\min_{v \in \mathcal{V}} P_V(v) \geq \frac{1}{M}$, it suffices to choose the sizes of $\mathcal{M}(v)$ such that $|\mathcal{M}(v)| = M P_V(v)$ for all $v \in \mathcal{V}$ and the random variable R such that for any $v \in \mathcal{V}$ there are $|\mathcal{M}(v)|$ values in $\mathcal{R}$ such that $P_{R|V}(r|v)$ is equal to $\frac{1}{|\mathcal{M}(v)|}$ and for the remaining values in $\mathcal{R}$ $P_{R|V}(r|v)$ is equal to 0 (if $MP_V(v)$ is not an integer, then it is only possible to get an approximate uniform distribution P_X). In this way we obtain $P_X(m) = \frac{P_V(v)}{|\mathcal{M}(v)|} = \frac{1}{M}$ for all $m \in \mathcal{M}(v)$.

Then any regular cipher guarantees $H(X|Y) = \log K$ but Theorem 54 tells us that $H(V|Y) < \log K$ if the condition (2.4.24) is not fulfilled. If (2.4.24) holds then $H(V|Y) = \log K$ for the cipher introduced in the proof of Theorem 54. From condition (2.4.24) follows in the described case

$$K \leq \frac{1}{\max_{v \in \mathcal{V}} P_V(v)} \leq \frac{M}{\rho(\mathcal{V})},$$

where $\rho(\mathcal{V}) \triangleq \frac{\max_{v \in \mathcal{V}} P_V(v)}{\min_{v \in \mathcal{V}} P_V(v)}$.

Shtarkov [25] concludes that the equality $H(V|Y) = \log K$ can be attained at the expense of an increase in M and hence, of implementation complexity. Therefore he compares the results achievable with and without randomization under the same complexity, i.e., for the same values of K and M.

Consider the following example where the letters in the output of a discrete memoryless source are splitted.

Suppose that the probabilities for the occurrence of all letters $u \in \mathcal{A}$ can be written as

$$P_{U_1}(u) = \gamma_u b^{-\omega} \quad \text{for some } \omega, \gamma_u \in \mathbb{N} \text{ with } 0 < \gamma_u < b^{\omega}.$$

(Recall that b is the size of the alphabet $\mathcal{B}$.) Then we can partition the set $\mathcal{B}^{\omega}$ of words of length ω over $\mathcal{B}$ into $a = |\mathcal{A}|$ disjoint sets $\mathcal{B}_u^{\omega}$, $u \in \mathcal{A}$, with $|\mathcal{B}_u^{\omega}| = \gamma_u$ (recall that $\sum_{u \in \mathcal{A}} \gamma_u b^{-\omega} = 1$). Given the letter $u \in \mathcal{A}$ as source output then we may replace it by any element of $\mathcal{B}_u^{\omega}$ with probability $\frac{1}{\gamma_u}$. We can do this independently n times ($n \in \mathbb{N}$) and define in this way the code

$$\phi : \mathcal{A}^n \times \mathcal{R} \to \mathcal{B}^{n\omega},$$

where we chose $\mathcal{V} \triangleq \mathcal{A}^n$ and $\mathcal{M} \triangleq \mathcal{B}^{n\omega}$.

By construction X has a uniform distribution on the set $\mathcal{M}$. Furthermore the resulting source over the alphabet $\mathcal{B}$ has independent and identically distributed random variables.

The source treated in the example at the end of the Section "Variable-to-Fixed Length Coding" of Sect. 2.4.4 allows such a form of randomization. In that case we have $\omega = 6$, $\gamma_0 = 59$ and $\gamma_1 = 5$. To get the same complexity as for the cases without randomization we should take the same values for M and K. As $M = 64$ we can only take $\mathcal{V} = \mathcal{A}$, i.e., $n = 1$. The values of $H(V|Y)$ for the cipher introduced in the proof of Theorem 54 are presented in Table 2.1 in the row (c). We see in any column of the table, i.e., for fixed K, that (under the same complexity) the randomization reduces the secrecy compared to an absence of the source coding and even more to the variable-to-fixed length coding.

Shtarkov [25] concludes that on the whole, one can reasonably believe, that the efficiency of the randomization has been overestimated but that there are no reasons to reject this approach completely.

2.5 Public-Key Cryptology

2.5.1 Introduction

In secret-key cryptology the cryptanalyst's task was to find out which of the possible keys $c_1, \ldots, c_K$ was used to encrypt the message. It was assumed that sender and receiver could agree on this key by communicating over a "secure" channel to which the cryptanalyst had no access. This assumption is often not realistic. In computer networks, for example, all users share the same net and there usually is no possibility to transmit messages over some private wire to which only the two communicating parties have access. Even if such a secure channel would exist, there is a further disadvantage of secret-key cryptology. Recall from the previous chapter that in order to really protect a message from being decrypted the amount of key space has to be as big as the amount of message space. So if we want to protect a message of length n bits, say, we have to transmit another n bits as the key. This, of course, will slow down the transmission of the message by a factor 2.

In their paper "New directions in cryptography" Diffie and Hellman [9] introduced the first public-key protocol, based on the discrete logarithm. In public-key cryptology communication over a secure channel is no longer necessary. There is only one key $c : \mathcal{M} \to \mathcal{M}$. We now drop the assumption that the cryptanalyst has unlimited computational power. It was already pointed out by Shannon in his pioneering paper that the complexity of encoding and decoding might be considered and Diffie and Hellman finally introduced the concept of a *one-way function*, i.e., a function, which is easy to evaluate but hard to invert. We shall later precise this notion. So if we use a one-way function as key c, then the encoding, i.e., the evaluation of $c(m)$ can be done rather fast, but in order to decrypt the transmitted message the cryptanalyst has to apply the inverse function c^{-1} to recover the original message $m = c^{-1}\big(c(m)\big)$ which is a task of much higher complexity and cannot be done in reasonable time. We shall present the protocol of Diffie and Hellman in order to get more insight.

The Diffie–Hellman Algorithm

(1) Person i chooses some $a_i \in \{1, 2, \ldots, p-1\}$ and stores the value $b_i = w^{a_i}$ in a public directory, accessible to everybody. p here is a large prime number and w some primitive element, i.e., the order of p in $GF(p)$ is $p-1$.

(2) If Persons i and j want to communicate, they calculate their common key

$$k_{ij} = b_i^{a_j} = w^{a_i \cdot a_j} = w^{a_j \cdot a_i} = b_j^{a_i} = k_{ji}$$

and encrypt and decrypt their message using this common key.

(3) In order to break the key, a third person has to know one of the numbers

$$a_i = \log_w b_i, a_j = \log_w b_j$$

(where $\log_w$ is the discrete logarithm to the base w in $\mathbb{Z}_p$).

The algorithm is already presented in such a form that it is clear how it will work in a multiuser system, e.g., in a computer network. Observe that there is only one key for communication between Persons i and j. For instance, they could split their message into blocks of length $\lceil \log_2 p \rceil$ and add k_{ij} to each of these blocks. If p is large enough, a third person will not be able to decipher the text. Additionally, every other user in the system has all the necessary information to calculate k_{ij}. He knows p and w and he also can deduce a_i and a_j from b_i and b_j, since $a_i \mapsto w^{a_i}$ is one-to-one.

However, in order to obtain a_i or a_j, a third person has to apply the discrete logarithm $\log_w b_i$ or $\log_w b_j$, which is a computationally hard task. The best known algorithm takes $O(\sqrt{p})$ steps. In contrast, Persons i and j have to exponentiate in order to obtain k_{ij}. This can be done in $O(\log p)$ steps using repeated squaring. The function $f(x) = w^x$ (in $GF(p)$) had been conjectured by Diffie and Hellman [9] to be a one-way function. Later Hellman and Pohlig [21] found that additionally $p-1$ must have a large prime factor.

Diffie and Hellman also introduced the concept of a *trapdoor one-way function*. This is a collection of functions $\{f_k\}_k$ with the properties that

(i) in knowledge of k there exist fast algorithms for the evaluation of f_k and f_k^{-1}.

(ii) when k is not known, then for almost all y it is hard to find the x with $f_k(x) = y$, even if the encoding procedure is known.

Diffie and Hellman did not give an example for a trapdoor one-way function. This was later done by Rivest, Shamir and Adleman. We shall now present the Rivest–Shamir–Adleman [22] (RSA) cryptosystem. The RSA-system is widely used today. The (conjectured) trapdoor one-way function here is obtained making use of the hardness of integer factorization.

The RSA—Public Key Cryptosystem

(1) Each person k selects two "large" prime numbers p and q and forms the product $n \triangleq p \cdot q$.
(2) Further, each person selects (at random) a "large" number d with the property that the greatest common divisor $gcd\big(d, (p-1)\cdot(q-1)\big) = 1$ and then computes its multiplicative inverse e, hence $e \cdot d \equiv 1 \mod (p-1)\cdot(q-1)$.
(3) The numbers e and n are published in a public directory.
(4) If another person wants to submit a message x to Person k, he encrypts it using the encoding functions

$$E_k(x) = x^e \mod n (=: y).$$

Person k can easily decrypt y by application of the decoding function

$$D_k(y) = y^d \mod n\big(= (x^e)^d = x^{e\cdot d} = x \mod n\big).$$

Again, it is obvious that the RSA-system is already constructed for multi-user networks. Since e and n are stored in a public directory, every other person can encrypt messages directed to Person k using the key E_k. Decoding is done very fast using the number d, which is only known to Person k. Anybody else has to find the prime factor p and q of n in order to obtain d. Now, there exist quite fast algorithms to find even large prime numbers, whereas factorization is a very hard computational task. This has not been proved, but under the assumption that there is a significant gap between the complexity of prime number generation and factorization a collection of functions $(E_k)_k$ as used in the RSA-system is a trapdoor one-way function.

Most of the cryptosystems we shall introduce in this chapter are based on the hardness of factorization. We shall discuss this in Sect. 2.5.3, where some prime number tests and the basic ideas of the best known factorization algorithms are presented. First, we need some background in elementary number theory, which is given in Sect. 2.5.2.

We introduced a one-way function as a function which is "easy" to evaluate but "hard" to invert. This is rather a heuristic approach and we did not say yet what we mean by easy and hard. We do not want to discuss this here, since it requires some background in Complexity Theory. However, we shall at least give the idea for those who are familiar with the notions. "f is easy to evaluate" means that there exists a probabilistic polynomial-time algorithm (Turing machine) that on input x outputs $f(x)$). "Hard to invert" analogously means that for all probabilistic polynomial-time algorithms A the probability that A finds the inverse for a given y is negligibly small.

The function presented in the Diffie–Hellman and RSA-cryptosystems have been conjectured to be one-way functions. However, this has not been proved. It is not even known if one-way functions exist at all. Computer Scientists say that the existence of a one-way function seems to be a stronger assumption than the famous $P \neq NP$, although it is widely believed that one-way functions exists.

Although the discrete logarithm and encoding functions based on integer factorization are often used in practice, from a theoretical point of view they are not quite satisfactory examples. It has not been shown that the inversion is really as hard as suggested. The only thing we know is that up to now the fastest known algorithms for the computation of the discrete logarithm and for integer factorization are much slower than repeated squaring (for exponentiation) and the best prime number tests, respectively. We shall discuss this briefly in Sect. 2.5.3 (factorization) and Sect. 2.5.4 (discrete logarithm).

On the other hand, there exist problems which are provably hard if we assume that $P \neq NP$, the NP-complete problems. Using an NP-complete problem as basic tool for the construction of an encoding (one-way) function might yield a cryptosystem which is secure—at least if we assume that $P \neq NP$. However most of the attempts to construct a cryptosystem based on some NP-complete problem, so far, have not been very satisfactory. We shall illustrate the difficulties which may arise, when the knapsack problem is used to encrypt messages, in Sect. 2.5.5.

In the two cryptosystems introduced by Diffie and Hellman, as in Shannon's model of secret-key cryptology, a message is encrypted in order to protect it against the cryptanalysts attempts to obtain the information contained in this message. In electronic communication further forms of protection may be required. We already saw in the chapter on authentication that the cryptanalyst could also have the possibility to replace a message. In order to prove the authenticity of a message, this message is often equipped with a signature—some extra bits of information, which prove to the receiver that the message really originated from the sender who encrypted it. There exist several public-key cryptosystems for digital signatures. Further, for many purposes it is required that a participant of a system has to prove his identity in order to get access. Think, e.g., of a password you have to enter in order to login into the computer or of a secret code for the credit card. If the person who has to verify the identity does not obtain any further information, the identity proof is said to be a zero-knowledge proof.

Digital signatures, identity proofs and further situations, for which public-key cryptosystems have been developed, will be discussed in Sect. 2.5.6.

2.5.2 *Number Theory*

In this section we shall present those results and facts from Number Theory which are important to understand the algorithms in the subsequent sections. We assume that the reader is familiar with basic notions such as prime number, greatest common divisor, congruences, group, ring, field, etc.

Euclidean Algorithm

The Euclidean algorithm yields the greatest common divisor of two natural numbers $a > b$, which we shall denote by $gcd(a, b)$. It proceeds as follows:

In the first step we divide the numbers a and b with remainder, i.e., we find non-negative integers t_0 and r_1 with $a = t_0 \cdot b + r_1$, where $0 \le r_1 < b$. This procedure is repeated with b and r_1 to obtain numbers t_1 and r_2 with $b = t_1 \cdot r_1 + r_2$ and $0 \le r_2 < r_1$. We continue with r_1 and r_2 until we finally find an r_m such that $r_{m-1} = t_m \cdot r_m + 0$ (since $0 < r_m < \cdots < r_2 < r_1 < b < a$, this algorithm really needs a finite number of m iterations).

Proposition 5 *The number r_m is the greatest common divisor $gcd(a, b)$.*

Proof We have to show that r_m divides a and b and that r_m is the largest number with this property. Since $r_{m-1} = t_m \cdot r_m$, r_m divides r_{m-1}. Of course, then r_m divides $r_{m-2} = t_{m-1} \cdot r_{m-1} + r_m = (t_m \cdot t_{m-1} + 1) \cdot r_m$. Inductively, r_m divides $r_{i-2} = t_{i-1} \cdot r_{i-1} + r_i$, since r_m is divisor of r_{i-1} and r_i, and hence r_m divides b and a. In order to show that r_m is really the greatest common divisor of a and b, we shall see that any d which divides a as well as b also has to divide r_m. To see this observe that d must divide $r_1 = t_0 b - a$, hence $r_2 = t_1 \cdot r_1 - b$ and finally (by induction) $r_m = t_{m-1} \cdot r_{m-1} - r_{m-2}$.

Proposition 6 *The greatest common divisor $gcd(a, b)$ can be written as $gcd(a, b) = u \cdot a + v \cdot b$ for some integers $u, v \in \mathbb{Z}$.*

Proof With $u_1 = 1$ and $v_1 = -t_0$ we have $r_1 = a - t_0 b = u_1 a + v_1 b$. Now assume that for some $u_k, v_k \in \mathbb{Z}$ it is $r_k = u_k a + v_k b$ $(k \le m - 1)$. Then

$$\begin{aligned} r_{k+1} &= r_{k-1} - t_k r_k = u_{k-1} a + v_{k-1} b - t_k (u_k a + v_k b) \\ &= (u_{k-1} - t_k \cdot r_k) a + (v_{k-1} - t_k \cdot r_k) b, \end{aligned} \tag{2.5.1}$$

and hence

$$u_{k+1} = u_{k-1} - t_k r_k, \, v_{k+1} = v_{k-1} - t_k \cdot r_k \in \mathbb{Z}.$$

With $u \triangleq u_m$ and $v \triangleq v_m$ the Proposition is proved. For a speed analysis of the Euclidean algorithm, recall that the Fibonacci numbers $\{F_n\}_{n=0}^{\infty}$ are defined by the recurrence $F_n = F_{n-1} + F_{n-2}$ with initial values $F_0 = 0$, $F_1 = 1$. It can be shown that $F_n = \frac{1}{\sqrt{5}} \left(\left(\frac{1+\sqrt{5}}{2} \right)^n + \left(\frac{1-\sqrt{5}}{2} \right)^n \right)$, especially, it turns out that $F_n \ge \left(\frac{1+\sqrt{5}}{2} \right)^{n-2}$. The proof is left as an exercise to the reader.

Proposition 7 (Lamé) *For positive integers $a > b$ the number of iterations to compute the greatest common divisor $gcd(a, b)$ via the Euclidean algorithm is at most $\lfloor \log_s a \rfloor - 2$, where $s = \frac{1+\sqrt{5}}{2}$.*

Proof For all $i = 1, \ldots, m$ it is $r_{i-2} = t_{i-1} \cdot r_{i-1} + r_i \ge r_{i-1} + r_i$ (since $t_{i-1} \ge 1$ and with the convention $r_{-1} \triangleq a$, $r_0 \triangleq b$). Since $\{r_i\}_i$ is a decreasing integer sequence with $r_m = gcd(a, b) \ge 1$, we see that $r_{i-2} \ge r_{i-1} + r_i$ must be larger than the $(i - m)$th Fibonacci number from which Proposition 3 follows. With Proposition 7 the Euclidean algorithm is a fast way to determine the greatest common divisor $gcd(a, b)$ of two non-negative integers a and b. It takes about $O(\log a)$ steps. The performance of the Euclidean algorithm can still be improved. Stein introduced a variant in which we get rid off the division with remainder, which is replaced by divisions by 2. This can be done much faster by processors.

In the design of cryptographic protocols the Euclidean algorithm is used to find the inverse of a given number $d \in \mathbb{Z}_n$. To see this, observe that d is invertible in $\mathbb{Z}_n$ if $gcd(d, n) = 1$. With Proposition 6 this means that $1 = u \cdot d + v \cdot n \equiv u \cdot d (\text{mod } n)$ and hence $u = d^{-1}$ in $\mathbb{Z}_n$.

Repeated Squaring

The reason for the speed in the encoding and decoding function of the Diffie–Hellman and of the RSA cryptosystems is that the determination of the inverse in $\mathbb{Z}_n$ and exponentiation can be done very fast. The inverse element is found using the Euclidean algorithm in $O(\log n)$ computation steps. We shall now present the repeated squaring algorithm, which computes the nth power of a given number in $O(\log n)$ steps.

Let

$$n = \sum_{i=0}^{t} a_i 2^i, a_i \in \{0, 1\}, t = \lfloor \log_2 n \rfloor$$

be the binary representation of n. Then

$$x^n = x^{a_0 + a_1 2 + \cdots + a_t 2^t} = x^{a_0} \cdot (x^2)^{a_1} \cdot (x^4)^{a_2} \cdot \cdots \cdot (x^{2^t})^{a_t}$$

with this product representation, it is clear what to do. Starting with x, we obtain $x, x^2, x^4, \dots, x^{2^t}$ by repeated squaring. This takes in total $t = \lfloor \log n \rfloor$ multiplications. Further, after each squaring, we look if the coefficient a_i is 0 or 1.

If $a_i = 0$ then x^{2^i} does not contribute to the product, if $a_i = 1$ then x^{2^i} occurs as a factor to the product $x^n = \prod_{\substack{i=1 \\ a_i=1}}^{t} x^{2^i}$.

So, to obtain x^n as product of the squares $(x^{2^i})_{i=1}^t$ we need at most another $t = \lfloor \log n \rfloor$ multiplications, such that the total number of multiplications is smaller than $2\lfloor \log n \rfloor$.

Euler's Totient Function

We denote by

$$\mathbb{Z}_n^* = \{x \in \mathbb{Z}_n : \exists y \in \mathbb{Z}_n \text{ such that } x \cdot y = 1\}$$

where multiplication is performed modulo n.

It can easily be verified that $\mathbb{Z}_n^*$ is a group. The order (number of elements) of $\mathbb{Z}_n^*$ is denoted by $\varphi(n)$. φ is called *Euler's totient function*. The proof of the following properties is left as an exercise to the reader.

Proposition 8 *Euler's φ-function has the following properties.*

(a) For all $x \in \mathbb{Z}_n^*$ it is $x^{\varphi(n)} \equiv 1 \mod n$
(b) $\sum_{d|n} \varphi(d) = n$

(c) *For a prime power* p^e, $e \in \mathbb{N}$, *it is* $\varphi(p^e) = p^{e-1}(p-1)$
(d) φ *is multiplicative, i.e.,* $\varphi(n_1 \cdot n_2) = \varphi(n_1) \cdot \varphi(n_2)$ *if* $gcd(n_1, n_2) = 1$
(e) $\varphi(n) = n \cdot \prod_{\substack{p|n \\ p\, prime}} \left(1 - \frac{1}{p}\right)$

If p is a prime number, then by (c) $\varphi(p) = p - 1$ and if $u = p \cdot q$ is the product of the different primes, then $\varphi(n) = (p-1) \cdot (q-1)$ by (e). Since by (a) $x^{\varphi(n)} \equiv 1 \mod n$, the condition $e \cdot d \equiv 1 \mod (p-1) \cdot (q-1)$ in Step 2 of the RSA-cryptosystem now becomes clear.

When p is a prime number it can be shown that the multiplicative group $\mathbb{Z}_p^*$ is cyclic, i.e., $\mathbb{Z}_p^* = \{1, x, x^2, x^3, \ldots, x^{p-1}\}$ is generated by some element x. We denote such an element as *primitive root.*

Proposition 9 *Let p be a prime number. In* $\mathbb{Z}_p^*$ *there are exactly* $\varphi(p-1)$ *primitive roots.*

Little Fermat

Fermat's "Little" Theorem is the central tool in the prime number tests we shall present in the next section.

Theorem 55 (Little Fermat) *Let p be a prime number. Then for any integer* $x \in \mathbb{Z}$ *not divisible by p*

$$x^{p-1} \equiv 1 \mod p.$$

Proof For any $y \in \mathbb{Z}$

$$(x+y)^p = \sum_{k=0}^{p} \binom{p}{k} x^k y^{p-k} \equiv x^p + y^p \mod p,$$

since $\binom{p}{k} \equiv 0 \mod p$ for $k = 1, \ldots, p-1$. So, especially $(x+1)^p \equiv x^p + 1 \mod p$.

By induction it is now clear that for all $x \in \mathbb{Z}$

$$x^p \equiv x \mod p$$

since with $x^p \equiv x \mod p$, also $(x+1)^p \equiv x^p + 1 \equiv x + 1 \mod p$. This is equivalent to

$$x(x^{p-1} - 1) \equiv 0 \mod p$$

and since by the assumption $x \not\equiv 0 \mod p$, Fermat's Little Theorem is proved.

Quadratic Residues

A number $x \in \mathbb{Z}_p^*$, p prime, is a *quadratic residue*, if there exists some $y \in \mathbb{Z}_p^*$ such that $y^2 \equiv x \mod p$. For $p = 7$ the quadratic residues in $\mathbb{Z}_p$ are 1, 2 and 4, whereas

3, 5 and 6 are non-residues. As this example suggests half of the elements in $\mathbb{Z}_p^* = \{1, \ldots, p-1\}$ are quadratic residues, more exactly

Proposition 10 *The squares in $\mathbb{Z}_p^*$ are a subgroup of $\mathbb{Z}_p^*$ with $\frac{p-1}{2}$ elements.*

Proof With $x^2 \cdot y^2 = (x \cdot y)^2$ and $(x^{-1})^2 \cdot x^2 = 1$ it is easy to verify that the squares form a subgroup. Since $\mathbb{Z}_p^*$ is cyclic it can be written as $\mathbb{Z}_p^* = \{1, w, w^2, \ldots, w^{p-1}\}$ for the generator w. Squares can only have an even exponent and indeed $|\{1, w^2, w^4, \ldots, (w^{\frac{p-1}{2}})^2\}| = \frac{p-1}{2}$.

In order to characterize, if a given $x \in \mathbb{Z}_p^*, p > 2$, is a quadratic residue the *Legendre symbol* $\left(\frac{x}{p}\right)$ is introduced, defined by

$$\left(\frac{x}{p}\right) = \begin{cases} +1, & \text{if } x \text{ is quadratic residue} \\ -1, & \text{else.} \end{cases}$$

The Legendre symbol defines a homomorphism from $\mathbb{Z}_p^*$ into $\{1, -1\}$, since $\left(\frac{x}{p}\right) \cdot \left(\frac{y}{p}\right) = \left(\frac{x \cdot y}{p}\right)$.

The Legendre symbol can be evaluated very fast using the following result.

Proposition 11 (Euler's lemma) *Let $p > 2$ be an odd prime number and $x \in \mathbb{Z}_p^*$. Then*

$$\left(\frac{x}{p}\right) \equiv x^{\frac{p-1}{2}} \mod p.$$

Proof By Fermat's Theorem the elements of $\mathbb{Z}_p^*$ are just the roots of the polynomial

$$z^{p-1} - 1 = (z^{\frac{p-1}{2}} - 1)(z^{\frac{p-1}{2}} + 1).$$

If x is a quadratic residue, then $x = y^2$ for some y and $x^{\frac{p-1}{2}} = y^{p-1} = 1$ by Fermat's Theorem.

If x is not a quadratic residue, then x must be a root of $(z^{\frac{p-1}{2}} + 1)$ (since there are exactly $\frac{p-1}{2}$ quadratic residues), hence $x^{\frac{p-1}{2}} = -1$.

With Euler's Lemma it is now easy to determine, whether a given $x \in \mathbb{Z}_p^*$ is a quadratic residue or not, just use repeated squaring to compute $x^{\frac{p-1}{2}}$.

We make use of this fact in order to present a fast probabilistic algorithm which finds a quadratic non-residue: Choose at random an $x \in \mathbb{Z}_p^*$ and compute $x^{\frac{p-1}{2}}$. If $x^{\frac{p-1}{2}} = -1$ we are done. Since exactly half of the elements in $\mathbb{Z}_p^*$ are quadratic non-residues, the probability that $x^{\frac{p-1}{2}} = -1$ is exactly $\frac{1}{2}$. So, on the average, after two attempts we are done. Note, that there is no deterministic algorithm known, which finds a quadratic non-residue this fast.

Once we know that x is a quadratic residue, we want to take the square root, i.e., to find a y with $y^2 = x$ in $\mathbb{Z}_p^*$ (of course with y also $p - y$ is square of x).

Proposition 12 *If x is quadratic residue modulo p and $\frac{p-1}{2}$ is odd, then*

$$y = x^{\frac{p+1}{4}}$$

and $p - y$ are the two square roots of x.

The proof is left as an exercise to the reader. Observe, that again we can apply repeated squaring in order to obtain a square root, if $\frac{p-1}{2}$ is odd. If this is not the case, there also exist fast algorithms, which solve this task. We do not want to discuss this here.

In cryptographic applications we are also interested in taking square roots in the ring $\mathbb{Z}_n$, when n is not a prime especially when $n = p \cdot q$ is the product of exactly two prime factors.

Proposition 13 *If $n = p \cdot q$, where p and q are distinct odd prime numbers, then there are exactly $\frac{(p-1)\cdot(q-1)}{4}$ quadratic residues in $\mathbb{Z}_n^*$, each of which has four distinct square roots.*

As an example consider $n = 15$. Here $\mathbb{Z}_n^* = \{1, 2, 4, 7, 8, 11, 13, 14\}$ and $x^2 = 1$ for $x = 1, 4, 11, 14$, whereas $x^2 = 4$ for $x = 2, 7, 8, 13$.

Let $n = p \cdot q$ as before and let y be a quadratic residue in $\mathbb{Z}_n^*$. Then x_1 and x_2 are said to be *essentially different square roots* of y if $x_1 \neq x_2$ and $x_1 \neq n - x_2$. So, for $n = 15$ in the above example 1 and 4 are essentially different square roots of 1, whereas 1 and 14 are not essentially different.

From the following proposition we can conclude that taking square roots in $\mathbb{Z}_n$, $n = p_1 \cdot p_2$ and factoring n are computationally equivalent tasks, in the sense that once one task is solved the other can be done with little extra effort.

Proposition 14 *If $n = p \cdot q$, where p and q are distinct odd primes and if x_1 and x_2 are essentially different square roots of some quadratic residue in $\mathbb{Z}_n^*$, then either*

$$gcd(x_1 + x_2, n) = p \;\; or \;\; gcd(x_1 + x_2, n) = q.$$

Proof Since x_1 and x_2 are square roots of the same element in $\mathbb{Z}_n^*, x_1^2 - x_2^2 \equiv 0 \mod n$ and hence $(x_1 - x_2)(x_1 + x_2) = t \cdot n = t \cdot p \cdot q$ for some integer t. Since x_1 and x_2 are essentially different, $n = p \cdot q$ cannot divide $x_1 - x_2$ or $x_1 + x_2$. So p divides one factor, either $(x_1 - x_2)$ or $(x_1 + x_2)$ and q divides the other one but not both, and hence either p or q (but not both) must divide $x_1 + x_2$.

With Proposition 14 it is clear that once we found two essentially different square roots, we can easily factor $n = p \cdot q$ using the Euclidean Algorithm. With the Chinese Remainder Theorem it can, on the other hand, be shown that if the prime factors p and q are known, then all four square roots of a quadratic residue can be found very fast. So taking square roots in $\mathbb{Z}_n$ and factoring $n = p \cdot q$ are of about the same computational complexity.

2.5.3 *Prime Number Tests and Factorization Algorithms*

Little Fermat, Pseudoprimes and Carmichael Numbers

The simplest way to factorize a given integer $n \in \mathbb{N}$ is to divide n by all numbers smaller than n. Indeed, we only have to check all numbers $m < \sqrt{n}$, since if $n = n_1 \cdot n_2$ is a product of two integers $n_1, n_2 > 1$, then one of the factors n_1 and n_2 must be smaller than $\sqrt{n}$. If none of these integers is a divisor of n, then n must be a prime. Hence with $O(\sqrt{n})$ computation steps we can determine if n is prime or not. Moreover, if n is not a prime, the above *trial division* algorithm will yield a prime factor.

We shall see in this section that the performance of factorization algorithms has not essentially been improved, whereas there are fast algorithms (at least probabilistic algorithms) known that determine if a number n is prime within running time $O(\log n)$.[4] This gap is exploited in the RSA cryptosystem.

The prime number tests are based on the Little Fermat, which states that if p is prime for all $b \in \mathbb{Z}_p^* = \{1, \dots, p-1\}$

$$b^{p-1} \equiv 1 \mod p.$$

So the Little Fermat yields a criterion for primality of an integer n which does not give any information about the prime factors of n. Just take a *base* $b \in \{1, \dots, n-1\}$ and check if $b^{n-1} \equiv 1 \mod n$. If this is not the case, then n cannot be prime. However, this "Fermat test" does not always work, since if n is not a prime there might exist bases b which pass the Fermat test. For instance $2^{340} \equiv 1 \mod 341$ but $3^{340} \equiv 54 \mod 341$.

We say in this case that n is *pseudoprime* to the base b. Even worse is, that there exist *Carmichael numbers*, which are pseudoprimes to every base b relatively prime to n (i.e., $gcd(b, n) = 1$). For instance $561 = 3 \cdot 11 \cdot 17$ is a Carmichael number. The Fermat test can be executed in $O(\log n)$ steps using repeated squaring. So, if we must only apply this to a small fraction of bases $b \in \{1, \dots, n-1\}$ in order to determine if n is prime, then we would have found a fast prime number test. Unfortunately it is not known if there are only finitely many Carmichael numbers, such that the Fermat test has to be executed for all bases.

---

[4] Remark by the editors: This statement is not up to date, because in the paper "M. Agrawal, N. Kayal, and N. Saxena, "PRIMES is in P", Annals of Mathematics, Vol. 160, No. 2, 781–793, 2004, the authors proved the asymptotic time complexity of the algorithm to be $\tilde{O}(\log^{12}(n))$. In other words, the algorithm takes less time than the twelfth power of the number of digits in n times a polylogarithmic (in the number of digits) factor. However, the upper bound proved in the paper was rather loose; indeed, a widely held conjecture about the distribution of the Sophie Germain primes would, if true, immediately cut the worst case down to $\tilde{O}(\log^{6}(n))$.

Probabilistic Prime Number Tests

Miller improved the Fermat test as follows. He proved that if n is prime and $n - 1 = r \cdot 2^k$, where r is odd and hence 2^k the highest power of 2 dividing n, then for every $b \in \{1, \dots, n-1\}$

$$b^r \equiv 1 \mod n \text{ or } b^{r \cdot 2^i} \equiv -1 \mod n \text{ for some } i \in \{1, \dots, k-1\}.$$

Again, if some base b does not pass the Miller test, then n must be a composite number. For the Miller test there is no analogon to the Carmichael numbers. More exactly, if n is an odd composite number, then the fraction of integers $b \in \{1, \dots, n\}$ which do not pass the Miller test is greater than $\frac{3}{4}$. This means that the probability that a randomly chosen $b \in \{1, \dots, n-1\}$ passes the test is smaller than $\frac{1}{4}$. If we choose t bases independently at random than the probability that all t numbers pass the Miller test for a composite number is smaller than $\frac{1}{4^t}$. If for a given n we find t randomly chosen numbers that pass the test, we say that p is a *probable prime*. We just described the probabilistic prime number test due to Rabin, which for a given degree of accuracy has running time $O(\log n)$. Note that the Miller test would yield a deterministic $O(\log^3 n)$ prime number test, if the generalized Riemann hypothesis would hold. In this case, for a composite number n, one would find a base b which does not pass the Miller test in the interval $\{2, 3, \dots, c \cdot \log^2 n\}$, where c is some universal constant not dependent on n. Hence the test would only have to be executed for the elements in this range.

Deterministic Prime Number Tests

The best known deterministic prime number tests[5] are based on factoring numbers related to the number n which has to be tested for primality. This is surprising, since we know that factoring is a hard task. However, the choice of the numbers which have to be factored is decisive.

Theorem 56 (Pocklington) *For an integer $n > 1$ let s be a divisor of $n - 1$. Suppose there is an integer b satisfying*

$$b^{n-1} \equiv 1 \mod n$$

$$gcd(b^{\frac{n-1}{q}} - 1, n) = 1 \textit{ for each prime } q \textit{ dividing } s.$$

Then for every prime factor p of n it is $p \equiv 1 \mod s$, and if $s > \sqrt{n} - 1$, then n is prime.

Pocklington's theorem yields a probabilistic prime number test analogous to the Rabin test, by random selection of several bases b for which the condition in the theorem is checked. There are similar tests using factors of $n+1, n^2+1, n^2+n+1$ or $n^2 - n + 1$. Note that a test based on Pocklington's theorem can only be fast if the

[5]See the Remark in the previous footnote.

factorization of s is easy, i.e., s only has small prime factors. If, e.g., $n - 1 = s_1 \cdot s_2$ where s_1 and s_2 are primes of about the same size, the test will be very slow. However the fastest prime number tests are based on similar arguments.

In the Jacobi-sum-test, the number s which is used for the single checks is no longer required to be a factor of $n - 1$, any product $s > \sqrt{n}$ can be used. So we can try to find an $s > \sqrt{n}$ which is the product $s = q_1 \dots q_r$ with the property that the least common multiple $t = \ell cm\{q_1 - 1, \dots, q_r - 1\}$ is small, i.e., the $q_i - 1$ have many factors in common. Odlyzko and Pomerance have shown that there is a positive constant c such that for every $n > e^e$ there exists an integer $t < (\log n)^{c \cdot \log \log \log n}$ such that the corresponding $s > \sqrt{n}$. Because a similar lower bound on t can be derived, it follows that the trial division step of this primality test requires slightly more than polynomially many steps, namely $(\log n)^{O(\log \log \log n)}$.

Another approach to overcome the difficulties in finding an appropriate number s is taken in the primality tests based on elliptic curves. Note that in the condition of Pocklington's theorem the number s is a divisor of $n - 1$ which is the order of the group $\mathbb{Z}_n^*$ if n is prime. Now to each prime p several groups over different elliptic curves are constructed. The group orders by a theorem of Hasse are between $p + 1 - 2\sqrt{p}$ and $p + 1 + 2\sqrt{p}$. Moreover, they are almost uniformly distributed in the interval $\{p + 1 - \sqrt{p}, \dots, p + 1 + \sqrt{p}\}$.

Now the groups are selected at random with the hope to find a group order T with a divisor s having a nice form.

Factorization Algorithms

The best factorization algorithms are rather slow compared to the best primality test. However, they show that in the construction of the RSA-cryptosystem and other schemes based on the hardness of factorization, one has to be very careful with the appropriate choice of the product $n = p \cdot q$.

In cryptographic applications, $n = p \cdot q$ is usually chosen as the product of two primes of about the same size $p \approx q$. In this case, one should first try the quadratic sieve method due to Pomerance. Lenstra developed a factorization algorithm based on elliptic curves. All these tests are not rigorously analyzed theoretically. However their performance in practice is good.

One should also take into account that a possible parallelization of a factorization algorithm might close the gap to primality tests a little bit. The RSA-129 (where the number n is a 129 digit number) was broken by factoring n using massive parallelization. The task was distributed worldwide via the Internet. A message encrypted with RSA-129 was presented in Scientific American 1977 as a "new kind of cipher that would take millions of years to break".

2.5.4 *The Discrete Logarithm*

Using repeated squaring $b = w^a$, $a \in \{0, \dots, n\}$ can be evaluated in $O(\log n)$ steps. The fastest known algorithm to find the discrete logarithm $a = \log_w b$ for a given b

(in an arbitrary multiplicative group) is due to Shanks. It has running time $O(\sqrt{n} \cdot \log \sqrt{n})$. The disadvantage is the enormous amount of storage space. However there are algorithms known, which are almost as fast and use less storage.

Shanks' algorithm consists of three stages.

(1) Select some $d \sim \sqrt{n}$. By Euclid's Algorithm there exist numbers Q and r such that $a = Qd + r$. The choice of d guarantees that all numbers involved (Q, d, r) have size not greater than $O(\sqrt{n})$.
(2) Make a table with entries $(x, \log_w x)$ for $\log_w x = 0, 1, \dots, d-1$ and sort this table on x.
(3) It is $b = w^a = w^{Qd+r}$ and hence $b(w^{-d})^Q = b(w^{n-d})^Q = w^r$. Now for $Q = 0, 1, 2, \dots$ compute $b(w^{n-d})^Q$ and compare the result with the entries in the table. Stop, when the result is equal to some x in the table. Then $r = \log_w x$ and $a = Qd + r$.

The most time-consuming task in this algorithm is the sorting of $O\sqrt{n}$ elements in the table in Step 2. This can be done using one of the best sorting procedures in time $O(\sqrt{n} \log \sqrt{n})$.

Note that taking logarithms can be done faster, when n is a composite number. In the Diffie–Hellman scheme this is the case, since $n = p - 1$, where p is prime. In order to keep the gap to the exponentiation algorithm large, n must then have a large prime factor. If this is not the case, $f(x) = w^x$ in $GF(p)$ is not a one-way function.

2.5.5 Knapsack Cryptosystems

We shall in this section discuss cryptosystems based on the knapsack problem. The knapsack problem is NP-complete and hence from a theoretical point of view such cryptosystems are quite attractive, since they are provably hard, as pointed out in the Introduction. However, in practice most of these cryptosystems have been broken.

The knapsack problem states as follows. For a given set of positive integers $a_1, \dots, a_n$ and s, determine if there is a subset of $\{a_1, \dots, a_n\}$ such that the sum of the a_i's in this subset is exactly s. In other words, do there exist variables $x_1, \dots, x_n \in \{0, 1\}$ such that

$$\sum_{i=1}^{n} x_i a_i = s.$$

The number s may be interpreted as the capacity of a knapsack. If the a_i's are the weights of certain goods, the question is, if it is possible to find a collection of these goods which exactly fills the knapsack.

If such a collection exists, the subset of the a_i's can be guessed and it is easy to verify that $\sum_{i=1}^{n} x_i a_i = s$ in linear time (using at most n additions). Hence there exists a non-deterministic algorithm which solves the knapsack problem in polynomial time.

A simple deterministic algorithm is to check all possible 2^n subsets for the condition. Of course, this takes an exponential number of steps. This naive way has not essentially been improved. The best known algorithm takes about $2^{\frac{n}{2}}$ operations. The idea is to form all sums

$$S_1 = \left\{ \sum_{i=1}^{\lfloor \frac{n}{2} \rfloor} x_i a_i, x_i \in \{0,1\} \right\}, \; S_2 = \left\{ \sum_{i=\lfloor \frac{n}{2} \rfloor + 1}^{n} x_i a_i, x_i \in \{0,1\} \right\},$$

sort each of the sets S_1 and S_2 and then try to find a common element. If such a common element exists,

$$\sum_{i=1}^{\lfloor \frac{n}{2} \rfloor} x_i a_i = s - \sum_{i=\lfloor \frac{n}{2} \rfloor + 1}^{n} x_i a_i \text{ and hence } \sum_{i=1}^{n} x_i a_i = s.$$

Like in Shanks' algorithm for the evaluation of the discrete logarithm, the speedup has to be paid with an enormous amount of storage space.

In a knapsack cryptosystem, a message $(x_1, \dots, x_n) \in \{0,1\}^n$ is encoded as

$$s = \sum_{i=1}^{n} a_i x_i$$

where the weights $\{a_1, \dots, a_n\}$ are stored in a public directory. The cryptanalyst then knows the $a_1, \dots, a_n$ from the public directory and the message s he intercepted. So he has all the necessary information to decode the cryptogram. However, in order to do so, he has to solve an NP-complete problem.

The problem is that also the receiver has to solve the knapsack problem. Without any additional information his task is as hard as the cryptanalyst's. To overcome this difficulty, we first consider knapsacks of a certain structure which are easy to attack. Namely, it is required that the coefficients $a_1, \dots, a_n$ form a *superincreasing* sequence, i.e., for all $i = 2, \dots, n$

$$a_i > \sum_{j=1}^{i-1} a_j.$$

A knapsack problem based on a superincreasing sequence can be solved inductively very fast. It is $x_n = 1$ exactly if $s > \sum_{i=1}^{n-1} a_i$. So after having determined x_n we are left with the smaller knapsack problem $s - x_n a_n = \sum_{i=1}^{n-1} x_i a_i$.

All public-key cryptosystems based on the knapsack problem use such a superincreasing sequence $b_1, \dots, b_n$, say, of coefficients. Of course, these coefficients cannot be published, since the cryptanalyst could easily decode the cryptogram in this case. The idea is to transform the superincreasing sequence $b_1, \dots, b_n$ to a sequence $a_1, \dots, a_n$ from which the cryptanalyst does not benefit. The a_i's are published and the message $(x_1, \dots, x_n)$ is encoded as $s = \sum x_i \cdot a_i$ using the public key. The cryptanalyst, hence, still has to solve a hard problem. The receiver, who can reconstruct the superincreasing sequence $b_1, \dots, b_n$, only has to solve an easy knapsack problem.

Merkle and Hellman [20] introduced the first knapsack cryptosystem. We shall now present the transformation they used.

The system consists of

(1) a superincreasing sequence $b_1, \dots, b_n$ with
$b_1 \approx 2^n$, $b_i > \sum_{j=1}^{i-1} b_j$ for $i = 2, \dots, n$, $b_n \approx 2^{2n}$,
(2) two positive integers, M and W such that
$M > \sum_{i=1}^{n} b_i$, $gcd(M, W) = 1$,
(3) a permutation $\pi : \{1, \dots, n\} \to \{1, \dots, n\}$.

The superincreasing sequence $b_1, \dots, b_n$ is transformed in two steps to a sequence $a_1, \dots, a_n$ of coefficients by

(a) $a'_i \equiv b_i \cdot W \mod M$
(b) $a_i = a'_{\pi(i)}$.

So first the b_i's are multiplied by W modulo M. Observe that $a'_i = 0$ cannot occur, since $gcd(M, W) = 1$ and $M > b_i$ for all i. Then the so obtained numbers are shuffled using the permutation π. The sequence $a_1, \dots, a_n$ is the public key. A message $(x_1, \dots, x_n) \in \{0, 1\}^n$ is hence encrypted as $s = \sum_{i=1}^{n} x_i a_i$.

The receiver has some information, which is not available to the cryptanalyst. Namely he knows the numbers M and W from which he can conclude to the superincreasing sequence $b_1, \dots, b_n$ as follows. He computes

$$\begin{aligned} C &\equiv s \cdot W^{-1} \mod M \\ &\equiv \sum_{i=1}^{n} x_i a_i W^{-1} \mod M \equiv \sum_{i=1}^{n} x_i a'_i W^{-1} \mod M \\ &\equiv \sum_{i=1}^{n} x_i b_{\pi(i)} \mod M \end{aligned}$$

by the encoding rules. So multiplication modulo M of the cryptogram s with W^{-1} leaves a knapsack based on a superincreasing sequence and this is an easy computational task for the receiver.

There also exists a refined version of the Merkle–Hellman system, where instead of the numbers (M, W) a sequence (M_k, W_k) is used to transform the superincreasing sequence iteratively. The Merkle–Hellman system has been broken by the following approach. By the encoding prescription it is $a_i \equiv b_{\pi(i)} W \mod M$ and hence $b_{\pi(i)} \equiv a_i W^{-1} \mod M$. So for some integer k_i it is $a_i W^{-1} - k_i M = b_{\pi(i)}$ and hence

$$\frac{W^{-1}}{M} - \frac{k_i}{a_i} = \frac{b_{\pi(i)}}{a_i M}.$$

This means that the quotients $\frac{k_i}{a_i}$ are close to $\frac{W^{-1}}{M}$, since M is large compared to the first b_i's, at least.

Shamir used this close approximation to obtain numbers W' and M' with $\frac{(W')^1}{M'}$ close to $\frac{W^{-1}}{M}$ from which a superincreasing sequence similar to $b_1, \ldots, b_n$ is obtained. Another attack using Diophantine approximation is due to Lenstra.

2.5.6 Further Cryptographic Protocols

As pointed out in the introduction, in multiuser computer-networks cryptographic protocols are needed not only for protecting a message from being deciphered. We already learned about Simmon's theory of authentication, where a message is protected from being replaced. This is often done by a *digital signature*. Further applications of cryptography are *proofs of identity*. For instance, you have to enter a code before using a credit card or a password is needed in order to login to a computer. Identity proofs are often required to be *zero-knowledge* interactive proofs, i.e., the verifier should obtain no more information from the prover except information that the verifier could produce alone, even if the verifier cheats.

Proof of Identity

The following interactive protocol for a proof of identity is due to Omura. It is based on the discrete logarithm. First, each user of a multiuser system chooses some x (from a finite field) and puts $y = w^x$ in a public directory. It is assumed that each user has a copy of this directory. The protocol then proceeds in three rounds of communication.

(1) The first message $M_1 = a$ sent by the person who wants to prove his identity is the index a of his position in the public directory.
(2) The verifier selects some number r and transmits in the second round the message $M_2 = w^r$.
(3) The prover rises M_2 to the power x_a (he has a copy of the public directory) and transmits $M_3 = M_2^{x_a} = w^{r \cdot x_a}$.
(4) Finally, the verifier computes $y_a^r = w^{x_a \cdot r}$ and compares the result with the last message M_3.

Observe that this is not a zero-knowledge proof of identity since the verifier may cheat by sending $M_2 = r$ (not the power w^r) as second message. In this case he learns $M_3 = r^{x_A}$ which he could not calculate himself (However, he still has to take the discrete logarithm to conclude to x_a, which is a difficult task. So this information does not help him so much).

We shall later on present a zero-knowledge proof of identity using quadratic residues. First we shall illustrate the idea by a method for executing a fair random experiment interactively (due to Rabin).

Coin-Flipping by Telephone

Two persons want to execute a fair random experiment. They are only connected by telephone and do not trust each other. So they have to simulate a coin-flipping by telephone. The simulation is based on the factorization of an integer $n = p \cdot q$, a product of two large primes formed by Person 1. Since factorization is a hard task, Person 2 is not able to find the prime factors. In the course of the protocol Person 1 now will give some information about the number n, which will allow Person 2 to factor n with probability $\frac{1}{2}$. So "head" just means that Person 2 can factor n, whereas "tail" corresponds to the event that Person 2 cannot factor n. The protocol proceeds as follows.

(1) As first message $M_1 = n$, Person 1 sends the number $n = p \cdot q$.
(2) Person 2 selects an element $x \in \mathbb{Z}_n^*$ and transmits as second message $M_2 = x^2$.
(3) Person 1 computes a square root y of M_2 and sends this as message $M_3 = y$.
(4) If now $y = x$ or $-x$ in $\mathbb{Z}_n^*$ Person 2 can factor n (cf. Sect. 2). Else, he cannot factor n. Observe that if he can factor n, he can also prove this to anyone.

The idea of finding a square root that allows to factor a composite number n with probability $\frac{1}{2}$ is also used in the following zero-knowledge proof of identity due to Fiat and Shamir (1986).

Fiat–Shamir Zero-Knowledge Proof of Identity

It is assumed that $n = p \cdot q$ is a product of two large prime factors which is publicly known. Further each user selects an element $x \in \mathbb{Z}_n^*$ and stores x^2 next to the index of his name in a public directory. Again the protocol consists of three rounds.

(1) First, Person 1 selects at random an element $r \in \mathbb{Z}_n^*$ and transmits as first message $M_1 = (a, r^2)$ the index of his name a and r^2.
(2) Person 2 randomly chooses a binary digit $b \in \{0, 1\}$ which he transmits as message $M_2 = b$.
(3) Person 1 sends the third message $M_3 = \begin{cases} r, & \text{if } b = 0 \\ r \cdot x_a, & \text{if } b = 1. \end{cases}$
(4) If $b = 0$, Person 2 checks that $M_3^2 = r^2$, which was sent in the first message. If $b = 1$, Person 2 checks that $M_3^2 = r^2 \cdot x_a^2$.

Why is this protocol a zero-knowledge proof. Observe that since $(r \cdot x_a) \cdot r^{-1} = x_a$, Person 1 can know both possible values for the third message M_3 only if he knows the secret x_a. Hence, the probability that a third person not knowing x_a is deceiving Person 2, is less than or equal to $\frac{1}{2}$. On the other hand, Person 2 does not obtain any further information. The number r was chosen at random, so the only thing transmitted from Person 1 to Person 2 in the course of the protocol is a random number (either r or $r \cdot x_1$) and its square. This could be generated by Person 2 himself.

Observe that, contrasting to the first proof-of-identity protocol presented, here the only message from Person 2 to Person 1 is a random number, such that it is not possible to cheat for him. By repetition of the Fiat–Shamir protocol k times, say, (giving Person 1 k secrets), the probability that a third person, who does not know these secrets, deceives, is smaller than 2^{-k} and can hence be made arbitrarily small by the appropriate choice of k.

Another well-known zero-knowledge protocol for a proof of identity is based on the graph-isomorphism problem, i.e., on the decision if two graphs are isomorphic. As the knapsack problem, the graph-isomorphism problem is NP-complete and hence the zero-knowledge protocol in this case is based on a provably hard problem. The Fiat–Shamir protocol, as the RSA-system depends on the hardness of factorization.

Digital Signatures

A signature is attached to a message in order to identify the producer of this message. Signatures may be *implicit* or *explicit*. An implicit signature is used when the message is written in a way that no one else can imitate. An example for an implicit signature is the encryption of a message with a secret key, since it is very improbable that a randomly chosen string will be accepted as a valid plain-text. However, the opponent could replace the cryptogram by an older valid cryptogram. In order to avoid such an attack, messages are usually equipped with a time stamp.

We will rather be concerned with explicit signatures. In this case the message has an inseparable mark attached that no one else can imitate.

Further, signatures may be *private* or *public*. In order to discover a private signature, one has to share a secret with the author of the message (for instance, the secret-key example of an implicit signature is also private). A public signature can be identified by anybody else.

Explicit signatures are often obtained using hashing functions. Reversible two-key cryptosystems automatically yield implicit public signatures.

In electronic banking *blind signatures* are important, i.e., the signer does not know what message he is signing but can later certify whether a message was signed by him or not.

A detailed discussion on digital signatures will be carried out in Chap. 4. An overview is given in the book [29].

References

1. R. Ahlswede, Remarks on Shannon's secrecy systems. Prob. Control Inf. Theory **11**(4), 301–318 (1982)
2. L.A. Bassalygo, Lower bounds for the probability of successful substitution of messages. Prob. Inf. Trans. **29**(2), 194–198 (1993)
3. L.A. Bassalygo, M.V. Burnashev, Estimate for the maximal number of messages for a given probability of successful deception. Probl. Inf. Trans. **30**(2), 129–134 (1994)
4. L.A. Bassalygo, M.V. Burnashev, Authentication, identification and pairwise separated measures. Problemy Peredachi Informacii (in Russian) **32**(1), 41–47 (1996)

5. R.E. Blahut, *Principles and Practice of Information Theory* (Addison-Wesley, Boston, 1987)
6. M.V. Burnashev, S. Verdu, Measures separated in L_1-metrics and ID-codes. Probl. Inf. Trans. **30**(3), 3–14 (1994)
7. D. Coppersmith, The data encryption standard (DES) and its strength against attacks. IBM J. Res. Dev. **38**(3), 243–250 (1994)
8. I. Csiszar, J. Körner, *Information Theory: Coding Theorems for Discrete Memoryless Systems* (Academic Press, Cambridge, 1981)
9. W. Diffie, M.E. Hellman, New directions in cryptography. IEEE Trans. Inf. Theory **22**(6), 644–654 (1976)
10. W. Feller, *An Introduction to Probability Theory and Its Applications*, 3rd edn. (Wiley, New York, 1968)
11. B. Fitingof, Z. Waksman, Fused trees and some new approaches to source coding. IEEE Trans. Inform. Theory **34**(3), 417–424 (1988)
12. E.N. Gilbert, F.J. Mac Williams, N.J.A. Sloane, Codes which detect deception. Bell Syst. Tech. J. **53**(3), 405–424 (1974)
13. M.E. Hellman, An extension of the shannon theory approach to cryptography. IEEE Trans. Inform. Theory **23**(3), 289–294 (1977)
14. R. Johannesson, A. Sgarro, Strengthening Simmons' bound on impersonation. IEEE Trans. Inform. Theory **37**(4), (1991)
15. D. Kahn, *The Codebreakers* (Mac Millan, New York, 1967)
16. D. Kahn, Modern cryptology. Sci. Am. 38–46 (1966)
17. F.J. MacWilliams, N.J.A. Sloane, *The Theory of Error Correcting Codes* (North-Holland, Amsterdam, 1977)
18. J.L. Massey, An introduction to contemporary cryptology, in *Contemporary Cryptology—the Science of Information Integrity*, ed. by G.J. Simmons (IEEE Press, New Jersey, 1992), pp. 1–39
19. U. Maurer, A unified and generalized treatment of authentication theory, in *Proceedings of the 13th Symposium on Theoretical Aspects of Computer Science (STACS '96)*, Lecture Notes in Computer Science (Springer, Heidelberg, 1996), pp. 387–398
20. R.C. Merkle, M.E. Hellman, Hiding information and signatures in trapdoor knapsacks, Secure communications and asymmetric cryptosystems, 197-215, in *AAAS Selected Symposium Series* (Westview, Boulder, 1982)
21. S. Pohlig, M. Hellman, An improved algorithm for computing logarithms in GF(p) and its cryptographic significance. IEEE Trans. Inform. Theory **24** (1978)
22. R. Rivest, A. Shamir, L.M. Adleman, A method for obtaining digital signatures and public-key cryptosystems. Commun. ACM **21**, 120–126 (1978)
23. A. Sgarro, Informational divergence bounds for authentication codes, advances in *Cryptology—Eurocrypt '89*, Lecture Notes in Computer Science (Springer, Heidelberg, 1990)
24. C.E. Shannon, Communication theory of secrecy systems. Bell Syst. Tech. J. **28**, 656–715 (1949)
25. Yu.M. Shtarkov, Some information-theoretic problems of discrete data protection. Prob. Inf. Trans. **30**(2), 135–144 (1994)
26. G.J. Simmons, Message authentication: a game on hypergraphs. Congressus Numerantium **45**, 161–192 (1984)
27. G.J. Simmons, Authentication theory/coding theory, advances in cryptology, in *Proceedings of the CRYPTO 84*, Lecture Notes in Computer Science, ed. by G.R. Blakley, D. Chaum (Springer, Heidelberg, 1985), pp. 411–431
28. G.J. Simmons, A survey of information authentication, in *Contemporary Cryptology—the Science of Information Integrity*, ed. by G.J. Simmons (IEEE Press, New Jersey, 1992), pp. 379–419
29. D.R. Stinson, *Cryptography—Theory and Practice, Discrete Mathematics and its Applications*, 3rd edn. (Chapman and Hall, London, 2006) (CRC, Florida)
30. B.P. Tunstall, Synthesis of Noiseless Compression Codes, Ph.D. Thesis, Georgia Institute of Technology, Atlanta, 1967

Chapter 3
The Mathematical Background of the Advanced Encryption Standard

3.1 Introduction

In 2001 *Rijndael* became the official new encryption standard named *Advanced Encryption Standard* (AES). It is the successor of the *Data Encryption Standard* (DES) and won the competition, started by the *National Institute for Standards and Technology* (NIST) in 1997, which we will briefly explain in Sect. 3.2. In this competition Rijndael, which was proposed by Joan Daemen and Vincent Rijmen [6], prevailed over the other proposals such as *Mars* by IBM [3], *RC6* by RSA Labs [19], *Serpent* by Ross Anderson et al. [1] and *Twofish* by Counterpane Inc. [20]. The goal of this section is to give comprehensive explanations about the design criteria of Rijndael and their specific realization.[1]

One of the AES requirements was that the submitted ciphers should be block ciphers, which are used for computer security such as online banking, smart cards, computer communication, etc. This means that input and output of the ciphers should be one-dimensional array of bits.

In Sect. 3.3 we will show that there exists a bijection from the set of all one-dimensional array of bits of length n to the set $GF(2)[x]|_n$ of all polynomials with coefficients in $GF(2)$ and degree less than n. Each of these polynomials and therewith each one-dimensional array of bits of length n represents an element of the *finite field* $GF(2^n)$. In this section we will define the addition and multiplication of the finite field $GF(2^8)$ and the finite ring $GF(2^{32})$ and show how byte-addition, byte-multiplication, 4-byte-column-addition and 4-byte-column-multiplication are realized in Rijndael. We will show that byte- and 4-byte-column-addition equal the bitwise *XOR* operation, which can be efficiently evaluated. Further on, we show that byte-multiplication, which equals the polynomial multiplication followed by the reduction via the modulo

[1] Rudolf Ahlswede was invited with his group to the "Seminar Rijndael" in June 2001 in Paderborn. There he noticed the very interesting mathematics of the new code. Therefore he decided to section about it. His student Christian Heup wrote his dipolma thesis about this topic.

A. Ahlswede et al. (eds.), *Hiding Data – Selected Topics*, Foundations in Signal Processing, Communications and Networking 12, DOI 10.1007/978-3-319-31515-7_3 operation of a so-called reduction polynomial, can be efficiently computed by the *xtime* operation, if at least one operand is 'small'.

In the following Sect. 3.4 we give some basic definitions of several *boolean functions*, which map a boolean vector onto another boolean vector. Since a boolean vector of length n equals an one-dimensional array of bits of length n, these functions can be used to describe the block cipher Rijndael. After that, this section finishes with the definitions of several types of *block ciphers*, such as the *iterated* block cipher, which consists of the repeated application of one and the same round function and the *key-iterated* block cipher, where every application of the round function is followed by an addition of a particular round key, which is derived from the given cipherkey.

Section 3.5 concentrates on the *design of Rijndael*, which arose from the cryptanalysis of the DES. There were two approaches to analyze DES, the *differential attack* developed by Biham and Shamir [2] and the *linear attack* developed by Matsui [13]. To attack a block cipher via the differential respectively linear attack an enemy has to find trails over all but a few rounds of the cipher with a high difference propagation probability respectively with a high correlation. We will define the differential and linear weight of differential and linear trails as the negative logarithm of their difference propagation probability and their correlation. To be secure against both attacks was the main security requirement for the AES candidates. In order to achieve this goal Rijndael was designed according to the *Wide Trail Strategy*, which was developed by Joan Daemen in his doctoral dissertation [5] in 1995 and offers design criteria for block ciphers, so that there exist no low-weighted differential or linear trails. The Wide Trail Strategy suggests that the round function should be decomposed into two different layers, a non-linear substitution layer γ, which operates on only a limited number of bits of the intermediate results, called *bundles*, with high minimum differential and linear weights in relation to the bundle size, and a linear diffusion layer λ, which increases minimum differential and linear weights round by round. With this $\gamma\lambda$ round structure we are able to eliminate any trails of given differential or linear weights by increasing the number rounds of the cipher.

After that, in Sect. 3.6, we give the exact *specifications* of the individual steps of Rijndael. Firstly, we show how the plaintext block is mapped on the state, which represents the intermediate results of Rijndael. Then we specify the non-linear layer *SubBytes*, which consists of the inverse mapping in $GF(2^8)$ followed by an affine mapping to avoid interpolation attacks [11], and the linear layer, which consists of *ShiftRows* and *MixColumns*, where *ShiftRows* shifts the individual bytes of each row of the state over its columns and *MixColumns* multiplies each column of the state by a fixed polynomial. In the next sections, we show that the round keys are added to the state by the simple bitwise *XOR* operation and derive the round keys from the cipherkey via the *Key Schedule*. Finally, we present the whole encryption and decryption as they are implemented in Rijndael, provide some facts about its complexity and show how the requirements of the Wide Trail Strategy are applied to the Rijndael cipher in order to make it secure against differential and linear cryptanalysis.

Section 3.7 treats an cryptanalytic attack, the *saturation attack*. This attack is a chosen-plaintext attack over up to six rounds of Rijndael, which means that it exploits its specific structure by encrypting properly chosen plaintexts in order to derive the unknown cipherkey. The set of chosen plaintexts are the so-called Λ-sets, which consists of 2^8 plaintexts in which all the 2^8 bytes at the same position of these plaintexts sum up to zero. The property of Rijndael, which is exploited by this attack, is that the steps *SubBytes*, *ShiftRows* and *AddRoundKey* do not destroy a Λ-set and, if the Λ-sets are properly chosen, the *MixColumns* step maintains a Λ-set two times. This means that all the bytes at the same positions of the state sum up to zero until the input of the third *MixColumns* step and since *MixColumns* is linear this property is still true to its output state and therefore remains until the input of the fourth *MixColumns* step. To obtain all bytes of one round key we then guess its value and verify its correctness by summing up all bytes at the same position of the input state of the fourth *MixColumns* step. If we obtain zero the guess was correct with some probability, and if we do not obtain zero our guess was wrong. If we have found one whole round key with this method, we are able to obtain the cipherkey by going the *Key Schedule* algorithm the other way round.

3.2 The AES Selection Process

The Data Encryption Standard (DES) describes the data encryption algorithm (DEA), which is an improvement of the algorithm Lucifer developed by IBM in the early 1970s. This standard is up to now the most widely spread encryption algorithm in the world. But since the development of the differential attack [2] and the linear attack [13], it is no longer considered to be secure enough for security-critical applications. For example the U.S. government is no longer allowed to use it. To gain a higher security level, triple-DES was invented, which consists of the threefold application of DES, but whose disadvantage is its efficiency.

In 1997 the National Institute for Standards and Technology (NIST) announced the start of a competition, whose goal was to find an encryption algorithm to become the Advanced Encryption Standard (AES).

The requirements to the submissions were that the algorithm should be a symmetric block cipher with 128-bit blocks and 128-, 192- and 256-bit keys, in contrast to DES, which uses 56-bit keys and to 3-DES, which uses 112-bit keys. Further on, the algorithm should offer at least as much security as 3-DES, but should be much more efficient and finally the algorithm should be available royalty-free world-wide, which includes that the security-testing would be realized by the world-wide cryptology community and be therefore much more reliable. Another innovation was that the submissions were international and not only reserved to American cryptographers.

In August 1998 the first AES conference was held and fifteen submissions were accepted, where the first ones were submitted by companies and the last ones by researchers:

CAST-256	by Entrust (CA)
Crypton	by Future Systems (KR)
E2	by NTT (JP)
Frog	by TecApro (CR)
Magenta	by Deutsche Telekom (DE)
Mars	by IBM (USA)
RC6	by RSA (USA)
SAFER+	by Cylink (USA)
Twofish	by Counterpane Inc. (USA)
DEAL	by Outerbridge, Knudsen (USA-DK)
DFC	by ENS-CNRS (FR)
HPC	by Schroeppel (USA)
LOKI97	by Brown (AU)
Rijndael	by Daemen, Rijmen (BE)
Serpent	by Anderson, Biham, Knudsen(UK-IL-DK)

At the second conference, which took place in Rome in 1999, the five finalists were selected:

- Mars (IBM)
- RC6 (RSA)
- Rijndael (Daemen, Rijmen)
- Serpent (Anderson, Biham, Knudsen)
- Twofish (Counterpane Inc.)

The other submissions withdrew because of security or efficiency problems.

The final AES conference was held in 2000 in New York.

All finalists offered adequate security, but Rijndael was selected because of its efficiency and its flexibility, which makes it usable on all kind of processors.

3.3 Finite Fields

In this section we will introduce the theory of finite fields, especially the finite field $GF(2^8)$. We will represent the elements of $GF(2^8)$ by polynomials of degree less than 8 and show how byte-addition and byte-multiplication are defined in Rijndael.

We start with the two basic definitions of a finite field and the characteristic of a finite field.

Definition 43 Let F be a set. A triple $< F, \oplus, \odot >$ is called a **finite field of order m**, denoted by $GF(m)$, if:

- $< F, \oplus >$ is an Abelian group, with **0** as the neutral element,
- $< F \backslash \{0\}, \odot >$ is an Abelian group, with **1** as the neutral element,
- Distributivity holds: $a \odot (b \oplus c) = a \odot b \oplus a \odot c \quad \forall\, a, b, c \in F$,
- $|F| = m < \infty$.

Definition 44 The **characteristic** of a finite field of order m, denoted by *char* $(GF(m))$, is defined by:

$$char(GF(m)) := \min_{l \in \mathbb{N}}\{l \mid \underbrace{\mathbf{1} \oplus \cdots \oplus \mathbf{1}}_{l\ times} = \mathbf{0}\}.$$

Now we come to some well known results, for example see [12], from the theory of finite fields, which we will need in the remaining section.

Theorem 57 *A finite field exists and has order m, if and only if m is a prime power, e.g.* $m = p^k$*, with* $p \in \mathcal{P}$ *and* $k \in \mathbb{N}_+$*, where* $\mathcal{P}$ *is the set of all primes.*

Theorem 58 *All finite fields of the same order are isomorphic, they differ only in the way of representing the elements.*

Theorem 59 *The characteristic of the finite field* $GF(p^k)$ *is p.*

From Theorems 57 and 58 it follows that for all $p \in \mathcal{P}$ and for all $k \in \mathbb{N}$ exists a unique finite field with p^k elements.

3.3.1 Polynomials Over a Field

From Theorem 58 it follows that besides the definition of addition and multiplication, we have to determine the representation of a finite field.

In Rijndael the polynomial representation is used, which means that every element of $GF(p^k)$ is represented by a polynomial of degree $k - 1$ and coefficients in $GF(p)$. This is done, because the definitions for addition and multiplication are quite intuitive with the polynomial representation.

Definition 45 A **polynomial over a field F** is an expression of the following form:

$$a(x) := a_{n-1}x^{n-1} + \cdots + a_1 x + a_0, \text{ with } a_i \in F.$$

Definition 46 Let F be a field. $F[x]$ is the set of all polynomials over F.

Definition 47 Let F be a field. $F[x]|_d$ is the set of all polynomials over F with degree less than d.

3.3.2 The Field $< F[x]|_d, \oplus, \odot >$

In Rijndael we are only interested in the case $F = GF(2^k)$. Since the construction of both addition and multiplication does not depend on the structure of the underlying field, we will do this in general for any field F.

In this section we will define the addition $\oplus$ and the multiplication $\odot$ in order to give $< F[x]|_d, \oplus, \odot >$ a field structure. To do this we have to choose a irreducible reduction polynomial in order to make the multiplication closed.

Definition 48 Let F be a field and $a(x), b(x) \in F[x]|_d$, then addition $c(x) := a(x) \oplus b(x)$ is defined by:

$$c_i = a_i + b_i \qquad \forall\, i \in \{0, \ldots, d-1\},$$

where $+$ is the addition in the field F.

Proposition 15 *$< F[x]|_d, \oplus >$ is an Abelian group.*

Proof The associativity and commutativity follow directly from the field structure of F. Now let $z(x) \in F[x]|_d$ be the polynomial with all its coefficients equal to the neutral element for addition in the field F. It follows that $z(x)$ is the neutral element for addition in $< F[x]|_d, \oplus >$ and it is denoted by **0**. For a given polynomial $a(x) \in F[x]|_d$, the polynomial $b(x) \in F[x]|_d$ with $b_i = -a_i$, where $-a_i$ is the additive inverse of a_i in F, is the additive inverse of $a(x)$ in $< F[x]|_d, \oplus >$. Finally, the above defined addition is closed, which means $a(x) \oplus b(x) \in F[x]|_d$, because, on the one hand, the addition in F is closed and from this it follows that all the coefficients $a_i + b_i$ are also in F. And on the other hand, $deg(a(x) \oplus b(x)) \leq max\{deg(a(x)), deg(b(x))\} < d$. □

As we can see this definition is equal to the known polynomial addition for any polynomial over a field F. From now on we will denote both the above defined addition and the normal polynomial addition with $\oplus$.

The definition for the multiplication $\odot$ is a bit more complicated, because we have to do a so-called *reduction* in order to make the multiplication closed.

It is known that the polynomial multiplication $\otimes$ is associative, commutative, distributive together with $\oplus$ and it has a neutral element $e(x)$, denoted by **1**, with $e_0 = 1_f$ and $e_i = 0_f$, for all $i \geq 1$, where 1_f is the neutral element for multiplication and 0_f is the neutral element for addition in the underlying field F. The problem is that the multiplication of polynomials is not closed over $F[x]|_d$, because $deg(a(x) \otimes b(x)) = deg(a(x)) + deg(b(x))$, which could certainly be bigger than $d - 1$.

To solve this problem we select a *reduction polynomial*.

Definition 49 Let F be a field and $d \in \mathbb{N}$.
A polynomial $m(x) \in F[x]$, with $deg(m(x)) = d$ and $m_d = 1_f$, is called a **reduction polynomial** in $F[x]|_d$.

Definition 50 Let $a(x), m(x) \in F[x]$.
$r(x)$ is called the **residue of a(x) modulo m(x)**,
written $a(x) = r(x) \pmod{m(x)}$, if and only if $\exists q(x), r(x) \in F[x]$, with $a(x) = q(x) \otimes m(x) \oplus r(x)$ and $deg(r(x)) < deg(m(x))$.

Claim *$q(x)$ and $r(x)$ from the above definition are unique.*

Proof Suppose there are $q(x), r(x), q'(x)$ and $r'(x)$, with:
$(q(x) \otimes m(x)) \oplus r(x) = a(x) = (q'(x) \otimes m(x)) \oplus r'(x)$
and $deg(r(x)) < deg(m(x)),\ deg(r'(x)) < deg(m(x))$
$\Rightarrow (q(x) \oplus (-q'(x)))m(x) = r'(x) \oplus (-r(x))$
$\Rightarrow q(x) = q'(x) \Rightarrow r(x) = r'(x)$,
because $deg(m(x)) > max\{deg(r(x)), deg(r'(x))\} \geq deg(r'(x) \oplus (-r(x)))$ □

Definition 51 Let F be a field and $a(x), b(x) \in F[x]|_d$, then the multiplication $\odot$ is defined by:

$$a(x) \odot b(x) := a(x) \otimes b(x) \pmod{m(x)},$$

where $m(x)$ is a reduction polynomial in $F[x]|_d$.

Together with the above definitions for the reduction polynomial and the residue modulo $m(x)$, it follows that the multiplication is closed over $F[x]|_d$.

But it is still a Ring, because not every element of $F[x]|_d$ needs to have a multiplicative inverse. The reason for this is that until now the only restrictions to the reduction polynomial are that its degree must equal d and $m_d = 1_f$. So if we choose the reduction polynomial $m(x) = m_1(x) \otimes m_2(x)$, with $d > deg(m_1(x)) > 0$ and $d > deg(m_2(x)) > 0$, then any polynomial of the form $a(x) = a_1(x) \otimes m_1(x)$ or $b(x) = b_1(x) \otimes m_2(x)$ has no multiplicative inverse. Because if we assume the opposite that, for example for $a(x) = a_1(x) \otimes m_1(x)$ there exists a multiplicative inverse, denoted by $a^{-1}(x)$, it would hold that:

$\exists q(x) \in F[x]$ with: $a(x) \otimes a^{-1}(x) = (q(x) \otimes m(x)) \oplus \mathbf{1}$
$\Leftrightarrow -(q(x) \otimes m(x)) \oplus (a(x) \otimes a^{-1}(x)) = \mathbf{1}$
$\Leftrightarrow (c_1(x) \otimes m_1(x)) \oplus (c_2(x) \otimes m_1(x)) = \mathbf{1}$,
where $c_1(x) = -q(x) \otimes m_2(x),\ c_2(x) = a_1(x) \otimes a^{-1}(x)$
$\Leftrightarrow m_1(x) \otimes (c_1(x) \oplus c_2(x)) = \mathbf{1}$

And this is a contradiction. Because if $c_1(x) \oplus c_2(x) \neq \mathbf{0}$, then $deg(m_1(x) \otimes (c_1(x) \oplus c_2(x))) > 0$ and $deg(\mathbf{1}) = 0$ and if $c_1(x) \oplus c_2(x) = \mathbf{0}$, it would follow that $\mathbf{0} = \mathbf{1}$.

The same holds for any polynomial of the form $b(x) = b_1(x) \otimes m_2(x)$. □

In order to obtain a field structure we have to choose a special kind of reduction polynomial, which does not have the property that it can be decomposed into two polynomials with degree bigger than zero. These polynomials are called *irreducible*.

Definition 52 Let F be a field. A polynomial $c(x) \in F[x]$ is called **irreducible**, if and only if there exist no two polynomials $a(x), b(x) \in F[x]$, with $c(x) = a(x) \otimes b(x)$ and $deg(a(x)) > 0$ and $deg(b(x)) > 0$.

Lemma 18 *If $m(x)$ is an irreducible reduction polynomial in $F[x]|_d$, then $gcd(a(x), m(x)) = \mathbf{1}$, for all $a(x) \in F[x]|_d$.*

Proof $a(x) = \mathbf{1}$: $gcd(a(x), m(x)) = gcd(\mathbf{1}, m(x)) = \mathbf{1}$.
$a(x) \neq \mathbf{1}$: The only possible divisor of $m(x)$ with degree less than d has degree 0. Since $m_d = 1_f$, it follows that the only possible divisors of $m(x)$ are $\mathbf{1}$ and $m(x)$ itself. Since $deg(m(x)) = d > deg(a(x))$, it follows that $gcd(a(x), m(x)) = \mathbf{1}$. □

Proposition 16 *If we choose the reduction polynomial to be irreducible it follows that* $< F[x]|_d, \oplus, \odot >$ *is a field.*

Proof Given two polynomials $a(x), m(x) \in F[x]$ the *Extended Euclidean Algorithm*, which is described in Sect. 3.8 determines uniquely $b(x), c(x) \in F[x]$, with $(a(x) \otimes b(x)) \oplus (m(x) \otimes c(x)) = gcd(a(x), m(x))$ and $deg(b(x)) < deg(m(x))$. In our case $a(x) \in F[x]|_d$ and $m(x)$ is an irreducible reduction polynomial in $F[x]|_d$.

It follows from Lemma 18 that $gcd(a(x), m(x)) = \mathbf{1}$.
$\Rightarrow (a(x) \otimes b(x)) \oplus (m(x) \otimes c(x)) = \mathbf{1}$
$\Rightarrow a(x) \otimes b(x) = \mathbf{1} \pmod{m(x)}$
$\Rightarrow a(x) \odot b(x) = \mathbf{1}$.
And since $deg(b(x)) < deg(m(x)) = d$, it follows $b(x) \in F[x]|_d$.

That means, by applying the *Extended Euclidean Algorithm* we can determine the unique multiplicative inverse for any given element of $F[x]|_d$. □

We showed that for any given field F we can construct a field $< F[x]|_d, \oplus, \odot >$ with $|F|^d$ elements. It follows that if $|F| < \infty$, also $| < F[x]|_d, \oplus, \odot > | = |F|^d < \infty$. This means that if F is a finite field, for example $F = GF(p^k), p \in \mathcal{P}$, then $< GF(p^k)[x]|_d, \oplus, \odot >$, is the finite field $GF(p^{kd})$. On the other hand, by starting with $GF(p)$, where $F = \{0, \ldots, p-1\}$, $\oplus$ is the addition modulo p and $\odot$ is the multiplication modulo p, we can obtain the elements of $GF(p^{kd})$ for all $k, d \in \mathbb{N}$, by constructing the polynomials over $GF(p)$ with degree less than kd.

The only thing left to do, is defining $\oplus_d$ and $\odot_d$. From this it immediately follows that both $GF(p)[x]|_{kd}$ and $GF(p^k)[x]|_d$ represent the elements of $GF(p^{kd})$. That means, with appropriate definitions for $\oplus$, $\odot$, $\oplus_d$ and $\odot_d$, it follows that:

$$< GF(p^k)[x]|_d, \oplus, \odot > = GF(p^{kd}) = < GF(p)[x]|_{kd}, \oplus_d, \otimes_d > .$$

3.3.3 Byte-Operations in Rijndael

In Rijndael we are only interested in the finite fields with characteristic 2, in particular in $GF(2^8)$ and $GF(2^{32})$. The reason for that is that Rijndael is an block-cipher which operates, on the one hand, on one-dimensional arrays of bits of length 8, called *bytes*, which represent the elements of $GF(2^8)$, and on the other hand, it deals with one-dimensional arrays of bytes of length 4, called *4-byte columns*, which represent the elements of $GF(2^{32})$.

As shown above, there are different ways to construct $GF(2^{kd})$. In the case of bytes, $GF(2^8)$ is constructed via $< GF(2)[x]|_8, \oplus, \odot >$ and in the case of 4-byte columns, $GF(2^{32})$ is constructed via $< GF(2^8)[x]|_4, \oplus, \odot >$.

The Finite Field GF(2^8)

In this subsection we will show how the addition and the multiplication work while operating on bytes. We will see that the addition is, by its definition, nothing more than the bitwise XOR-operation and that the multiplication can be done efficiently by applying the *xtime*-operation.

The set of all possible bytes, denoted by $\mathcal{B}$, has 2^8 elements. From Theorem 57 it follows that we can use this set to represent the elements of $GF(2^8)$. Following Sects. 3.3.1 and 3.3.2 we can also represent the elements of $GF(2^8)$ by all possible polynomials of degree less than 8 with coefficients in $GF(2)$. By applying Theorem 58 it follows that there has to exist a bijection $\varphi : \mathcal{B} \to GF(2)[x]|_8$. Since every bit of a byte is either 0 or 1, this bijection is quite natural and defined as follows.

Definition 53 For a given byte $\beta = \beta_7\beta_6 \dots \beta_0 \in \mathcal{B}$, where the β_i's are bits, $\varphi(\beta)$ is defined via:

$$\varphi(\beta) := b(x) \in GF(2)[x]|_8, \qquad \text{with } b_i = \beta_i.$$

From now on we will write $\beta = \beta_7\beta_6 \dots \beta_0 \sim b_7x^7 + b_6x^6 + \dots + b_0 = b(x)$ or "the byte β *corresponds* to the polynomial $b(x)$", if $\varphi(\beta) = b(x)$.

For example $10110101 \sim x^7+x^5+x^4+x^2+1$. In some cases it is more convenient to write a byte not in the binary notation but in the hexadecimal notation. For example the hexadecimal notation of 10110101 is "B5". We will always use quotes, if we mean the hexadecimal notation.

Byte-Addition ⊕

We will now give an example for the byte-addition in Rijndael and we will show that it is in fact a very simple byte-level operation, which can be evaluated by computer-hardware very fast.

$$\begin{aligned}
&\text{"B5"} \oplus \text{"6C"}\\
&= 10110101 \oplus 01101100\\
&\sim (x^7 + x^5 + x^4 + x^2 + 1) \oplus (x^6 + x^5 + x^3 + x^2)\\
&= x^7 + x^6 + (1 + 1)x^5 + x^4 + x^3 + (1 + 1)x^2 + 1\\
&= x^7 + x^6 + x^4 + x^3 + 1\\
&\sim 11011001\\
&= \text{"D9"}
\end{aligned}$$

As we can see, this is the same as the simple bitwise exclusive-or-operation, which is the following for given bits β_1 and β_2:

$$XOR(\beta_1, \beta_2) := \begin{cases} 0, & \text{if } \beta_1 = \beta_2 \\ 1, & \text{otherwise} \end{cases}$$

From now on we will denote both the addition of bytes and XOR by $\oplus$.

Remark 30 Since the characteristic of $GF(2^8)$ is 2, every element is its own additive inverse.

Byte-Multiplication $\odot$

In order to define the multiplication of $GF(2^8)$ we have to choose an irreducible reduction polynomial $m(x)$ in $GF(2)[x]|_8$.

In Rijndael $m(x) := x^8 + x^4 + x^3 + x + 1 \sim 100011011 =$ "11B" is chosen to be this reduction polynomial.

Example 8 "57" $\odot$ "83"
$= 01010111 \odot 10000011$
$\sim (x^6 + x^4 + x^2 + x + 1) \odot (x^7 + x + 1)$
$= (x^6 + x^4 + x^2 + x + 1) \otimes (x^7 + x + 1) \pmod{m(x)}$
$= (x^{13} + x^{11} + x^9 + x^8 + x^7) \oplus (x^7 + x^5 + x^3 + x^2 + x) \oplus (x^6 + x^4 + x^2 + x + 1)$
$(\text{mod} m(x))$
$= (x^{13} + x^{11} + x^9 + x^8 + x^6 + x^5 + x^4 + x^3 + 1) \pmod{m(x)}$
$= (x^7 + x^6 + 1) \pmod{(x^8 + x^4 + x^3 + x + 1)} \sim 11000001 =$ "C1"

The disadvantage of multiplication compared to addition is the fact that there is no obvious simple byte-operation, as there is the XOR-operation for addition. But any monomial of a polynomial over $GF(2)$ is either 0, 1 or it is a power of x. Since, as we will show, the multiplication by $x \sim$ "02" can be done efficiently, also the multiplication of any monomial can be done efficiently, by an iterated application of the multiplication of x. In order to obtain the whole polynomial, we only have to do an XOR of all the monomials.

Multiplication by x

Let $b(x) \in GF(2)[x]|_8$.

From the definition of the multiplication $\odot$ it follows that:

$$b(x) \odot x = b(x) \otimes x = b_7x^8 + b_6x^7 + \cdots + b_1x^2 + b_0x \pmod{x^8 + x^4 + x^3 + x + 1}.$$

If $b_7 = 0$:
$b(x) \odot x = b_6x^7 + \cdots b_1x^2 + b_0x$
In this case the multiplication by x is a left-shift of the bits over one bit, where the last bit of the result is filled up with the zero bit.

If $b_7 = 1$:
$b(x) \odot x$
$= x^8 + b_6x^7 + \cdots b_1x^2 + b_0x \pmod{x^8 + x^4 + x^3 + x + 1}$
$= (x^8 + b_6x^7 + \cdots b_1x^2 + b_0x) \oplus (x^8 + x^4 + x^3 + x + 1)$
$= b_6x^7 + b_5x^6 + b_4x^5 + (b_3 \oplus 1)x^4 + (b_2 \oplus 1)x^3 + b_1x^2 + (b_0 \oplus 1)x + 1$

In this case the multiplication by x is a left-shift of the bits over one bit, followed by a bitwise XOR with "1B".

So in both cases the multiplication by x consists only of simple byte-operations, a left-shift and an optional XOR. We will denote the multiplication of $b(x)$ by x with $xtime(b)$.

We will show now, by the example of "57" $\odot$ "13", how the multiplication of two bytes is done via the multiplication of "02" $\sim x$.

The first step is to obtain the product of "57" with all the monomials of "13". Since "13" $\sim x^4 + x + 1$, it suffices to apply *xtime* four times to obtain "10" $\sim x^4$.

$$\begin{aligned}
\text{“57”} &= 01010111 \\[2ex]
\text{“57”} \odot \text{“02”} &= xtime(\text{“57”}) \\
&= 10101110 = \text{“AE”} \\[2ex]
\text{“57”} \odot \text{“04”} &= xtime(\text{“AE”}) \\
&= 01011100 \oplus 00011011 \\
&= 01000111 = \text{“47”} \\[2ex]
\text{“57”} \odot \text{“08”} &= xtime(\text{“47”}) \\
&= 10001110 = \text{“8E”} \\[2ex]
\text{“57”} \odot \text{“10”} &= xtime(\text{“8E”}) \\
&= 00011100 \oplus 00011011 \\
&= 00000111 = \text{“07”}
\end{aligned}$$

The second step is then to add all the obtained monomials in order to get the final result.

$$\begin{aligned}
\text{“57”} \odot \text{“13”} &= \text{“57”} \odot (\text{“10”} \oplus \text{“02”} \oplus \text{“01”}) \\
&= \text{“07”} \oplus \text{“AE”} \oplus \text{“57”} \\
&= 00000111 \\
&\oplus 10101110 \\
&\oplus 01010111 \\
&= 11111110 = \text{“FE”}.
\end{aligned}$$

We have seen that the byte-multiplication can be done efficiently if it is done by an iterated application of *xtime*. The efficiency depends on the smaller operand, "13" in the above example. The bigger this smaller operand is, the more often *xtime* has to be applied and the byte-multiplication via *xtime* becomes less efficient.

In the subsection "The *MixColums* Step" of Sect. 3.6.3 we will see that in the only case Rijndael uses byte-multiplication, one operand will always be small.

The Finite Ring $< \mathbf{GF(2^8)[x]|_4, \oplus, \bullet} >$

In this subsection we will introduce addition and multiplication of 4-byte-columns.

A *4-byte column* is a one-dimensional array of bytes. The set $\mathcal{C}$ of all possible 4-byte columns has $(2^8)^4 = 2^{32}$ elements and therefore can be used to represent the elements of $GF(2^{32})$. But also $GF(2^8)[x]|_4$ represents the elements of $GF(2^{32})$, so a bijection $\psi : \mathcal{C} \to GF(2^8)|_4$ has to exist. Since every byte represents an element of $GF(2^8)$, this bijection is defined as follows:

Definition 54 For a given 4-byte column $\gamma = \gamma_3\gamma_2\gamma_1\gamma_0 \in \mathcal{C}$, with $\gamma_i \in \mathcal{B}$ for $i \in \{0, \dots, 3\}$, $\psi(\gamma)$ is defined via:

$$\psi(\gamma) := c(x) \in GF(2^8)|_4, \qquad \text{with } c_i = \gamma_i.$$

4-Byte-Column-Addition $\oplus$

As shown before the addition of a 4-byte-column consists of the addition of the coefficients in $GF(2^8)$. Since this addition is only a bitwise XOR of the individual bits, again the addition of a 4-byte-column equals a bitwise XOR, not only over the bits of one byte, but over all the bits of the two 4-byte-columns. Therefore we will denote the 4-byte-column-addition also with $\oplus$.

4-Byte-Column-Multiplication $\bullet$

In order to get a closed multiplication we have to choose a reduction polynomial. For the multiplication of 4-byte-columns $l(x) := x^4 + 1 \in GF(2^8)[x]|_4$ was chosen. In $GF(2^k)$ this polynomial satisfies the "Freshman's Dream", which means that $x^4+1 = (x+1)^4$, and from this it follows that $l(x)$ is *not* irreducible. This property holds for every polynomial $x^{2^a} + 1 \in GF(2^k)[x]|_d$, where $k \in \mathbb{N}$ and $2^a < d$.

Proof Following Theorem 59, the characteristic of $GF(2^k)$ is 2. Further on, it holds that $(x+1)^{2^a} = \sum_{i=0}^{2^a} \binom{2^a}{i} x^i$, where all the binomial coefficients, except the first and the last, are even and therefore every addend, except the first and last, each sums up to zero. □

This definition for the reduction polynomial gives $< GF(2^8)|_4, \oplus, \bullet >$ not a field structure but a ring structure, which means that not every element needs to have a multiplicative inverse.

In particular an element a of $< GF(2^8)|_4, \oplus, \bullet >$ has an inverse, if and only if its corresponding polynomial $a(x)$ is not of the form $a_1(x) \otimes (x+1)$.

We have shown before, in Sect. 3.3.2 that if $a(x) = a_1(x) \otimes (x+1)$, an inverse element for $a(x)$ does not exist and if $a(x)$ is not of this form, it follows that $gcd(a(x), l(x)) = \mathbf{1}$ and the Extended Euclidean Algorithm determines a unique inverse element.

But this fact is not important for Rijndael, because in Rijndael the 4-byte-column-multiplication is done by a fixed polynomial $c(x)$, with $gcd(c(x), l(x)) = \mathbf{1}$ and so the multiplication by a fixed polynomial will be invertible.

Multiplication by Fixed Polynomial

The reason for the choice of $l(x)$ is that with this choice the multiplication with a fixed polynomial can be written as a matrix multiplication by a circulant matrix and therewith can be efficiently computed.

Let $c(x) = c_3x^3 + c_2x^2 + c_1x + c_0 \in GF(2^8)[x]|_4$ be the fixed polynomial and $a(x) = a_3x^3 + a_2x^2 + a_1x + a_0 \in GF(2^8)[x]|_4$ be another polynomial. The coefficients of $d'(x) := c(x) \otimes a(x)$ are:

$$\begin{array}{lllllll}
d'_0 = & c_0 \odot a_0 & & & & & \\
d'_1 = & c_1 \odot a_0 & \oplus & c_0 \odot a_1 & & & \\
d'_2 = & c_2 \odot a_0 & \oplus & c_1 \odot a_1 & \oplus & c_0 \odot a_2 & \\
d'_3 = & c_3 \odot a_0 & \oplus & c_2 \odot a_1 & \oplus & c_1 \odot a_2 & \oplus \; c_0 \odot a_3 \\
d'_4 = & & & c_3 \odot a_1 & \oplus & c_2 \odot a_2 & \oplus \; c_1 \odot a_3 \\
d'_5 = & & & & & c_3 \odot a_2 & \oplus \; c_2 \odot a_3 \\
d'_6 = & & & & & & \quad\; c_3 \odot a_3
\end{array}$$

Now we come to the claim, which is the basis for the choice of $l(x) = x^4 + 1$.

Claim $x^j = x^{j \bmod 4} \pmod{(x^4+1)}$.

Proof Let $j = 4q + r$, with $0 \leq r < 4$.

$$\begin{aligned}
x^j = \quad x^{4q+r} &= x^{4(q-1)+r}\,(x^4+1) + x^{4(q-1)+r} \\
x^{4(q-1)+r} &= x^{4(q-2)+r}\,(x^4+1) + x^{4(q-2)+r} \\
&\;\vdots \\
x^{4+r} &= x^r(x^4+1) + x^r
\end{aligned}$$

$\Rightarrow x^{4q+r} = \left(\sum_{i=1}^{q} x^{4(q-i)+r}\right)(x^4+1) + x^r$

$\Rightarrow x^j = x^r \pmod{(x^4+1)}$, with $r = j \bmod 4$. □

With this we get the following coefficients for
$d(x) := c(x) \bullet a(x) = c(x) \otimes a(x) \pmod{(x^4+1)}$:

$$\begin{aligned}
d_0 &= c_0 \odot a_0 \oplus c_3 \odot a_1 \oplus c_2 \odot a_2 \oplus c_1 \odot a_3 \\
d_1 &= c_1 \odot a_0 \oplus c_0 \odot a_1 \oplus c_3 \odot a_2 \oplus c_2 \odot a_3 \\
d_2 &= c_2 \odot a_0 \oplus c_1 \odot a_1 \oplus c_0 \odot a_2 \oplus c_3 \odot a_3 \\
d_3 &= c_3 \odot a_0 \oplus c_2 \odot a_1 \oplus c_1 \odot a_2 \oplus c_0 \odot a_3
\end{aligned}$$

So if we write this system of equations as a matrix-multiplication, we see that we obtain a circulant matrix:

$$\begin{pmatrix} d_0 \\ d_1 \\ d_2 \\ d_3 \end{pmatrix} = \begin{pmatrix} c_0 & c_3 & c_2 & c_1 \\ c_1 & c_0 & c_3 & c_2 \\ c_2 & c_1 & c_0 & c_3 \\ c_3 & c_2 & c_1 & c_0 \end{pmatrix} \begin{pmatrix} a_0 \\ a_1 \\ a_2 \\ a_3 \end{pmatrix}.$$

Since the 4-byte-column-multiplication equals, as shown, the iterated application of the byte-multiplication $\odot$ and the byte-addition $\oplus$, it can be evaluated efficiently if the coefficients of the fixed polynomial are small.

3.4 A Key-Iterated Block Cipher

In this section we will give some important definitions about boolean functions and introduce the *key-iterated block cipher*.

3.4.1 Boolean Functions

Firstly, we give the definition of a *boolean vector*, which is, as we will see, the input and the output of Rijndael.

Definition 55 A **boolean vector** b **of length n** is a vector, whose entries are bits:

$$b \in GF(2)^n.$$

A boolean vector of length n is also called a **one-dimensional array of bits of length n**.

Rijndael is a cipher, which operates on bytes. We have seen in the last section, that any boolean vector of length n represents an element of the finite field $GF(2^n)$. We will now define a *boolean function*, which operates on the finite field $GF(2^n)$.

Definition 56 A **boolean function** ϕ is a mapping, which maps a boolean vector to another boolean vector:

$$\phi : GF(2)^n \rightarrow GF(2)^m.$$

Definition 57 A **boolean transformation** χ is a boolean function, which maps a boolean vector to another boolean vector of the same length:

$$\chi : GF(2)^n \rightarrow GF(2)^n.$$

We say, it operates on an **n-bit-state**.

If the boolean transformation χ is invertible it is called a **series boolean permutation**.

We come now to three special boolean functions. The *bricklayer function*, the *transposition* and the *iterative boolean function*. As we will see, Rijndael is an iterative boolean permutation, which concatenates several boolean round functions. These boolean round functions are again iterative boolean permutations, which consist of three individual boolean permutations, namely a non-linear bricklayer permutation, a transposition and a linear bricklayer permutation.

Definition 58 A **bricklayer function** is a boolean function that can be decomposed into several boolean functions, operating independently on subsets of bits of the input vector.

If these boolean functions are linear, they are called **diffusion boxes**, or **D-boxes**, and if they are non-linear, they are called **substitution boxes**, or **S-boxes**.

Definition 59 A **bricklayer transformation (permutation)** is a bricklayer function, which can be decomposed into several boolean transformations (permutations).

Definition 60 A **transposition** is a boolean permutation, for which the binary output vector has the same hamming weight like the input vector.

Definition 61 An **iterative boolean function** $\psi : GF(2)^{n_0} \rightarrow GF(2)^{n_k}$ is the concatenation of $k \in \mathbb{N}$ boolean functions $\psi_i : GF(2)^{n_i} \rightarrow GF(2)^{n_{i+1}}$, for $i \in \{0, \ldots, k-1\}$:

$$\psi := \psi_{k-1} \circ \cdots \circ \psi_0.$$

Definition 62 An **iterative boolean transformation (permutation)** is the concatenation of boolean transformations (permutations).

3.4.2 A Key-Iterated Block Cipher

The input of Rijndael consists of the *plaintext* and the *cipherkey*, the output is the encrypted plaintext, called the *ciphertext*.

Rijndael is a *symmetric* cipher, which means that the same cipherkey is used for both encryption and decryption.

A *block cipher* is a cipher, where the plaintext is a one-dimensional array of bits of an arbitrary length, which is divided into several *blocks*, which are again one-dimensional array of bits, but all of the same given length N_B. All of these blocks are encrypted separately and in the same way, which means with the same algorithm and the same cipherkey. After the encryption of the blocks, the derived ciphertext-blocks, which are still of the same length N_B, are stuck together in order to obtain the whole ciphertext.

From now on, we will speak of only one block to be the input of a block cipher and therefore the output is also only one ciphertext block. Of course, every ciphertext block has to be uniquely decryptable into the same plaintext block from which it was encrypted in order to make a cipher work. In other words, the encryption has to be invertible. Since the plaintext block is a one-dimensional array of bits of length N_B, it follows that a block cipher is a boolean permutation which operates on an N_B-bit state.

An *iterated block cipher* consists of several *rounds* and in every round a *round transformation* is applied to the block. Each round transformation does not change the length of its input vector and since the whole cipher has to be invertible, every single round transformation has to be invertible, too.

Similar to the whole cipher, the round transformations are also boolean permutations, which operate on N_B-bit states. The individual round transformations depend on individual *roundkeys*, which are derived form the cipherkey. So if we denote the ith round transformation with ρ_i, the ith roundkey with κ_i, the whole block cipher with B and the number of rounds with N_r, an iterated block cipher can be written as follows:

$$B = \rho_{N_r}[\kappa_{N_r}] \circ \rho_{N_r-1}[\kappa_{N_r-1}] \circ \cdots \circ \rho_1[\kappa_1].$$

In a *key-alternating block cipher* each key-depended round transformation can be decomposed into a key-independent round transformation ρ_i and an addition of the roundkey, which is denoted by $\sigma[\kappa_i]$, and an additional key-addition of the 0th roundkey, which is applied before the first round transformation ρ_1. With this notation a key-iterated block cipher can be written in the following form:

$$B = \sigma[\kappa_{N_r}] \circ \rho_{N_r} \circ \sigma[\kappa_{N_r-1}] \circ \rho_{N_r-1} \circ \cdots \circ \sigma[\kappa_1] \circ \rho_1 \circ \sigma[\kappa_0].$$

A *key-iterated block cipher* is a key-alternating block cipher, where all rounds, except perhaps the first or the last, use the same round transformation ρ:

$$B = \sigma[\kappa_{N_r}] \circ \rho \circ \sigma[\kappa_{N_r-1}] \circ \rho \circ \cdots \circ \sigma[\kappa_1] \circ \rho \circ \sigma[\kappa_0].$$

3.5 The Wide Trail Strategy

In this section we will introduce the *Wide Trail Strategy*, which was developed by Joan Daemen in his doctoral dissertation [5]. The first section introduces linear trails. In a linear attack of a key-iterated block cipher a attacker needs to find linear trails over all but a few rounds with a high correlation. In the second section we come to differential trails, which are needed in differential cryptanalysis. In the third section the properties of linear and differential trails, derived in the first both sections, are used to design a key-iterated block cipher, which is secure against both attacks.

3.5.1 *Linear Trails*

Correlation

In this first subsection we will explain what is meant by a *correlation between two binary boolean functions*.

Definition 63 A **parity** is a binary boolean function $p : GF(2)^n \to GF(2)$, with:

$$p(a) = \bigoplus_{j \in \mathcal{J}_p} a_j, \text{ where } \mathcal{J}_p \subset \{0, \ldots, n-1\}.$$

Definition 64 The **selection pattern** u of a parity p is a boolean vector, with:

$$u_i = \begin{cases} 1, & \text{if} \quad i \in \mathcal{J}_p \\ 0, & \text{if} \quad i \notin \mathcal{J}_p \end{cases}$$

It follows that $p(a) = u^T a := u_0 a_0 \oplus \cdots \oplus u_{n-1} a_{n-1}$.

From now on we will write "$u^T\_$", if we speak of a parity p.

Definition 65 The **correlation** $C(f, g)$ between two binary boolean functions $f : GF(2)^n \to GF(2)$ and $g : GF(2)^n \to GF(2)$ is defined as:

$$C(f, g) := 2 \cdot Prob(f = g) - 1,$$

where $Prob(f = g) := \frac{1}{2^n} \cdot \#\{i \in \{0, \dots, 2^n - 1\} | f(a_i) = g(a_i), a_i \in GF(2)^n\}$

It follows that $C(f, g) \in [-1, 1]$.

Two binary boolean functions f, g are said to be *correlated*, if $C(f, g) \neq 0$.

We will now show that any binary boolean function can be written in terms of its input parities and the correlation between itself and these input parities. To do this we have to show that any binary boolean functions can be understood as an element of the vector space $< \mathbb{R}^{2^n}, +, . >$.

Definition 66 The **real-valued counterpart** $\hat{f} : GF(2)^n \rightarrow \mathbb{R}$ of a binary boolean function f is defined as:

$$\hat{f}(a_i) := (-1)^{f(a_i)}, \quad \forall\, a_i \in GF(2)^n \text{ and } i \in \{0, \dots, 2^n - 1\}.$$

The real-valued counterpart $\hat{f}$ of a binary boolean function f can be seen as an element of the vector space $< \mathbb{R}^{2^n}, +, . >$, where $\hat{f}$ is represented and defined by the vector $\begin{pmatrix} \hat{f}(a_0) \\ \vdots \\ \hat{f}(a_{2^n-1}) \end{pmatrix}$ $(a_j \in GF(2)^n, j \in \{0, \dots, 2^n - 1\})$.

$\hat{f}(a_j)$ is then the jth component of this vector.

We denote the above vector by $\hat{f}$ and since $\hat{f}$ determines f uniquely, we will say that the vector $\hat{f}$ represents the binary boolean function f.

From the definitions of the inner product and the norm in $< \mathbb{R}^{2^n}, +, . >$ follow directly the definitions of the inner product and the norm of two binary boolean functions.

Definition 67 The **inner product** of two binary boolean functions f and g is defined as:

$$< f, g > := < \hat{f}, \hat{g} > = \sum_{i=0}^{2^n-1} \hat{f}(a_i)\hat{g}(a_i).$$

Definition 68 The **norm** of a binary boolean function f is defined as:

$$||f|| := ||\hat{f}|| = \sqrt{< \hat{f}, \hat{f} >}.$$

It follows that: $||\hat{f}|| = 2^{\frac{n}{2}}$, since $\hat{f}(a_i)\hat{f}(a_i) = 1$, for all $i \in \{0, \dots, 2^n - 1\}$.

Proposition 17 *For two binary boolean function f, g, it holds that:*

$$C(f, g) = \frac{< \hat{f}, \hat{g} >}{||\hat{f}|| \cdot ||\hat{g}||}.$$

Proof

$$\begin{aligned}
\frac{<\hat{f},\hat{g}>}{||\hat{f}||\cdot||\hat{g}||} &= 2^{-n}\sum_{i=0}^{2^n-1}\hat{f}(a_i)\hat{g}(a_i) \\
&= 2^{-n}(\sum_{f(a_i)=g(a_i)} 1 - \sum_{f(a_i)\neq g(a_i)} 1) \\
&= 2^{-n}(\#\{i \in \{0,\dots,2^n-1\}|f(a_i)=g(a_i), a_i \in GF(2)^n\} \\
&\quad -(2^n - \#\{i \in \{0,\dots,2^n-1\}|f(a_i)=g(a_i), a_i \in GF(2)^n\}) \\
&= 2^{-n}(2\cdot\#\{i \in \{0,\dots,2^n-1\}|f(a_i)=g(a_i), a_i \in GF(2)^n\} - 2^n) \\
&= 2\cdot Prob(f=g)-1 \\
&= C(f,g)
\end{aligned}$$

□

In other words, the correlation between two binary boolean functions is the angle between their representing vectors in $< \mathbb{R}^{2^n}, +, . >$.

Proposition 18 *The representing vectors of the parities form an orthogonal basis in* $< \mathbb{R}^{2^n}, +, . >$.

Proof For any two parities $u^T\_$ and $v^T\_$ it holds that:

$$\begin{aligned}
< (-1)^{u^T_}, (-1)^{v^T_} > &= \sum_{i=0}^{2^n-1}(-1)^{u^Ta_i}(-1)^{v^Ta_i} \\
&= \sum_{i=0}^{2^n-1}(-1)^{u^Ta_i\oplus v^Ta_i} \\
&= \sum_{i=0}^{2^n-1}(-1)^{(u\oplus v)^Ta_i}
\end{aligned}$$

Since we sum up over all a_i's, the sum contains exactly 2^{n-1} 1's and 2^{n-1} (-1)'s, if $u \oplus v \neq 0$, and therefore sums up to 0.

And if $u \oplus v = 0$, every addend equals 1 and we obtain 2^n.

We have shown that all the 2^n parities are pairwise orthogonal and therefore form an orthogonal basis in $< \mathbb{R}^{2^n}, +, . >$. □

This means that the representing vector $\hat{f}$ of every binary boolean function f can be written as the linear combination of the parity vectors:

$$\hat{f} = \sum_u \lambda_u(-1)^{u^T\_}.$$

The next proposition shows that the coefficients λ_u equal the correlation $C(f, u^T\_)$ between the binary boolean function f and the parity $u^T\_$, which means that a binary boolean function f can be completely determined by the correlations between itself and its input parities $u^T\_$.

Proposition 19 *For all $i \in \{0, \ldots, 2^n - 1\}$, it holds that:*

$$\hat{f}(a_i) = \sum_u C(f, u^T\_)(-1)^{u^T a_i}.$$

Proof

$$\sum_u C(f, u^T\_)(-1)^{u^T a_i} = 2^{-n} \sum_u \Big(\sum_{j=0}^{2^n-1} \hat{f}(a_j)(-1)^{u^T a_j} \Big) (-1)^{u^T a_i}$$

$$= 2^{-n} \sum_u \sum_{j=0}^{2^n-1} \hat{f}(a_j)(-1)^{u^T a_j}(-1)^{u^T a_i}$$

$$= 2^{-n} \sum_u \sum_{j=0}^{2^n-1} \hat{f}(a_j)(-1)^{u^T (a_j \oplus a_i)}$$

$$= 2^{-n} \sum_u \Big(\hat{f}(a_i) + \sum_{j \neq i} \hat{f}(a_j)(-1)^{u^T (a_j \oplus a_i)}\Big)$$

$$= \hat{f}(a_i) + 2^{-n} \sum_u \sum_{j \neq i} \hat{f}(a_j)(-1)^{u^T (a_j \oplus a_i)}$$

$$= \hat{f}(a_i) + 2^{-n} \sum_{j \neq i} \underbrace{\sum_u \hat{f}(a_j)(-1)^{u^T (a_j \oplus a_i)}}_{=0}$$

$$= \hat{f}(a_i)$$ □

As a special case it holds for an output parity $w^T f$ of an binary boolean function and for every a_i that:

$$w^T f(a_i) = \sum_u C(w^T f, u^T\_)(-1)^{u^T a_i}.$$

Definition 69 For a given binary boolean function f we define, according to [10, 18], the **spectrum** $F(u)$ of f by:

$$F(u) := C(f, u^T\_).$$

Correlation Matrices

Up to now we have only considered binary boolean functions.

We come now to the more general case of boolean transformations, which can be represented by their *correlation matrix*.

Reminding of Definition 57 a boolean transformation χ is a boolean function, operating on a n-bit state:

$$\chi : GF(2)^n \to GF(2)^n.$$

A boolean transformation can be decomposed into n binary boolean functions:

$$\chi_i : GF(2)^n \to GF(2), \text{ for } i \in \{0, \ldots, n-1\}.$$

These binary boolean functions χ_i can be represented by the vector

$$\hat{\chi}_i = \begin{pmatrix} \hat{\chi}_i(a_0) \\ \vdots \\ \hat{\chi}_i(a_{2^n-1}) \end{pmatrix}$$

and it holds that:

$$\hat{\chi}_i(a_j) = \sum_u C(\chi_i, u^T\_)(-1)^{u^T a_j}.$$

Let $X_i(u) = C(\chi_i, u^T\_)$ be the spectrum of χ_i.

By applying Lemma 19 we will obtain that the spectrum W_χ of the output parity $w^T\chi$ of χ is:

$$W_\chi(u) = C(w^T\chi, u^T\_) = \bigotimes_{u_i=1} X_i(u).$$

Definition 70 The $2^n \times 2^n$ **correlation matrix** $\mathcal{C}^\chi$ of a boolean function χ is defined via its input parities $u^T\_$ and its output parities $w^T\chi$ in the following way:

$$\mathcal{C}^\chi = (C^\chi_{w,u}), \text{ with}$$
$$C^\chi_{w,u} := C(w^T\chi, u^T\_).$$

It can be proved in the same way like in the proof of Proposition 19 that it holds for every a_i:

$$(-1)^{w^T\chi(a_i)} = \sum_u C^\chi_{w,u}(-1)^{u^T a_i}.$$

Hence, each row of the correlation matrix expresses an output parity of a boolean transformation with respect to its input parities.

Definition 71 The **linear weight** $\mathrm{w}_l^\chi(w, u)$ of a correlation $C^\chi_{w,u}$ between an output parity $w^T\chi$ and an input parity $u^T\_$ of a boolean transformation χ is defined via:

$$\mathrm{w}_l^\chi(w, u) := -log_2(C^\chi_{w,u}).$$

We will now consider two special cases of boolean transformations, iterative boolean transformations and bricklayer transformations, which we will need in the remaining section.

Proposition 20 *Let* $\pi = \pi_1 \circ \pi_0 : GF(2)^n \to GF(2)^n$ *be an iterative boolean transformation, with* $\pi_i : GF(2)^n \to GF(2)^n$*. Further on, let* C^{π_i} *be the* $2^n \times 2^n$ *correlation matrix of* π_i*. Then it holds for the* $2^n \times 2^n$ *correlation matrix* C^{π} *of* π *that:*

$$C^{\pi} = C^{\pi_1} \times C^{\pi_0}.$$

Proof We have for all a_i:

$$\begin{aligned}(-1)^{w^T\pi(a_i)} &= \sum_v C^{\pi_1}_{w,v}(-1)^{v^T\pi_0(a_i)} \\ &= \sum_v C^{\pi_1}_{w,v} \sum_u C^{\pi_0}_{v,u}(-1)^{u^T a_i} \\ &= \sum_u \big(\sum_v C^{\pi_1}_{w,v} C^{\pi_0}_{v,u}\big)(-1)^{u^T a_i}.\end{aligned}$$

From this follows:

$$C^{\pi} = C^{\pi_1} \times C^{\pi_0}.$$

□

From this proposition follows for $\pi = \pi_1 \circ \pi_0$ that the correlation between an output parity $w^T\pi$ and an input parity $u^T\_$ of π is given by:

$$C(w^T\pi, u^T\_) = \sum_v C(w^T\pi_1, v^T\_)C(v^T\pi_0, u^T\_). \tag{3.5.1}$$

By an iterated application of Proposition 20 we obtain the following fact for boolean transformations consisting of more than two boolean transformations.

Proposition 21 *For* $k \in \mathbb{N}$ *let* $\pi = \pi_{k-1} \circ \cdots \circ \pi_0 : GF(2)^n \to GF(2)^n$ *be an iterative boolean transformation, with* $\pi_i : GF(2)^n \to GF(2)^n$*.*

Further on, let C^{π_i} *be the* $2^n \times 2^n$ *correlation matrix of* π_i*.*

Then it holds for the $2^n \times 2^n$ *correlation matrix* C^{π} *of* π *that:*

$$C^{\pi} = C^{\pi_{k-1}} \times \cdots \times C^{\pi_0}.$$

A bricklayer transformation h can be decomposed into $k \in \{2, \ldots, n\}$ boolean transformations h_i, for $i \in \{0, \ldots, k-1\}$, which operate independently on different bits of the n-bit state. We will only consider bricklayer transformations, whose boolean transformations h_i operate on the same number n_h of independent bits, but the results can be applied to all bricklayer transformations.

We have $n = kn_h$ and denote the ith n_h bits of an n-bit state a by $a_{(i)}$.
With this notation we have:

$$b = h(a) \Leftrightarrow b_{(i)} = h_i(a_{(i)}), \text{ with } i \in \{0, \dots, k-1\}.$$

We can write the h_i's in the following form:

$$h_i = (h_i^0, \dots, h_i^{n-1}), \text{ with } h_i^j = 0, \text{ if } j \notin \{in_h, \dots, (i+1)n_h - 1\}.$$

It follows that:

$$h = \bigoplus_{i=0}^{k-1} h_i.$$

We will show that the correlation of a bricklayer transformation can be derived from correlations of its underlying boolean transformations. To do this we have to revert to binary boolean functions. As we have seen a bricklayer transformation can be written as a XOR of its underlying boolean transformations. This yields to the following lemma.

Lemma 19 *For two binary boolean functions f and g, with spectra $F(u)$ and $G(v)$, let $h := f \oplus g$. Then it holds for the spectrum $H(w)$ of h:*

$$H(w) = F(u) \otimes G(v) := \sum_v F(v \oplus w)G(v).$$

Proof Firstly, we show that the real-valued counterpart of the XOR $\oplus$ of two binary boolean functions is the product of the individual real-valued counterparts.

$$\widehat{f \oplus g} = (-1)^{f \oplus g} = (-1)^f(-1)^g = \hat{f} \cdot \hat{g}.$$

Now it holds for every a_i that:

$$\begin{aligned}
\hat{f}(a_i)\hat{g}(a_i) &= \sum_u F(u)(-1)^{u^T a_i} \sum_v G(v)(-1)^{v^T a_i} \\
&= \sum_u \sum_v F(u)G(v)(-1)^{(u \oplus v)^T a_i} \\
&= \sum_w \Big(\sum_v F(v \oplus w)G(v)\Big)(-1)^{w^T a_i}
\end{aligned}$$

From Proposition 19 and Definition 69 it follows that:

$$H(w) = \sum_v F(v \oplus w)G(v).$$

□

As mentioned on p. 28, from this lemma follows that the spectrum W_χ of the output parity $w^T\chi$ of χ is:

$$W_\chi(u) = C(w^T\chi, u^T\_) = \bigotimes_{u_i=1} X_i(u).$$

The individual boolean transformations of a bricklayer transformation operate independently on subsets of the input vector. This fact simplifies the above lemma. For this we preliminarily define the *support space* of a binary boolean function.

Definition 72 Let f be a binary boolean function and $F(u)$ its spectrum.
The subspace of $GF(2)^n$ generated by the selection patterns u, with $F(u) \neq 0$, is called the **support space** $\mathcal{V}_f$ of f.

The following property holds for the support space of the XOR of two binary boolean functions.

Lemma 20 *Let f and g be two binary boolean functions with support spaces $\mathcal{V}_f$ and $\mathcal{V}_g$ and let $h = f \oplus g$.*
Then it holds for the support space $\mathcal{V}_h$ of h:

$$\mathcal{V}_h = \mathcal{V}_{f \oplus g} \subseteq \mathcal{V}_f \oplus \mathcal{V}_g.$$

Proof Let $w \in \mathcal{V}_{f\oplus g}$, then it follows by Definition 72 and Lemma 19 that:

$$H(w) = \sum_v F(v \oplus w)G(v) \neq 0.$$

Further on, it holds that:

$$\begin{aligned} 0 \neq \sum_v F(v \oplus w)G(v) &= \sum_{v \in \mathcal{V}_g} F(v \oplus w)G(v) \\ &= \sum_{v \in \mathcal{V}_g \wedge (v \oplus w) \in \mathcal{V}_f} F(v \oplus w)G(v) \end{aligned}$$

From this it follows that there exist $v \in \mathcal{V}_g$ and $u = v \oplus w \in \mathcal{V}_f$, with $w = u \oplus v$ and this yields to:

$$w \in \mathcal{V}_f \oplus \mathcal{V}_g.$$ □

The independence of the individual boolean transformations can be translated into terms of the support space.

Definition 73 Two binary boolean functions f and g are called **disjoint**, if it holds for their support spaces $\mathcal{V}_f$ and $\mathcal{V}_g$:

$$\mathcal{V}_f \cap \mathcal{V}_g = \{0\}.$$

We are now able to simplify Lemma 19.

Lemma 21 *Let f and g be two disjoint binary boolean functions with spectra $F(u)$ and $G(v)$ and let $h = f \oplus g$.*

Then there exist unique $u \in \mathcal{V}_f$ and $v \in \mathcal{V}_g$, with $w = u \oplus v \in \mathcal{V}_h$ and it holds for the spectrum $H(w)$ of h:

$$H(w) = F(u)G(v), \text{ where } w = u \oplus v.$$

Proof Lemma 20 states that each $w \in \mathcal{V}_h$ can be written as the XOR of $u \in \mathcal{V}_f$ and $v \in \mathcal{V}_g$. Suppose there exist $u, u' \in \mathcal{V}_f$ and $v, v' \in \mathcal{V}_g$ and it holds that:

$$u \oplus v = w = u' \oplus v' \Leftrightarrow u \oplus u' = v \oplus v'$$

Since $u \oplus u' \in \mathcal{V}_f$, $v \oplus v' \in \mathcal{V}_g$ and $\mathcal{V}_f \cap \mathcal{V}_g = \{0\}$, it follows:

$$u \oplus u' = v \oplus v' = 0 \Rightarrow u = u' \wedge v = v'.$$

For the spectrum of h it holds:

$$H(w) = \sum_{v \in \mathcal{V}_g \wedge u \in \mathcal{V}_f} F(u)G(v).$$

Since, as shown above, $u \in \mathcal{V}_f$ and $v \in \mathcal{V}_g$ are unique, it follows:

$$H(w) = F(u)G(v), \text{ with } w = u \oplus v.$$ □

With this lemma we are able to show how the correlation matrix of a bricklayer transformation can be derived from the correlation matrices of its underlying boolean transformations.

Proposition 22 *Let h be a bricklayer transformation consisting of $k \in \{2, \ldots, n\}$ boolean transformations h_i, for $i \in \{0, \ldots, k-1\}$. Further on, let $C^{h_i}_{w_{(i)},u_{(i)}}$ be the correlation between the output parities ${w_{(i)}}^T h_i$ and the input parities ${u_{(i)}}^T\_$ of h_i, where $w_{(i)}, u_{(i)} \in GF(2)^{n_h}$. It holds for the correlation $C^h_{w,u}$ between the output parities $w^T h$ and the input parities $u^T\_$ of h:*

$$C^h_{w,u} = \prod_{i=0}^{k-1} C^{h_i}_{w_{(i)},u_{(i)}} \quad \Leftrightarrow \quad \mathrm{w}^h_l(w, u) = \sum_{i=0}^{k-1} \mathrm{w}^{h_i}_l(w_{(i)}, u_{(i)}),$$

where $w = (w_{(0)}, \ldots, w_{(k-1)})$ and $u = (u_{(0)}, \ldots, u_{(k-1)})$.

Proof Like we have done above with the individual h_i's, we write the individual $w_{(i)}$'s in the following form:

$$w_{(i)} = (w_{(i)}^0, \ldots, w_{(i)}^{n-1}), \text{ with } w_{(i)}^j = 0, \text{ if } j \notin \{in_h, \ldots, (i+1)n_h - 1\}.$$

If we do the same with the individual $u_{(i)}$'s, we obtain:

$$w = \bigoplus_{i=0}^{k-1} w_{(i)} \text{ and } u = \bigoplus_{i=0}^{k-1} u_{(i)}.$$

From this follows:

$$w^T h = \bigoplus_{i=0}^{k-1} {w_{(i)}}^T h_i.$$

Now we denote the spectrum of $w^T h$ by $W_h(u)$ and the spectra of ${w_{(i)}}^T h_i$ by $W_{h_i}(u_{(i)})$. Further on, we denote the support spaces of ${w_{(i)}}^T h_i$ by $\mathcal{V}_i$.

From the structure of h_i and $w_{(i)}$ it follows that:

$$\mathcal{V}_i \cap \mathcal{V}_j = \{0\}, \text{ for all } i \neq j \text{ and } i, j \in \{0, \ldots, k-1\}.$$

We are now in the situation of Lemma 21 and an iterated application of this lemma yields to:

$$W_h(u) = \prod_{i=0}^{k-1} W_{h_i}(u_{(i)}).$$

Since, by definition, $W_h(u) = C_{w,u}^h$ and $W_{h_i}(u_{(i)}) = C_{w_{(i)},u_{(i)}}^{h_i}$, this proves the proposition. □

Linear Trails

We will now define a *linear trail* and the *weight* of a linear trail and finish this subsection with the *Theorem of Linear Trail Composition*.

Let ρ be an iterative boolean transformation, operating on a n-bit state:

$$\rho = \rho_{r-1} \circ \cdots \circ \rho_0.$$

It follows by Proposition 21 for the correlation matrix C^ρ of ρ:

$$C^\rho := C^{\rho_{r-1}} \times \cdots \times C^{\rho_0},$$

where C^{ρ_i} is the correlation matrix of the boolean transformation ρ_i.

Definition 74 **A linear trail** $\mathcal{U}_\rho$ over an iterative boolean transformation ρ with r rounds consists of a sequence of $(r+1)$ selection patterns $u^{(i)}$:

$$\mathcal{U}_\rho = \left(u^{(0)}, \ldots, u^{(r)}\right),$$

for which each of the r steps $(u^{(i)}, u^{(i+1)})$ ($i \in \{0, \ldots, r-1\}$) has a correlation given by:

$$C^{\rho_i}_{u^{(i+1)},\ u^{(i)}} = C({u^{(i+1)}}^T \rho_i, {u^{(i)}}^T \_) \neq 0.$$

Definition 75 The **correlation contribution** C_p of a linear trail $\mathcal{U}_\rho$ is defined via:

$$C_p(\mathcal{U}_\rho) := \prod_{i=0}^{r-1} C^{\rho_i}_{u^{(i+1)},\ u^{(i)}}.$$

Definition 76 The **weight $\mathrm{w}_l(\mathcal{U}_\rho)$ of a linear trail** $\mathcal{U}_\rho$ is defined by:

$$\mathrm{w}_l(\mathcal{U}_\rho) := -log_2(C_p(\mathcal{U}_\rho)).$$

It follows that the weight of a linear trail is the sum of the linear weights of the correlations of its steps:

$$\mathrm{w}_l(\mathcal{U}_\rho) = \sum_{i=0}^{r-1} \mathrm{w}_l^{\rho_i}(u^{(i+1)},\ u^{(i)}). \tag{3.5.2}$$

Definition 77 Let $\rho = \rho_{r-1} \circ \cdots \circ \rho_0$ be an iterative boolean transformation with r rounds. The set containing all linear trails $\mathcal{U}_\rho$, with $u^{(0)} = u$ and $u^{(r)} = w$ is denoted by:

$$U_{w,u}.$$

From Definitions 74 and 75 and the iterated application of (3.5.1) follows:

Theorem 60 (Theorem of linear trail composition, [5]) *The correlation $C^\rho_{w,u}$ between an output parity $w^T\rho$ and an input parity $u^T\_$ of an iterative boolean transformation ρ with r rounds is given by:*

$$C^\rho_{w,u} = C(w^T\rho, u^T\_) = \sum_{\mathcal{U}_\rho \in U_{w,u}} C_p(\mathcal{U}_\rho).$$

3.5.2 *Differential Trails*

Differential Propagation Probability

Consider a boolean transformation χ, operating on a n-bit state, and two n-bit vectors a_i and a_j, with $a_i \oplus a_j = a'$.

Let $b_i = \chi(a_i)$, $b_j = \chi(a_j)$ and $b' = b_i \oplus b_j$.

We say that the difference pattern a' propagates to the difference pattern b' through χ with a particular probability, which is called the *difference propagation probability* and defined as follows.

Definition 78 Given two difference patterns a' and b', the **difference propagation probability** $Prob^{\chi}(a', b')$ of χ is defined via:

$$Prob^{\chi}(a', b') := 2^{-n} \sum_{i=0}^{2^n-1} \delta(b' \oplus \chi(a_i \oplus a') \oplus \chi(a_i)),$$

where $\delta(x) := \begin{cases} 1, & \text{if } x = 0 \\ 0, & \text{if } x \neq 0 \end{cases}$ is the Kronecker delta function.

Definition 79 The **differential weight** $\mathbf{w}_d^{\chi}(a', b')$ of a difference propagation (a', b') through χ is defined via:

$$\mathrm{w}_d^{\chi}(a', b') := -log_2(Prob^{\chi}(a', b')).$$

Proposition 23 *For given difference patterns a' and b' it holds that:*

$$Prob^{\chi}(a', b') = 2^{1-n}k, \textit{ for } k \in \{0, \ldots, 2^{n-1}\}.$$

Proof Suppose we have found an $i \in \{0, \ldots, 2^n - 1\}$, with:

$$b' = \chi(a_i \oplus a') \oplus \chi(a_i).$$

Then we have for $j \in \{0, \ldots, 2^n - 1\}, j \neq i$ and $a_j = a_i \oplus a'$:

$$\chi(a_j \oplus a') \oplus \chi(a_j) = \chi(a_i) \oplus \chi(a_i \oplus a') = b'. \qquad \square$$

For given difference patterns a' and b', $Prob^{\chi}(a', b')$ is the fraction of the set of all n-bit vectors, for which a' propagates to b'.

We denote the set of all a_i's, for which $b' = \chi(a_i \oplus a') \oplus \chi(a_i)$ by:

$$\mathcal{M} := \{a_i \in GF(2)^n |\ b' = \chi(a_i \oplus a') \oplus \chi(a_i)\}$$

and obtain:

$$\#\mathcal{M} = 2^n Prob^{\chi}(a', b') = 2k, \text{ for } k \in \{0, \ldots, 2^{n-1}\}.$$

If $k = 0$, we say that a' and b' are *incompatible* and from now on we will only consider the case $k \neq 0$.

We will now consider a special case of boolean transformations, the bricklayer transformations. A bricklayer transformation h consists of $k \in \{2, \ldots, n\}$ boolean transformations h_i, operating on individual bits of the n-bit state a. Let the boolean transformation h_i operate on n_i bits, with $n_0 + \cdots + n_{k-1} = n$, and without loss of generality we assume that h_0 operates on the bits from position 0 to position $n_0 - 1$, denoted by $a^{(0)}$, and for $i \in \{1, \ldots, k-1\}$ we assume that h_i operates on the bits from position $\sum_{j=0}^{i-1} n_j$ to position $\left(\sum_{j=0}^{i} n_j\right) - 1$, denoted by $a^{(i)}$. Hence, $a = (a^{(0)}, \ldots, a^{(k-1)})$.

With the above notation we can prove the following proposition.

Proposition 24 *Let h be a bricklayer transformation consisting of $k \in \{2, \ldots, n\}$ boolean transformations h_i, for $i \in \{0, \ldots, k-1\}$. Further on, let $Prob^{h_i}(a^{(i)'}, b^{(i)'})$ be the probability that $a^{(i)'}$ propagates to $b^{(i)'}$ through h_i.*

Then it holds for the difference propagation probability of h:

$$Prob^h(a', b') = \prod_{i=0}^{k-1} Prob^{h_i}(a^{(i)'}, b^{(i)'}) \quad \Leftrightarrow \quad \mathrm{w}_d^h(a', b') = \sum_{i=0}^{k-1} \mathrm{w}_d^{h_i}(a^{(i)'}, b^{(i)'}).$$

Proof Let:

$$\mathcal{M} := \{a_j \in GF(2)^n |\, b' = h(a_j \oplus a') \oplus h(a_j)\}$$

and for all $i \in \{0, \ldots, k-1\}$, let:

$$\mathcal{M}_i := \{a_j^{(i)} \in GF(2)^{n_i} |\, b^{(i)'} = h(a_j^{(i)} \oplus a^{(i)'}) \oplus h(a_j^{(i)})\}.$$

Since $\mathcal{M} = \mathcal{M}_0 \times \cdots \times \mathcal{M}_{k-1}$, it follows:

$$\#\mathcal{M} = \prod_{i=0}^{k-1} \#\mathcal{M}_i.$$

We obtain:

$$\begin{aligned} Prob^h(a', b') &= 2^{-n}\#\mathcal{M} = 2^{-n} \prod_{i=0}^{k-1} \#\mathcal{M}_i = \prod_{i=0}^{k-1} 2^{-n_i}\#\mathcal{M}_i \\ &= \prod_{i=0}^{k-1} Prob^{h_i}(a^{(i)'}, b^{(i)'}) \end{aligned}$$

□

Differential Trails

Let $\rho = \rho_{r-1} \circ \cdots \circ \rho_0$ be an iterative boolean transformation with r rounds. We will now define a *differential trail* and the *weight of a differential trail*.

Definition 80 A **differential trail** $\mathcal{Q}_\rho$ over an iterative boolean transformation ρ with r rounds consists of a sequence of (r + 1) difference patterns q_i:

$$\mathcal{Q}_\rho := (q_0, \dots, q_r),$$

for which each of the r steps (q_i, q_{i+1}), for $i \in \{0, \dots, r-1\}$, has a differential weight given by:

$$\mathrm{w}_d^{\rho_i}(q_i, q_{i+1}).$$

Definition 81 The **weight $\mathrm{w}_d(\mathcal{Q}_\rho)$ of a differential trail** $\mathcal{Q}_\rho$ is defined via:

$$\mathrm{w}_d(\mathcal{Q}_\rho) := \sum_{i=0}^{r-1} \mathrm{w}_d^{\rho_i}(q_i, q_{i+1}).$$

With this definitions we are ready to define the *difference propagation probability* of an iterative boolean transformation over r rounds.

Definition 82 Let $\rho = \rho_{r-1} \circ \cdots \circ \rho_0$ be an iterative boolean transformation with r rounds. The set containing all differential trails $\mathcal{Q}_\rho$, with $q_0 = a'$ and $q_r = b'$ is denoted by:

$$Q_{a',b'}.$$

Definition 83 The **difference propagation probability** $Prob^\rho(a', b')$ of an iterative boolean transformation ρ over r rounds is defined as:

$$Prob^\rho(a', b') := \sum_{\mathcal{Q}_\rho \in Q_{a',b'}} 2^{-\mathrm{w}_d(\mathcal{Q}_\rho)}.$$

3.5.3 *The Wide Trail Strategy*

In linear cryptanalysis the attacker needs to know a correlation over all but a few rounds with a high amplitude and in differential cryptanalysis he needs to know an input difference, which propagates to an output difference with a high probability.

The approach of the *Wide Trail Strategy* is to design a key-iterated block cipher, which combines security and efficiency. By security we mean that there do not exist any low weighted linear or differential trails.

The $\gamma\lambda$ Round Structure

Each round transformation ρ consists of two layers γ and λ, where γ is a non-linear bricklayer permutation and λ is a linear permutation, which provides a *high diffusion*:

$$\rho = \lambda \circ \gamma.$$

This structure is called a $\gamma\lambda$ *round transformation*.

The Construction of γ

The first layer γ is a non-linear bricklayer permutation, which means that it consists of n_γ invertible S-boxes S_{γ_i}, operating independently on different bits of the state.

The *first construction step* is that all the S_{γ_i}'s operate on the same number m of bits. This restricts the block length n to be $n_\gamma \cdot m$.

Definition 84 Let $a \in GF(2)^n$ be an n-bit state.

The ith **bundle** $a_{(i)}$ of a is defined via:

$$a_{(i)} := (a_{im}, \dots, a_{(i+1)m-1}), \text{ for } i \in \{0, \dots, n_\gamma - 1\}.$$

This partition of the n-bit state according to γ, is called the **bundle partition** of γ.

The *second construction step* is that S_{γ_i} operates on the ith bundle $a_{(i)}$:

$$b_{(i)} = S_{\gamma_i}(a_{(i)}).$$

From Proposition 22 follows that the linear weight of a correlation between an output and an input parity of γ is the sum of the linear weights of the correlations between the particular output and input parities of S_{γ_i}. And from Proposition 24 follows that the differential weight of an difference propagation of two difference patterns through γ is the sum of the differential weights of the difference propagations of the particular difference patterns through S_{γ_i}.

Definition 85 Let $u = (u_{(0)}, \dots, u_{(n_\gamma-1)})$ be a selection pattern, according to the bundle partition of γ. A bundle $u_{(i)}$ of u is called **active**, if:

$$u_{(i)} \neq 0.$$

Definition 86 Let $q = (q_{(0)}, \dots, q_{(n_\gamma-1)})$ be a difference pattern, according to the bundle partition of γ. A bundle $q_{(i)}$ of q is called **active**, if:

$$q_{(i)} \neq 0.$$

Definition 87 If we consider a linear trail $\mathcal{U}_\rho$ over an iterated block cipher ρ, we call a bundle $a_{(i)}$ of the input state a of a particular round **active**, if $u_{(i)}$ is active, where u is the input selection pattern of this round.

If we consider a differential trail $\mathcal{Q}_\rho$ over an iterated block cipher ρ, we call a bundle $a_{(i)}$ of the input state a of a particular round **active**, if $q_{(i)}$ is active, where q is the input difference pattern of this round.

Definition 88 The **bundle weight $\mathbf{w_b(u)}$ of a selection pattern u** is the number of active bundles in u.

Definition 89 The **bundle weight $\mathrm{w_b}(\mathbf{q})$ of a difference pattern q** is the number of active bundles in q.

Definition 90 The **bundle weight $\mathrm{w_b}(\mathbf{a})$ of a state a** is the number of active bundles in a.

Let us consider a linear trail. From the above definitions follows that:

$$\mathrm{w}_b(u) = \mathrm{w}_b(a),$$

if u is the input selection pattern and a is the input state of the same round.

The same holds for differential trails:

$$\mathrm{w}_b(q) = \mathrm{w}_b(a),$$

if q is the input difference pattern and a is the input state of the same round.

If the input selection pattern $u_{(i)}$ is zero it follows that the output selection $w(i)$ is zero, because otherwise:

$$C(w_{(i)}^T S_{\gamma_i}, 0) = 2^{-m} \sum_{j=0}^{2^m-1} (-1)^{w_{(i)}^T S_{\gamma_i}(a_j)} = 0$$

and from Proposition 22 follows that in this case $C(w^T\gamma, u^T\_\ ) = 0$, which is a contradiction to the definition of linear trails.

We obtain:

$$C(0,0) = 2^{-m} \sum_{j=0}^{2^m-1} 1 = 1$$

and hence:

$$\mathrm{w}_l^{S_{\gamma i}}(0,0) = 0. \tag{3.5.3}$$

Similarly, if the input difference pattern $a_{(i)}{}'$ is zero it follows that $b_{(i)}{}'$ is zero and therewith:

$$Prob^{S_{\gamma i}}(0,0) = 2^{-m} \sum_{j=0}^{2^m-1} \delta(S_{\gamma_i}(a_j) \oplus S_{\gamma_i}(a_j)) = 1$$

and hence:

$$\mathrm{w}_d^{S_{\gamma i}}(0,0) = 0. \tag{3.5.4}$$

Definition 91 The **bundle weight $\mathrm{w_b}(\mathcal{U}_\rho)$ of a linear trail $\mathcal{U}_\rho$** is the sum of the bundle weights of the input states of the individual rounds.

Definition 92 The **bundle weight $\mathrm{w_b}(\mathcal{Q}_\rho)$ of a differential trail $\mathcal{Q}_\rho$** is the sum of the bundle weights of the input states of the individual rounds.

Let us assume that the round transformation ρ consist only of the non-linear bricklayer permutation γ and consider a linear (differential) trail $\mathcal{U}_\rho = (u^{(0)}, \dots, u^{(r)})$ ($\mathcal{Q}_\rho = (q_0, \dots, q_r)$) over r rounds.

Applying Eq. (3.5.2), Proposition 22 and Eq. (3.5.3) we obtain:

$$\begin{aligned} \mathrm{w}_l(\mathcal{U}_\rho) &= \sum_{i=0}^{r-1} \mathrm{w}_l^{\rho_i}(u^{(i+1)},\ u^{(i)}) = \sum_{i=0}^{r-1} \sum_{j=0}^{n_\gamma - 1} \mathrm{w}_l^{S_{\gamma_j}}(u_{(j)}^{(i+1)},\ u_{(j)}^{(i)}) \\ &\geq \mathrm{w}_b(\mathcal{U}_\rho) \cdot \min_{i \in \{0,\dots,r-1\}, j \in \{0,\dots,n_\gamma - 1\}} \mathrm{w}_l^{S_{\gamma_j}}(u_{(j)}^{(i+1)},\ u_{(j)}^{(i)}) \end{aligned} \tag{3.5.5}$$

Analogous we obtain by Eq. (3.5.4), Definition 81 and Proposition 24:

$$\mathrm{w}_d(\mathcal{Q}_\rho) \geq \mathrm{w}_b(\mathcal{Q}_\rho) \cdot \min_{i \in \{0,\dots,r-1\}, j \in \{0,\dots,n_\gamma - 1\}} \mathrm{w}_d^{S_{\gamma_j}}(q_{(j)i},\ q_{(j)i+1}). \tag{3.5.6}$$

From this follows the *third construction step*, which is to find a S-box S_γ with good non-linearity properties and use this on all n_γ bundles $a_{(i)}$.

By good non-linearity properties we mean that the minimum of $\mathrm{w}_l^{S_\gamma}(u_{(j)}^{(i+1)},\ u_{(j)}^{(i)})$ and $\mathrm{w}_d^{S_\gamma}(q_{(j)i},\ q_{(j)i+1})$ should be high.

In [16] Kaisa Nyberg gave several examples for S-boxes with good non-linearity properties.

With this construction step Eqs. (3.5.5) and (3.5.6) become:

$$\mathrm{w}_l(\mathcal{U}_\rho) \geq \mathrm{w}_b(\mathcal{U}_\rho) \cdot \min_{i \in \{0,\dots,r-1\}, j \in \{0,\dots,n_\gamma - 1\}} \mathrm{w}_l^{S_\gamma}(u_{(j)}^{i+1},\ u_{(j)}^{i}) \tag{3.5.7}$$

and:

$$\mathrm{w}_d(\mathcal{Q}_\rho) \geq \mathrm{w}_b(\mathcal{Q}_\rho) \cdot \min_{i \in \{0,\dots,r-1\}, j \in \{0,\dots,n_\gamma - 1\}} \mathrm{w}_d^{S_\gamma}(q_{(j)i},\ q_{(j)i+1}). \tag{3.5.8}$$

Equations (3.5.7) and (3.5.8) provide two possibilities to increase the lower bounds of linear and differential trails. The first is, to construct a S-box with a high minimum linear and differential weight, but both minimum weights are upper bounded by the number of bits on which the S-box operates. This would mean we have to increase the bundle size m. This has a high implementation cost and hence this disagrees with the efficiency approach of the Wide Trail Strategy. The second possibility is to extend the round transformation ρ by the linear diffusion step λ, which increases the bundle weight of linear and differential trails.

Branch Numbers

All the discussions in this subsection are done with respect to the bundle partition given by γ.

λ is a linear boolean permutation $\lambda : GF(2)^n \rightarrow GF(2)^n$, with $\lambda(a) = \mathsf{M}a$, where M is a binary $n \times n$ matrix.

For an output selection pattern w we have:

$$w^T \lambda(a) = w^T \mathsf{M} a = (\mathsf{M}^T w)^T a.$$

It follows for the elements of the correlation matrix $\mathcal{C}^\lambda$ of λ:

$$\begin{aligned}\mathcal{C}^\lambda_{w,\,u} &= 2^{-n} \sum_{i=0}^{2^n-1} (-1)^{(\mathsf{M}^T w)^T a_i} (-1)^{u^T a_i} = 2^{-n} \sum_{i=0}^{2^n-1} (-1)^{((\mathsf{M}^T w) \oplus u)^T a_i} \\ &= \delta((\mathsf{M}^T w) \oplus u).\end{aligned}$$

Definition 93 The **linear branch number** $\mathcal{B}_l(\varphi)$ of a boolean permutation φ is defined by:

$$\mathcal{B}_l(\varphi) := \min_{w,u,\mathcal{C}^\varphi_{w,u} \neq 0} \{\mathrm{w}_b(u) + \mathrm{w}_b(w)\}.$$

If the boolean permutation is linear and denoted by λ, the branch number is defined via:

$$\mathcal{B}_l(\lambda) := \min_{u \neq 0} \{\mathrm{w}_b(u) + \mathrm{w}_b(\mathsf{M}^T u)\}.$$

Definition 94 The **differential branch number** $\mathcal{B}_d(\varphi)$ of a boolean permutation φ is defined by:

$$\mathcal{B}_d(\varphi) := \min_{a, b \neq a} \{\mathrm{w}_b(a \oplus b) + \mathrm{w}_b(\varphi(a) \oplus \varphi(b))\}.$$

If the boolean permutation is linear and denoted by λ, the branch number is defined via:

$$\mathcal{B}_d(\lambda) := \min_{a' \neq 0} \{\mathrm{w}_b(a') + \mathrm{w}_b(\mathsf{M} a')\}.$$

The remaining discussions of this subsection are valid both for linear and differential branch numbers so that we denote both $\mathcal{B}_l$ and $\mathcal{B}_d$ by $\mathcal{B}$ and speak of a pattern, instead of a selection or difference pattern.

Since the output pattern corresponding to an input pattern with a single non-zero bundle has at least one and at most n_γ non-zero bundle(s), it holds for the branch number $\mathcal{B}(\lambda)$ of a linear permutation:

$$2 \leq \mathcal{B}(\lambda) \leq n_\gamma + 1. \tag{3.5.9}$$

We have derived the following properties:

- from the symmetry of the Definitions 93 and 94 follows:

$$\mathcal{B}(\varphi) = \mathcal{B}(\varphi^{-1}).$$

- a pattern is not affected by a key addition and hence its bundle weight is not affected
- a bricklayer permutation operates independently on individual bundles and therefore cannot turn an active bundle into a non-active bundle and vice versa. Hence, it does not affect the bundle weight
- if ρ is a $\gamma\lambda$ round transformation it follows:

$$\mathcal{B}(\rho) = \mathcal{B}(\lambda).$$

Let us consider a key-iterated block cipher over two rounds with a $\gamma\lambda$ round transformation ρ. The bundle weight of a two-round trail is the number of active bundles at the input of the first and at the input of the second round. The state of the input of of the second round is the XOR of the output of the first round and a round key. With the above properties we obtain the following theorem.

Theorem 61 (Two-Round Propagation Theorem, [5]) *For a key-iterated block cipher over two rounds with a $\gamma\lambda$ round structure, it follows for any two-round trail $\mathcal{T}_{\gamma\lambda}$:*

$$\mathrm{w}_b(\mathcal{T}_{\gamma\lambda}) \geq \mathcal{B}(\lambda).$$

The Construction of λ

According to Theorem 61, one possibility to obtain high lower bounds on the bundle weight of multiple round trails would be to construct the linear diffusion layer λ as a linear boolean permutation with a high branch number. Similar to large S-boxes this has a high implementation cost and hence contradicts to the efficiency approach of the Wide Trail Strategy.

Instead, the Wide Trail Strategy suggests the construction of a key-iterated block cipher, whose linear diffusion layer λ consists of a sequence of two steps:

- θ: a linear bricklayer permutation, which offers a high local diffusion. The D-boxes of θ operate independently on *columns*, which consists of bundles with respect to the bundle partition of γ.
- π: a transposition, which provides a high *dispersion*. Dispersion means that bundles, which are in the same column are moved to different columns.

The Construction of θ

The diffusion step θ is a linear bricklayer permutation, which consists of n_θ D-boxes D_{θ_j} operating independently on different bundles with respect to the bundle partition of γ.

The *first construction step* of θ is that each of the D-boxes operates on the same number n_ξ of bundles. This restricts the number n_γ of bundles to be $n_\theta \cdot n_\xi$ and hence the block size n to be $n_\theta \cdot n_\xi \cdot m$.

Definition 95 Let $c \in GF(2)^n$ which has been partitioned into bundles $c_{(i)}$, for $i \in \{0, \ldots, n_\gamma - 1\}$, with respect to the bundle partition of γ.

The jth **column** $\mathbf{c}^{(j)}$ is defined by:

$$c^{(j)} := (c_{(jn_\xi)}, \ldots, c_{((j+1)n_\xi-1)}), \text{ for } j \in \{0, \ldots, n_\theta - 1\}.$$

Similar to the construction of γ, the *second construction step* is that the D-box D_{θ_j} operates on column $c^{(j)}$. Since the D-boxes are linear, they can be written as a $n_\xi \times n_\xi$ matrix $\mathsf{D}^\theta_{(j)}$:

$$d^{(j)} = D_{\theta_j}(c^{(j)}) = \mathsf{D}^\theta_{(j)} c^{(j)}. \tag{3.5.10}$$

The measure for diffusion is the branch number $\mathcal{B}(\theta)$. Since the output state of θ corresponding to the input state with one active bundle in one column active has at least one and at most n_ξ active bundles, it follows:

$$2 \leq \mathcal{B}(\theta) \leq n_\xi + 1. \tag{3.5.11}$$

The *third construction step* is then to find a D-box with the maximum branch number $n_\xi + 1$ and once it is found this one is used on every column.

We can now define the diffusion step θ.

Definition 96 The **linear bricklayer permutation** $\theta : GF(2)^{mn_\xi} \to GF(2)^{mn_\xi}$ is defined by:

$$d = \theta(c) \quad :\Leftrightarrow \quad d^{(j)} = \mathsf{D}^\theta c^{(j)}, \text{ for all } j \in \{0, \ldots, n_\theta - 1\},$$

where D^θ is a $n_\xi \times n_\xi$, with entries in $GF(2)^m$.

The **inverse** permutation θ^{-1} is defined via:

$$c = \theta^{-1}(d) \quad :\Leftrightarrow \quad c^{(j)} = \mathsf{D}^{\theta^{-1}} d^{(j)}, \text{ for all } j \in \{0, \ldots, n_\theta - 1\},$$

where $\mathsf{D}^{\theta^{-1}}$ is a $n_\xi \times n_\xi$, with entries in $GF(2)^m$ and $\mathsf{D}^{\theta^{-1}} \times \mathsf{D}^\theta = \mathsf{I}_{n_\xi}$.

Since $n_\xi = n_\theta^{-1} \cdot n_\gamma$, the implementation cost of a linear diffusion step with branch number $n_\xi + 1$ is much lower than the cost of one with branch number $n_\gamma + 1$.

We can now adopt Theorem 61 and obtain the following proposition:

Proposition 25 *For a key-iterated block cipher over two rounds, in which the first round transformation has a $\gamma\theta$ structure, it holds for any two-round trail $\mathcal{T}_{\gamma\theta}$:*

$$\mathrm{w}_b(\mathcal{T}_{\gamma\theta}) \geq N\mathcal{B}(\theta),$$

where N is the number of active columns at the input of the second round.

Proof We can apply Theorem 61 to each column separately. Each active column at the input of the second round imposes that the same column was active at the input of the first round and hence there are at least $\mathcal{B}(\theta)$ active bundles in that column in both input states together. □

The Construction of π

We will now define the transposition π and introduce the *diffusion optimality*, which means that π offers the highest possible dispersion.

Definition 97 The **bundle transposition** $\pi : GF(2)^n \rightarrow GF(2)^n$ is defined as:

$$b = \pi(a) \quad :\Leftrightarrow b_{(i)} = a_{(p(i))},$$

where $p(i)$ is a permutation of the bundle partition of γ. The **inverse** bundle transposition π^{-1} is defined by:

$$a = \pi(b) \quad :\Leftrightarrow a_{(i)} = b_{(p^{-1}(i))}.$$

Since θ and π together provide an inter-column action it is no longer sufficient to concentrate only on the branch number but also on the *column branch number*. To do so, we, firstly, define the *column weight* of a pattern.

Definition 98 A column $c^{(j)}$ is called **active** if at least one of its bundles $c_{(jn_\xi)}$, $\ldots, c_{((j+1)n_\xi-1)}$ is active.

Definition 99 The **column weight $w_c(a)$ of a (selection or difference) pattern a** is the number of active columns in the pattern a.

Definition 100 The **column branch number $\mathcal{B}^c(\Theta)$** of a linear boolean permutation Θ is defined as:

$$\mathcal{B}^c(\Theta) := \min_{a \neq 0}\{w_c(a) + w_c(\Theta(a))\}.$$

We will show now that if π is properly chosen the column branch number of $\pi \circ \theta \circ \pi$ equals the branch number of θ.

Definition 101 The bundle transposition π is called **diffusion optimal**, if and only if all bundles, which were in the same column of the input state of π are in different columns of its output state.

From Definition 97 follows that if π is diffusion optimal, π^{-1} also is diffusion optimal.

Further on, Definition 98 imposes that the number n_θ of columns has to be at least as big as the number n_ξ of bundles in each column. This restricts the block size n in the following way:

$$n_\theta \geq n_\xi \Rightarrow n = n_\theta \cdot n_\xi \cdot m \geq n_\xi^2 \cdot m. \tag{3.5.12}$$

Proposition 26 *If the bundle transposition π is diffusion optimal and the diffusion step θ has a maximum branch number $\mathcal{B}(\theta)$, it holds for $\Theta := \pi \circ \theta \circ \pi$:*

$$\mathcal{B}^c(\Theta) = \mathcal{B}(\theta).$$

Proof Let a denote the input state of Θ, d denote its output state and b and c its intermediate states, with:

$$b = \pi(a), c = \theta(b) = \theta(\pi(a)) \text{ and } d = \pi(c) = \pi(\theta(\pi(a))) = \Theta(a).$$

Firstly, we assume that $\mathrm{w}_b(a) = 1$ and hence $\mathrm{w}_c(a) = 1$.

From this follows that there exists exactly one active column $b^{(j)}$ in b, with:

$$\mathrm{w}_b(b^{(j)}) = 1.$$

The property that θ has a maximum branch number $\mathcal{B}(\theta)$ induces that there exists exactly one column $c^{(j)}$ in c, with:

$$\mathrm{w}_b(c^{(j)}) = \mathcal{B}(\theta) - 1.$$

Since π is diffusion optimal, all the $\mathcal{B}(\theta) - 1$ active bundles in $c^{(j)}$ are mapped to different columns of d and this yields to:

$$\mathrm{w}_c(d) = \mathcal{B}(\theta) - 1.$$

It follows that $\mathrm{w}_c(a) + \mathrm{w}_c(d) = \mathcal{B}(\theta)$ and hence:

$$\mathcal{B}^c(\Theta) \leq \mathcal{B}(\theta).$$

Secondly, we will show that $\mathcal{B}^c(\Theta) \geq \mathcal{B}(\theta)$, for all $0 \neq a \in GF(2)^n$.

For all $a \neq 0$ holds $\mathrm{w}_c(a) \geq 1$ and hence $\mathrm{w}_c(b) \geq 1$.

For any active column $b^{(j)}$ in b it follows that $c^{(j)}$ is active, too, and:

$$\mathrm{w}_b(b^{(j)}) + \mathrm{w}_b(c^{(j)}) = \mathcal{B}(\theta).$$

If $b^{(j)}$ and hence $c^{(j)}$ would be the single active columns in b and c it would follow by the diffusion optimality of π and π^{-1} that:

$$\mathrm{w}_c(d) = \mathrm{w}_b(c^{(j)}) \text{ and } \mathrm{w}_c(a) = \mathrm{w}_b(b^{(j)}).$$

But if the number of active columns in b and c is greater than 1 it could occur that:

$$\mathrm{w}_c(d) > \mathrm{w}_b(c^{(j)}) \text{ and } \mathrm{w}_c(a) > \mathrm{w}_b(b^{(j)}).$$

Altogether, we have:

$$\mathrm{w}_c(a) + \mathrm{w}_c(d) \geq \mathrm{w}_b(b^{(j)}) + \mathrm{w}_b^{(j)} = \mathcal{B}(\theta).$$ □

Now we are able to prove the final statement of the Wide Trail Strategy.

Theorem 62 *For a key-iterated block cipher B with a $\gamma\pi\theta$ round transformation, diffusion optimal π and where θ has a maximum branch number $\mathcal{B}(\theta)$, it holds for any four-round trail $\mathcal{T}_{\gamma\pi\theta}$:*

$$\mathrm{w}_b(\mathcal{T}_{\gamma\pi\theta}) \geq \mathcal{B}(\theta)^2.$$

Proof Consider four rounds of a key-iterated block cipher with a $\gamma\pi\theta$ round transformation:

$$B_4 := \sigma[\kappa_4]\circ(\theta\circ\pi\circ\gamma)\circ\sigma[\kappa_3]\circ(\theta\circ\pi\circ\gamma)\circ\sigma[\kappa_2]\circ(\theta\circ\pi\circ\gamma)\circ\sigma[\kappa_1]\circ(\theta\circ\pi\circ\gamma)\circ\sigma[\kappa_0],$$

where $\sigma[\kappa_i]$ is the round key addition of round key κ_i.

Since the non-linear bricklayer permutation γ and the key addition σ have no impact on the bundle weight of the trail, we write:

$$\begin{aligned} B_4 &= (\theta \circ \pi) \circ (\theta \circ \pi) \circ (\theta \circ \pi) \circ (\theta \circ \pi) \\ &= \theta \circ \pi \circ \theta \circ (\pi \circ \theta \circ \pi) \circ \theta \circ \pi. \end{aligned}$$

We denote the input state of B_4 by a and according to that:

$$a' := \pi(a),\ b := \theta(a'),\ b' := \pi(b),\ c := \theta(b'),\ c' := \pi(c) \text{ and } d := \theta(c').$$

We have to show that:

$$\mathrm{w}_b(a) + \mathrm{w}_b(b) + \mathrm{w}_b(c) + \mathrm{w}_b(d) \geq \mathcal{B}(\theta)^2.$$

Since π does not change the number of active bundles and θ does not change the number of active columns, it holds inter alia:

(i) $\mathrm{w}_b(a') = \mathrm{w}_b(a)$
(ii) $\mathrm{w}_b(c') = \mathrm{w}_b(c)$
(iii) $\mathrm{w}_c(d) = \mathrm{w}_c(c')$

Since $c' = \pi(\theta(\pi(b)))$, it follows from (3.5.12) and (iii):

$$\mathrm{w}_c(d) + \mathrm{w}_c(b) \geq \mathcal{B}(\theta). \tag{3.5.13}$$

Further on, by applying (3.5.12) and (i), we obtain:

$$\mathrm{w}_b(a) + \mathrm{w}_b(b) \geq \mathrm{w}_c(b) \cdot \mathcal{B}(\theta)$$

and from (3.5.12) and (ii):

$$\mathrm{w}_b(c) + \mathrm{w}_b(d) \geq \mathrm{w}_c(d) \cdot \mathcal{B}(\theta)$$

Together we have:

$$w_b(a) + w_b(b) + w_b(c) + w_b(d) \geq (w_c(b) + w_c(d)) \cdot \mathcal{B}(\theta)$$

and hence with (3.5.13):

$$w_b(a) + w_b(b) + w_b(c) + w_b(d) \geq \mathcal{B}(\theta)^2$$

□

With this final theorem Eqs. (3.5.7) and (3.5.8) become:

$$w_l(\mathcal{U}_\rho) \geq B(\theta)^2 \cdot \min_{i\in\{0,\dots,r-1\},j\in\{0,\dots,n_\gamma-1\}} w_l^{S_\gamma}(u_{(j)}^{i+1},\ u_{(j)}^{i}) \tag{3.5.14}$$

and:

$$w_d(\mathcal{Q}_\rho) \geq B(\theta)^2 \cdot \min_{i\in\{0,\dots,r-1\},j\in\{0,\dots,n_\gamma-1\}} w_d^{S_\gamma}(q_{(j)i},\ q_{(j)i+1}). \tag{3.5.15}$$

To construct a key-iterated block cipher, which resists linear and differentials attacks, we have to give it a $\gamma\pi\theta$ round transformation, where S_γ operates on only a small number of bits with a high minimum linear and differential weight, π is diffusion optimal and θ has the maximum possible branch number. It follows from Theorem 62 that Eqs. (3.5.14) and (3.5.15) hold for any four-round trail. To obtain a given security level we only have to increase the number of rounds, which will increase the bundle weight of any trails over all but a few rounds of the cipher.

3.6 The Specifications of Rijndael

In this section we will explain exact specifications of Rijndael and show how the individual steps of the round transformations work.

Rijndael is a key-iterated block cipher. It was developed to work for different values of the block length N_B and the cipherkey length N_C. Both are either 128, 192 or 256 bits. The number of rounds N_r depends on N_B and N_C and is defined as follows:

$$N_r := \begin{cases} 10, & \text{if } \max\{N_B, N_C\} = 128 \\ 12, & \text{if } \max\{N_B, N_C\} = 192 \\ 14, & \text{if } \max\{N_B, N_C\} = 256 \end{cases}$$

The design of Rijndael was derived from the Wide Trail Strategy and therefore it has a linear and a non-linear layer. The non-linear layer consists of one step, the *SubBytes* step, and the linear layer consists of two steps, the *ShiftRows* step and the *MixColumns* step. Each round transformation is followed by a *Key Addition* with the particular roundkey which is derived from the cipherkey via the *Key Schedule*.

The last section of this section covers the decryption, which has the nice added feature that it can be done in mainly the same way as the encryption.

3.6.1 The Input, the Output, and the State

As explained in Sect. 3.4 the inputs of Rijndael are the plaintext block, which is an one-dimensional array of bits of length N_B and the cipherkey, a one-dimensional array of bits of length N_C. The plaintext block is also a one-dimensional array of bytes of length $\frac{1}{8}N_B$, which is denoted by $p_0p_1 \dots p_{\frac{1}{8}N_B-1}$. Similarly the output, which is the ciphertext block, is a one-dimensional array of bytes of length $\frac{1}{8}N_B$, denoted by $c_0c_1 \dots c_{\frac{1}{8}N_B-1}$.

All steps of Rijndael, this means the round transformations and all their individual steps and the roundkey addition, operate on the intermediate results, called the *states*. Each state can be seen as a two-dimensional array of bytes with four rows and $N_b := \frac{1}{32}N_B$ columns. Figure 3.1 shows the state for $N_B = 128$.

The very first step of Rijndael is the mapping of the plaintext block $p_0 \dots p_{\frac{1}{8}N_B-1}$ to the state. This is done via the following equation:

$$a_{ij} = p_{i+4j}, \quad \text{for } 0 \le i < 4 \text{ and } 0 \le j < N_b.$$

This means that the state is 'filled up' column by column from the upper left to the lower right with the individual bytes of the plaintext block.

After the last step of the encryption the final state is mapped on the ciphertext block via:

$$c_i = a_{i \bmod 4, \lfloor \frac{i}{4} \rfloor}, \quad \text{for } i \in \{0, \dots, 4N_b - 1\}.$$

So the state is 'released' into the ciphertext block again column by column from the upper left to the lower right.

Fig. 3.1 The state for $N_B = 128$

$a_{0,0}$	$a_{0,1}$	$a_{0,2}$	$a_{0,3}$
$a_{1,0}$	$a_{1,1}$	$a_{1,2}$	$a_{1,3}$
$a_{2,0}$	$a_{2,1}$	$a_{2,2}$	$a_{2,3}$
$a_{3,0}$	$a_{3,1}$	$a_{3,2}$	$a_{3,3}$

3.6.2 *The Non-linear Layer*

The *SubBytes* Step

The *SubBytes* step is a bricklayer permutation, so it can be decomposed into several boolean permutations, which operate independently on subsets of the state. Since it is a non-linear permutation, these boolean permutations are called S-boxes. For simplicity all the S-boxes are, in fact, one and the same boolean permutation, so we have only one S-box, which is denoted by S_{RD}. The subsets on which this S-box operates, are the individual bytes a_{ij} of the state, which is visualized in Fig. 3.2.

Design Criteria

There are three design criteria, which were considered in the development of the *SubBytes* step. The first is, of course, that it should offer a high *shape non-linearity*, the second is that it should be *algebraic complex* and the third criterion is the *simplicity*, which means that it should be easy to describe and have an efficient computability.

In his work [16], Kaisa Nyberg gave the following four criteria, which a *substitution step*, like the S-box, should satisfy:

- high non-linearity
- resistance against linear cryptanalysis
- resistance against differential cryptanalysis
- efficient construction and computability

He also gave several alternatives of functions, which satisfy the above criteria. For Rijndael the following of these alternatives was chosen:

$$g : GF(2^8) \to GF(2^8)$$
$$g(a) = a^{-1},$$

In this equation a^{-1} is the multiplicative inverse of a in $GF(2^8)$, with $m(x) = x^8 + x^4 + x^3 + x + 1$ as the irreducible reduction polynomial.

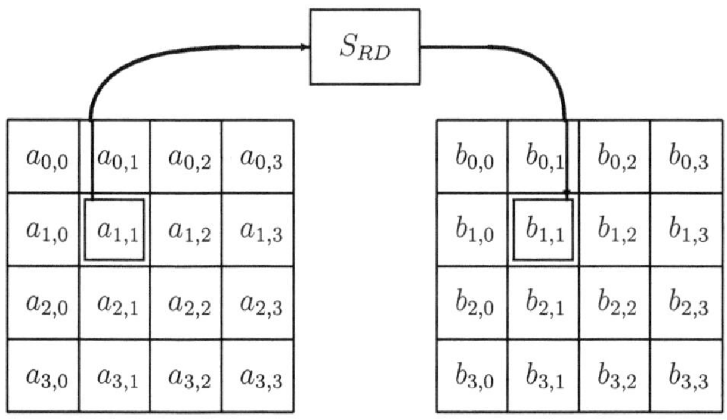

Fig. 3.2 The subsets on which the S-box operates

The disadvantages of this choice for the S-box are, on the one hand, the fact that $g(\text{"00"}) = \text{"00"}$ and on the other hand, this function has a very simple algebraic expression, since $a^{-1} = a^{254}$ in $GF(2^8)$. This fact would offer vulnerability against the *interpolation attack* [11], which was developed by Thomas Jakobsen and Lars R. Knudsen.

To get rid of these two disadvantages we combine the non-linear permutation g with the affine permutation f, which is defined as follows:

$$f : GF(2^8) \to GF(2^8)$$
$$f(a) = b$$
$$\Updownarrow$$
$$\begin{pmatrix} b_0 \\ b_1 \\ b_2 \\ b_3 \\ b_4 \\ b_5 \\ b_6 \\ b_7 \end{pmatrix} = \begin{pmatrix} 1 & 0 & 0 & 0 & 1 & 1 & 1 & 1 \\ 1 & 1 & 0 & 0 & 0 & 1 & 1 & 1 \\ 1 & 1 & 1 & 0 & 0 & 0 & 1 & 1 \\ 1 & 1 & 1 & 1 & 0 & 0 & 0 & 1 \\ 1 & 1 & 1 & 1 & 1 & 0 & 0 & 0 \\ 0 & 1 & 1 & 1 & 1 & 1 & 0 & 0 \\ 0 & 0 & 1 & 1 & 1 & 1 & 1 & 0 \\ 0 & 0 & 0 & 1 & 1 & 1 & 1 & 1 \end{pmatrix} \begin{pmatrix} a_0 \\ a_1 \\ a_2 \\ a_3 \\ a_4 \\ a_5 \\ a_6 \\ a_7 \end{pmatrix} \oplus \begin{pmatrix} 1 \\ 1 \\ 0 \\ 0 \\ 0 \\ 1 \\ 1 \\ 0 \end{pmatrix}$$

Since f is an affine permutation, it does not effect the non-linearity of g. Moreover f was chosen in such a way that S_{RD} has no fixed points ($S_{RD}(a) \oplus a = \text{"00"}$) or opposite fixed points ($S_{RD}(a) \oplus a = \text{"}FF\text{"}$). This was only done as a precautionary measure, since, up to now, there are no attacks known which exploit the existence of (opposite) fixed points.

By applying the *Lagrange interpolation technique* we obtain the following expression for S_{RD}:

$$\begin{aligned} S_{RD}(a) &= \text{"05"}\,a^{254} \oplus \text{"09"}\,a^{253} \oplus \text{"}F9\text{"}\,a^{251} \oplus \text{"25"}\,a^{247} \\ &\oplus \text{"}F4\text{"}\,a^{239} \oplus \text{"01"}\,a^{223} \oplus \text{"}B5\text{"}\,a^{191} \oplus \text{"}8F\text{"}\,a^{127} \oplus \text{"63"}. \end{aligned}$$

Together with the linear layer, consisting of *ShiftRows* and *MixColumns*, this expression offers sufficient security against the interpolation attack.

There is one other fact about the affine permutation f. As we see, the matrix which defines f is a circulant (8×8)-matrix. So if we go the other way round as in the subsection "The Finite Ring $< GF(2^8)[x]|_4, \oplus, \bullet >$" of Sect. 3.3.3, where we showed that the multiplication by a fixed polynomial in $< GF(2^8)|_4, \oplus, \bullet >$ can be written as a circulant (4×4)-matrix, then we can show that f can be seen as a multiplication by a fixed polynomial $c(x)$, with $m'(x) = x^8 + 1$ as the reducible reduction polynomial, followed by an addition with $d(x) = x^6 + x^5 + x + 1$. Since in this case, the first row of the matrix would be $(c_0 \;\; c_7 \;\; c_6 \;\; c_5 \;\; c_4 \;\; c_3 \;\; c_2 \;\; c_1)$, this fixed polynomial has to be $c(x) = x^4 + x^3 + x^2 + x + 1$.

Altogether f can be written as:

$$f(a) = ((x^4+x^3+x^2+x+1) \otimes (a_7x^7+\cdots+a_0)) \oplus (x^6+x^5+x+1) \pmod{x^8+1}.$$

If we denote the multiplication modulo $m'(x)$ of two polynomials, which are elements of $GF(2)[x]|_8$, by $\odot'$, it follows that the triple
$< GF[2][x]|_8, \oplus, \odot' >$ forms a Ring.

Definition 102 The **SubBytes step** is a bricklayer permutation, which consists of the $\frac{1}{8}N_B$-fold application of the S-box $S_{RD} : GF(2^8) \rightarrow GF(2^8)$, operating on the individual bytes α of the input state, which is defined by:

$$S_{RD}(\alpha) := f(g(\alpha)),$$

where $g(\alpha) \odot \alpha = \mathbf{1}$ and
$f(\beta) = ((x^4 + x^3 + x^2 + x + 1) \odot' b(x)) \oplus (x^6 + x^5 + x + 1)$, where $\beta \sim b(x)$.

InvSubBytes

The inverse operation of *SubBytes* is called *InvSubBytes* and is obtained by the application of the inverse permutation of f, called f^{-1}, followed by g, because g is the inverse operation and therewith self-inverse.

For f^{-1}, it must hold that $f^{-1}(f(a)) = a(x) \sim a \in \mathcal{B}$, for all $a \in \mathcal{B}$, where $\mathcal{B}$ is the set of all bytes. Additionally, it should be of the same form like f, which means that for suitable choices for the constant polynomials $c'(x)$ and $d'(x)$: $f^{-1}(b) = (c'(x) \odot' b(x)) \oplus d'(x)$.

Together the following must hold for all $a(x) \sim a \in \mathcal{B}$:

$$\begin{aligned} & f^{-1}(f(a)) = a(x) \\ \Leftrightarrow\ & (c'(x) \odot' ((c(x) \odot' a(x)) \oplus d(x))) \oplus d'(x) = a(x) \\ \Leftrightarrow\ & (c'(x) \odot' c(x) \odot' a(x)) \oplus (c'(x) \odot' d(x)) \oplus d'(x) = a(x) \\ \Leftrightarrow\ & c'(x) = c^{-1}(x) \pmod{x^8 + 1} \quad \wedge \quad d'(x) = c^{-1}(x) \odot' d(x). \end{aligned}$$

Since $c(x)$ is coprime to $x^8 + 1$, $c^{-1}(x)$ exists and therewith $c'(x)$ and $d'(x)$ are well-defined. By applying the *Extended Euclidean Algorithm* we can determine $c^{-1}(x) = x^6 + x^3 + x$ and it follows that:
$f^{-1}(b) = ((x^6 + x^3 + x) \odot' b(x)) \oplus (x^2 + 1)$. Again $f^{-1}(b) = a$ can be written as a multiplication by a circulant (8×8)-matrix followed by an addition with $d' \sim d'(x)$:

$$\begin{pmatrix} a_0 \\ a_1 \\ a_2 \\ a_3 \\ a_4 \\ a_5 \\ a_6 \\ a_7 \end{pmatrix} = \begin{pmatrix} 0&0&1&0&0&1&0&1 \\ 1&0&0&1&0&0&1&0 \\ 0&1&0&0&1&0&0&1 \\ 1&0&1&0&0&1&0&0 \\ 0&1&0&1&0&0&1&0 \\ 0&0&1&0&1&0&0&1 \\ 1&0&0&1&0&1&0&0 \\ 0&1&0&0&1&0&1&0 \end{pmatrix} \begin{pmatrix} b_0 \\ b_1 \\ b_2 \\ b_3 \\ b_4 \\ b_5 \\ b_6 \\ b_7 \end{pmatrix} \oplus \begin{pmatrix} 1 \\ 0 \\ 1 \\ 0 \\ 0 \\ 0 \\ 0 \\ 0 \end{pmatrix}$$

Definition 103 The **InvSubBytes step** is a bricklayer permutation, which consists of the $\frac{1}{8}N_B$-fold application of the S-box $S_{RD}^{-1} : GF(2^8) \rightarrow GF(2^8)$, operating on the individual bytes β of the input state, which is defined by:

$$S_{RD}^{-1}(\beta) := g(f^{-1}(\beta)),$$

where $g(\alpha) \odot \alpha = \mathbf{1}$ and
$f^{-1}(\beta) = ((x^6 + x^3 + x) \odot' b(x)) \oplus (x^2 + 1)$, where $\beta \sim b(x)$.

3.6.3 *The Linear Layer*

The *ShiftRows* Step

The *ShiftRows* step is a byte transposition. It consists only of a cyclically left-shift of the bytes of each row, where the bytes in row i are shifted over C_i bytes. The only thing remaining is the choice of the four constants $C_0, \ldots, C_3$.

Design Criteria

The two design criteria for the *ShiftRows* step are:

- simplicity
- diffusion optimality

The simplicity criterion means nothing more than that one of the C_i's equals zero. In Rijndael C_0 is zero.

The *ShiftRows* step is diffusion optimal, if all bytes, which were in the same column of the input state, are mapped into different columns of the output state. To achieve this criterion, all the C_i's have to be different modulo N_b. The C_i's are obtained from the following table:

N_B	C_0	C_1	C_2	C_3
128	0	1	2	3
192	0	1	2	3
256	0	1	3	4

Definition 104 Let a_{ij} be the byte in row i and column j of the input state s_{input} and b_{ij} the byte in row i and column j of the output state s_{output} of *ShiftRows SR*.

SR is then defined by:

$$SR(s_{input}) = s_{output},$$

with $b_{ij} := a_{i,(j+C_i) \bmod N_b}$ and the $C_i's$ are obtained from the above tabular.

Figure 3.3 shows the *ShiftRows* step for $N_B = 128$.

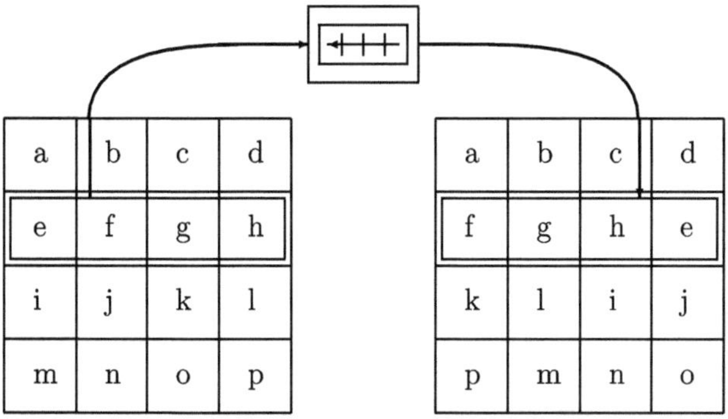

Fig. 3.3 The ShiftRows step for $N_B = 128$

InvShiftRows

The inverse of *ShiftRows*, which is denoted by *InvShiftRows*, is, of course, the byte transposition, which cyclically shifts the bytes of row i over C_i bytes to the right.

Definition 105 Let b_{ij} be the byte in row i and column j of the input state t_{input} and a_{ij} the byte in row i and column j of the output state t_{output} of *InvShiftRows* SR^{-1}.
SR^{-1} is then defined by:

$$SR^{-1}(t_{input}) = t_{output},$$
$$\text{with } a_{ij} := b_{i,(j-C_i) \bmod N_b}, \text{ with the same } C_i^{\prime}s \text{ like before.}$$

The *MixColumns* Step

The *MixColumns* step is a bricklayer permutation. It can be decomposed into several linear boolean permutations which are called *D-boxes* according to Definition 58. In fact, like for the *SubBytes* step, in *MixColumns* there is only one D-box, denoted by D_{RD}, operating independently on each of the N_b 4-byte columns of the state. This D-box consists of the multiplication by a fixed polynomial $c(x) \in GF(2^8)[x]|_4$, with $l(x) = x^4 + 1$ as the reducible reduction polynomial.

Figure 3.4 shows the application of the D-box for $N_B = 128$.

Design Criteria

In order to define D_{RD} we have to choose the fixed polynomial $c(x)$. Of course, $c(x)$ has to coprime to $l(x) = x^4 + 1 = (x + 1)^4$, which leads to the criterion that the decomposition of $c(x)$ must not include the factor $x + 1$.

Further on, it should have an efficient computability.

Let $c(x) = c_3x^3 + c_2x^2 + c_1x + c_0 \in GF(2^8)[x]|_4$ be the fixed polynomial and $a(x) = a_3x^3 + a_2x^2 + a_1x + a_0 \in GF(2^8)[x]|_4$ be a 4-byte column of the input state of *MixColumns*. As we have seen in subsection "The Finite Ring $< GF(2^8)[x]|_4, \oplus, \bullet >$" of Sect. 3.3.3 the multiplication of $c(x)$ and $a(x)$ modulo $l(x)$ can be written as a matrix multiplication. This means that the coefficients $c_i, a_i \in$

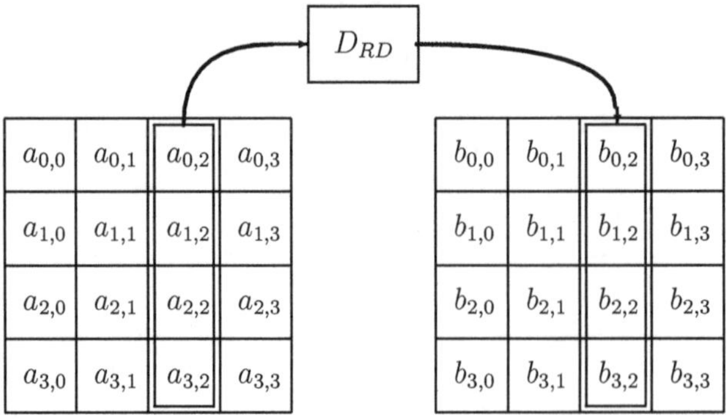

Fig. 3.4 The application of the D-box for $N_B = 128$

$GF(2^8)$ are multiplied by the application of the multiplication, which was defined in subsection "The Finite Field $GF(2^8)$" of Sect. 3.3.3.

In the same subsection it was shown that this multiplication can be done efficiently by the application of *xtime*, if the coefficients of $c(x)$ are 'small'.

From this it follows that the criterion of efficient computability can be translated into the requirement that the coefficients of $c(x)$ are 'small'.

Since also the inverse operation of *MixColumns* should be efficiently computable, the criterion has to be extended in such a way that also the coefficients of the fixed polynomial $d(x) \in GF(2^8)[x]|_4$, by which a 4-byte-column in *InvMixColumns* is multiplied, have to be 'small'.

In Rijndael a coefficient of the fixed polynomials $c(x), d(x) \in GF(2^8)[x]|_4$ is said to be 'small' if it is less than "10".

The last design criterion is that the coefficients of $c(x)$ are chosen in such a way that the branch number of *MixColumns* is 5 which is the maximum branch number.

Definition 106 The **MixColumns step** is a bricklayer permutation, which consists of the N_b-fold application of the D-box D_{RD}, operating independently on the individual 4-byte-columns α of the input state, which is defined by:

$$D_{RD}(\alpha) := c(x) \bullet a(x),$$
$$\text{where } \psi(\alpha) = a(x) \text{ and } c(x) := \text{“03”}\,x^3 + x^2 + x + \text{“02”}.$$

Following subsection "The Finite Ring $< GF(2^8)[x]|_4, \oplus, \bullet >$" of Sect. 3.3.3, $b(x) = b_3x^3 + b_2x^2 + b_1x + b_0 = (\text{“03”}\,x^3 + x^2 + x + \text{“02”}) \bullet (a_3x^3 + a_2x^2 + a_1x + a_0)$ can be written as the multiplication by the following circulant matrix:

$$\begin{pmatrix} b_0 \\ b_1 \\ b_2 \\ b_3 \end{pmatrix} = \begin{pmatrix} 02 & 03 & 01 & 01 \\ 01 & 02 & 03 & 01 \\ 01 & 01 & 02 & 03 \\ 03 & 01 & 01 & 01 \end{pmatrix} \begin{pmatrix} a_0 \\ a_1 \\ a_2 \\ a_3 \end{pmatrix}.$$

There is one other interesting way of rewriting *MixColumns*, denoted by MC. Let a_{ij} denote the byte in row i and column j of the input state and b_{ij} denote the byte in row i and column j of the output state.

It follows: $b_{ij} = MC(a_{ij}) :=$ "02" $\odot\, a_{ij} \oplus$ "03" $\odot\, a_{i+1,j} \oplus a_{i+2,j} \oplus a_{i+3,j}$, where the $(i+k)$'s are taken modulo 4, for $k \in \{1, 2, 3\}$.

InvMixColumns

InvMixColumns is also a bricklayer permutation, consisting of one D-box D_{RD}^{-1}, operating on each of the N_b 4-byte-columns of the input state. Again D_{RD}^{-1} is the multiplication by a fixed polynomial $d(x) \in GF(2^8)[x]|_4$ modulo $l(x) = x^4 + 1$.

It must hold that:

$$c(x) \bullet d(x) = \mathbf{1}.$$

By applying the *Extended Euclidean Algorithm* we obtain:

$$d(x) = \text{“0B”}\, x^3 + \text{“0D”}\, x^2 + \text{“09”}\, x + \text{“0E”}.$$

Definition 107 The **InvMixColumns step** is a bricklayer permutation which consists of the N_b-fold application of the D-box D_{RD}^{-1}, operating independently on the individual 4-byte-columns β of the input state, which is defined by:

$$D_{RD}^{-1}(\beta) := d(x) \bullet b(x),$$
$$\text{where } \psi(\beta) = b(x) \text{ and } d(x) := \text{“0B”}\, x^3 + \text{“0D”}\, x^2 + \text{“09”}\, x + \text{“0E”}.$$

Again, $a(x) = d(x) \bullet b(x)$ can be written as the multiplication by the following circulant matrix:

$$\begin{pmatrix} a_0 \\ a_1 \\ a_2 \\ a_3 \end{pmatrix} = \begin{pmatrix} 0E & 0B & 0D & 09 \\ 09 & 0E & 0B & 0D \\ 0D & 09 & 0E & 0B \\ 0B & 0D & 09 & 0E \end{pmatrix} \begin{pmatrix} b_0 \\ b_1 \\ b_2 \\ b_3 \end{pmatrix}.$$

And *InvMixColumns*, denoted by MC^{-1}, can be written as:
$a_{ij} = MC^{-1}(b_{ij}) :=$ "0E" $\odot\, b_{ij} \oplus$ "0B" $\odot\, b_{i+1,j} \oplus$ "0D" $b_{i+2,j} \oplus$ "09" $b_{i+3,j}$, where the b_{ij}'s are the individual bytes of the input state and the a_{ij}'s are the individual bytes of the output state of *InvMixColumns*.

3.6.4 The AddRoundKey Step

We have seen in Sect. 3.4 that Rijndael is a key-iterated block cipher.

Up to now we have defined how the plaintext block is mapped on the state and back on the ciphertext block and how the individual steps of the round transformation operate on the state. So the only things left to do, are to define the *AddRoundKey* step and how the individual RoundKeys are derived from the cipherkey.

Let us suppose we have generated all the required RoundKeys rk_i. Since there is an additional *AddRoundKey* step before the first round, we will need $(N_r + 1)$ different RoundKeys, all of the length N_B.

Definition 108 The **AddRoundKey step** of round i, with $i \in \{0, 1, \ldots, N_r\}$, is a bitwise XOR of its input state and the ith RoundKey rk_i.

Since the XOR operation on bits is self-inverse, it follows that the *AddRoundKey* step is also self-inverse so that the *AddRoundKey* step is used in both encryption and decryption.

3.6.5 The Key Schedule

The *Key Schedule* consists of two different parts. The first part is the *Key Expansion*, where the cipherkey is expanded into the *ExpandedKey* and the second part is the *RoundKey Selection*, where the ExpandedKey is decomposed into the individual RoundKeys rk_i. Since we need $(N_r + 1)$ RoundKeys, all of length N_B, the length of the ExpandedKey has to be $N_B(N_r + 1)$ bits.

The Key Expansion

The *Key Expansion* is a boolean function, consisting of several rounds $N_{KE} := \frac{1}{N_k} N_b(N_r + 1)$, where $N_k := \frac{1}{32} N_C$, with the cipherkey as its input and the ExpandedKey, consisting of all the RoundKeys, as its output. Of course, this is a security-critical part of Rijndael so that the Key Expansion was based on the following design criteria:

- non-linearity
- diffusion
- no symmetry in the rounds

The non-linearity criterion is obtained by the application of a S-box, which is in fact the same S-box S_{RD} like in the *SubBytes* step. In order to offer diffusion, a shift over the columns is applied and finally different constants for each round are used to delete the symmetry of each round.

The ExpandedKey is a two-dimensional array of four rows and $N_b(N_r + 1)$ columns. The first round of the KeyExpansion is different than the other rounds, since there only the cipherkey is used to fill up the first columns of the ExpandedKey and in the other rounds the already filled up columns are used to fill up the remaining columns.

We will visualize how the KeyExpansion works for $N_C = 128$.

In the *first round* the cipherkey is mapped into the first $N_k = 4$ columns of the ExpandedKey. This is done in the same way like the plaintext block is mapped into the state. Let $z = z_0 z_1 \ldots z_{\frac{1}{8}N_C - 1}$ be the cipherkey, where the z_i's are bytes and k_{ij} be the byte in row i and column j of the ExpandedKey (Fig. 3.5).

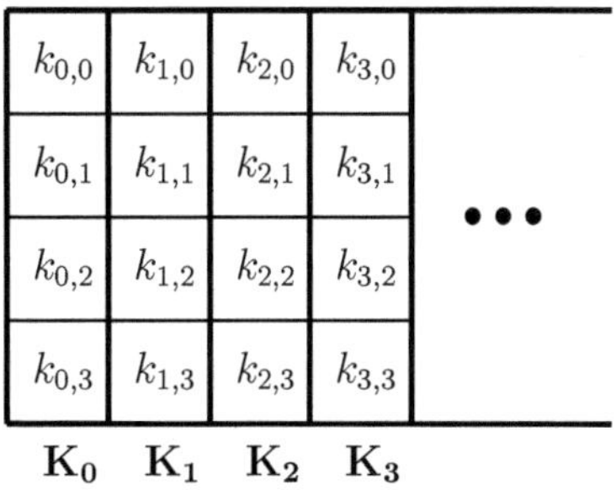

Fig. 3.5 The first round

Then:

$$k_{ij} = z_{i+4j}, \text{ for } 0 \leq i < 4 \text{ and } 0 \leq j < N_k$$

The *second* to the *last round* all have the same structure and can be divided into two different cases (Figs. 3.6 and 3.7).

If $j \neq 0 \mod N_k$, then the jth column K_j is the XOR of the previous column K_{j-1} and column K_{j-N_k}, written: $K_j = K_{j-N_k} \oplus K_{j-1}$.

And if $j = 0 \mod N_k$, then the jth column K_j is a XOR of column K_{j-N_k} and the previous column K_{j-1}, after function F was applied to K_{j-1}. This is written in the following form: $K_j = K_{j-N_k} \oplus F(K_{j-1})$.

The function F is the iterated application of the following parts.

Firstly, each byte of K_{j-1} is transformed via S_{RD}, then K_{j-1} is shifted over one byte to the top and, lastly, the round constant $RC(m) := x^{m-1}$, for $m \in \{2, \ldots, N_{KE}\}$, is added via the bitwise XOR-operation.

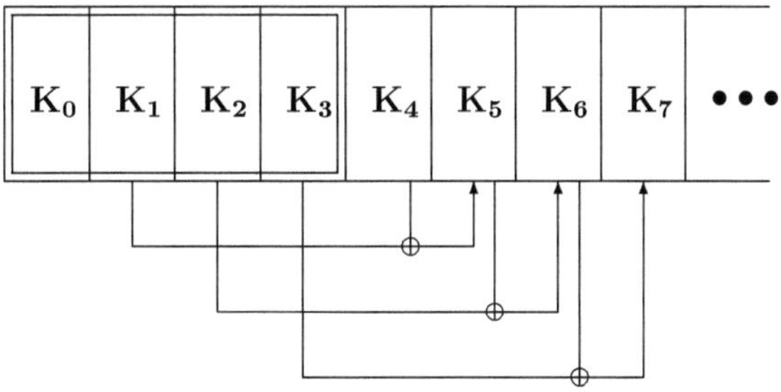

Fig. 3.6 First case of the other rounds

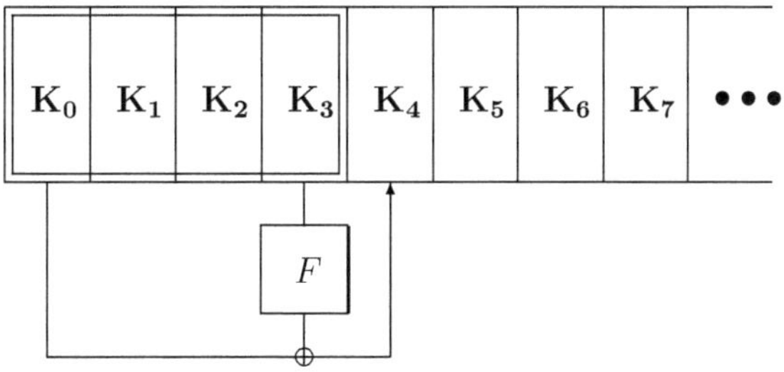

Fig. 3.7 Second case of the other rounds

Altogether we have:

$$k_{ij} = k_{i,j-N_k} \oplus S_{RD}(k_{i+1,j-1}) \oplus RC\left(\frac{j}{N_k}\right),$$

where k_{ij} is the ith byte of column j.

The RoundKey Selection

Finally the RoundKey rk_j, for $j \in \{0, \ldots, N_r\}$, of the jth round is given by the columns $K_{N_b \cdot j}, \ldots, K_{N_b \cdot (j+1)-1}$.

Figure 3.8 visualizes the RoundKey selection for $N_C = 192$ and $N_B = 128$.

3.6.6 *Encryption*

This encryption is written in pseudo-code, which means that both input and `output` are arguments of the individual functions.

For example *Rijndael*(plaintext, cipherkey, `ciphertext`) means that the arguments of the whole cipher are the plaintext, the cipherkey and the `ciphertext`, where `ciphertext` is an empty argument and obtains its value during the execution of the function *Rijndael*. For some functions like *AddRoundKey*, *Round*, *FinalRound* and the individual steps of each round there is no particular `output` given. The output of these functions is always the state.

For given N_B, N_C and N_r the encryption is done in the following way:

Rijndael(plaintext, cipherkey, `ciphertext`)
{
  PlainToState(plaintext, `state`);
  KeySchedule(cipherkey, `roundkeys[i]`);
  AddRoundKey(state, roundkeys[0]);
  for (i = 1, i < N_r, i++) {

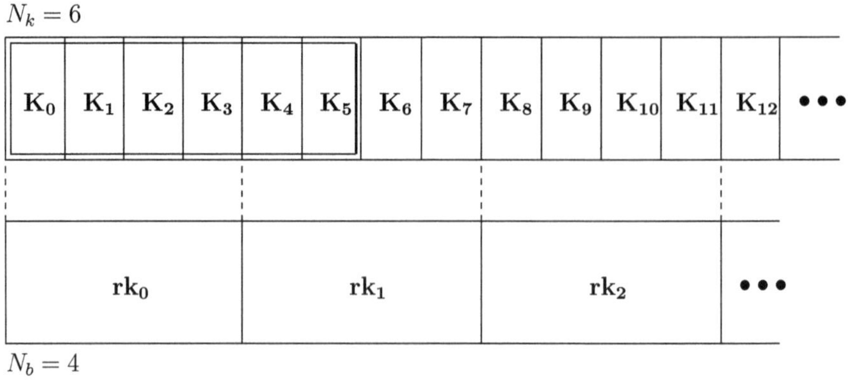

Fig. 3.8 The RoundKey selection for $N_C = 192$ and $N_B = 128$

```
    Round(state, roundkeys[i]);}
  FinalRound(state, roundkeys[N_r]);
  StateToCipher(state, ciphertext);
}
```

with:

```
KeySchedule(cipherkey, roundkeys[i])
{
  KeyExpansion(cipherkey, expkey);
  RoundKeySelection(expkey, roundkeys[i]);
}


Round(state, roundkeys[i])
{
  SubBytes(state);
  ShiftRows(state);
  MixColumns(state);
  AddRoundKey(state, roundkeys[i]);
}


FinalRound(state, roundkeys[N_r])
{
  SubBytes(state);
  ShiftRows(state);
  AddRoundKey(state, roundkeys[N_r]);
}
```

The variables of the *Rijndael* cipher and its individual functions are:

- plaintext, ciphertext:
  one-dimensional arrays of bytes of length $\frac{1}{8}N_B$
- cipherkey:
  one-dimensional array of bytes of length $\frac{1}{8}N_C$
- state:
  two-dimensional array of bytes with 4 rows and $\frac{1}{32}N_B$ columns
- expkey:
  one-dimensional array of bytes of length $\frac{1}{8}(N_r + 1)N_B$
- roundkeys[i]:
  one-dimensional array of round keys of length $N_r + 1$, where roundkeys[i] is the ith round key

3.6.7 Decryption

There are two ways in which the decryption can be done. The first is the *straight-forward* decryption, where the decryption is done by applying the operations exactly the other way round.

Table 3.1 shows this for a three-round Rijndael.

The other way is called the *equivalent* decryption, where the decryption is done in mainly the same way as the encryption. This can be done because of the following properties of the individual steps of Rijndael.

Since *InvSubBytes* operates on each byte of the state independently and *InvShiftRows* is a shift of the rows of the state and has no effect on the values of the individual bytes, these two steps can be interchanged.

In order to interchange *InvMixColumns* and *AddRoundKey* we have to take advantage of the linear structure of *InvMixColumns*.

From the linearity of *InvMixColumns* it follows that:

$$InvMixColumns(a \oplus rk_i) = InvMixColumns(a) \oplus InvMixColumns(rk_i)$$

It follows that if the RoundKey rk_j is changed into $InvMixCoulmns(rk_j)$ then *InvMixColumns* and *AddRoundKey* can be interchanged, too.

Table 3.2 shows the equivalent decryption for a three-round Rijndael.

The advantage of the equivalent decryption is that it can be done by the same algorithm as the encryption, where only the kes schedule has to be adapted. This is especially important if the cipher is constructed in hardware, since we are able to encrypt and decrypt with the same hardware.

Table 3.1 The straight-forward decryption for a three-round Rijndael

Encryption	Decryption
$AddRoundKey(rk_0)$	$AddRoundKey(rk_3)$
SubBytes	*InvShiftRows*
ShiftRows	*InvSubBytes*
MixColumns	
$AddRoundKey(rk_1)$	$AddRoundKey(rk_2)$
SubBytes	*InvMixColumns*
ShiftRows	*InvShiftRows*
MixColumns	*InvSubBytes*
$AddRoundKey(rk_2)$	$AddRoundKey(rk_1)$
SubBytes	*InvMixColumns*
ShiftRows	*InvShiftRows*
	InvSubBytes
$AddRoundKey(rk_3)$	$AddRoundKey(rk_0)$

Table 3.2 The equivalent decryption for a three-round Rijndael

Encryption	Decryption
AddRoundKey(rk_0)	*AddRoundKey*(rk_3)
SubBytes	*InvSubBytes*
ShiftRows	*InvShiftRows*
MixColumns	*InvMixColumns*
AddRoundKey(rk_1)	*AddRoundKey*(*InvMixColumns*(rk_2))
SubBytes	*InvSubBytes*
ShiftRows	*InvShiftRows*
MixColumns	*InvMixColumns*
AddRoundKey(rk_2)	*AddRoundKey*(*InvMixColumns*(rk_1))
SubBytes	*InvSubBytes*
ShiftRows	*InvShiftRows*
AddRoundKey(rk_3)	*AddRoundKey*(rk_0)

3.6.8 Complexity

First we will calculate the complexity of the individual steps of Rijndael. The measure of the complexity is how often the S-box S_{RD} and the XOR-operation on bytes are applied.

SubBytes:
In the *SubBytes* step the S-box S_{RD} is applied to each of the $\frac{1}{8}N_B = 4N_b$ bytes of the state so that its complexity is:

$$4N_b \ S_{RD}\text{'s}.$$

ShiftRows:
The *ShiftRows* step consists only of a shift on byte-level and therefore does not contribute to the complexity of the cipher.

MixColumns:
If we denote one column of the input state of the *MixColumns* step with (a_0, a_1, a_2, a_3) and the corresponding column of the output state with (b_0, b_1, b_2, b_3), the *MixColumns* step *MC* can be written as follows:

$$\begin{aligned} b_i = MC(a_i) &= \text{“02”} \odot a_i \oplus \text{“03”} \odot a_{i+1} \oplus a_{i+2} \oplus a_{i+3} \\ &= xtime(a_i) \oplus xtime(a_{i+1}) \oplus a_{i+1} \oplus a_{i+2} \oplus a_{i+3} \end{aligned}$$

It follows that each application of the D-box of the *MixColumns* step consists of four applications of the XOR-operations on bytes and two applications of *xtime*. In subsection “The Finite Field $GF(2^8)$” of Sect. 3.3.3 we have seen that the *xtime* operation consists either only of a left-shift of bits or of a left-shift of bits followed by one XOR-operation on bytes. Since the shift operation does not contribute to the complexity, we assume that the *xtime* operation equals one XOR-operation on bytes.

The D-box is applied to each of the N_b columns so that the whole *MixColumns* step has a complexity of:

$$6N_b \text{ XORs}.$$

AddRoundKey:
The *AddRoundKey* step is the N_B-fold application of the bitwise XOR of the state and the particular RoundKey, which corresponds to the $(\frac{1}{8}N_B = 4N_b)$-fold application of the XOR-operation on bytes and therefore its complexity is:

$$4N_b \text{ XORs}.$$

Table 3.3 shows the complexity of each of the individual steps of Rijndael.

We can now calculate the complexity of the whole Rijndael cipher.

As shown in Sect. 3.6.6 Rijndael consists of the *KeyExpansion*, the initial *AddRoundKey* step, the *Round* and the *FinalRound*.

KeyExpansion:
In subsection "The Key Expansion" of Sect. 3.6.5 we have seen that the *KeyExpansion* consists of
N_{KE} rounds, where $N_{KE} = \frac{1}{N_k}N_b(N_r + 1)$ and $N_k = \frac{1}{32}N_C$.

In the first round no calculation is done, since there only the cipherkey is mapped into the first N_k columns and in the following rounds each column K_j of the ExpandedKey is derived from the previous columns.

If $j \neq 0 \pmod{N_k}$, $K_j = K_{j-N_k} \oplus K_{j-1}$, which corresponds to four XOR-operations on bytes.

If $j = 0 \pmod{N_k}$, $K_j = K_{j-N_k} \oplus F(K_{j-1})$.

The map F consists of four applications of the S-box S_{RD}, one shift and four XOR-operations on bytes, from which it follows that in this case four applications of S_{RD} and eight XOR-operations on bytes are done.

It follows that each round, besides the first, consists of four applications of S_{RD} and $4(N_k + 1)$ XOR-operations on bytes and therewith the complexity of the whole *KeyExpansion* is:

$$4(\tfrac{1}{N_k}N_b(N_r + 1) - 1)\, S_{RD}\text{'s}$$
$$\text{and } 4(N_k + 1)(\tfrac{1}{N_k}N_b(N_r + 1) - 1) \text{ XORs}.$$

Table 3.3 Complexity of the individual steps of Rijndael

Step	S_{RD}	XOR
SubBytes	$4N_b$	–
ShiftRows	–	–
MixColumns	–	$6N_b$
AddRoundKey	–	$4N_b$

initial *AddRoundKey* step:

As seen above the initial *AddRoundKey* step has a complexity of:

$$4N_b \text{ XORs.}$$

Round:
Each *Round* consists of all the previously calculated steps so that its complexity is:

$$4N_b\, S_{RD}\text{'s and } 10N_b \text{ XORs.}$$

FinalRound:
In the *FinalRound* the *MixColumns* step is omitted, which leads to a complexity of:

$$4N_b\, S_{RD}\text{'s and } 4N_b \text{ XORs.}$$

The complexity for the Rijndael cipher with block length $N_B = 32N_b$ bits and cipherkey length $N_C = 32N_k$ bits over N_r rounds is:

$$4(\tfrac{1}{N_k}N_b(N_r+1)-1)+4N_bN_r\,S_{RD}\text{'s}$$
$$\text{and } 4(N_k+1)(\tfrac{1}{N_k}N_b(N_r+1)-1)+10N_b(N_r-1)+8N_b \text{ XORs.}$$

3.6.9 Security

Rijndael has been designed according to the Wide Trail Strategy with the following properties for:

- the bundle size m:
  $m = 8$
- the column size n_θ:
  $n_\theta = 4$
- the non-linear bricklayer permutation γ:
  $\gamma =$ *SubBytes*, whose S-box S_{RD} has been selected from [16] so that its minimum linear weight is at least 3 and its minimum differential weight is at least 6.
- the byte transposition π:
  $\pi =$ *ShiftRows*, which is diffusion optimal.
- the linear bricklayer permutation θ:
  $\theta =$ *MixColumns*, where coefficients of the fixed polynomial $c(x)$ has been chosen in such a way that the branch number of *MixColumns* is 5, the maximum possible branch number

From Eqs. (3.5.14) and (3.5.15) follows that the minimum weight for any linear trail over four rounds is at least 75 and the minimum weight for any differential trail is at least 150. Hence any eight round linear (differential) trail has a weight of at least 150 (300).

The authors of [6] "consider this sufficient to resist differential and linear attacks".

3.7 Cryptanalysis

In this section we introduce the *saturation attack*. The saturation attack is an attack by the authors of Rijndael themselves, which exploits the specific structure of the round transformation, to launch an attack of up to six rounds of Rijndael.

3.7.1 The Saturation Attack

This attack is based on the *Square attack*, developed by Lars Knudsen, which was designed to attack the block cipher *Square* [7]. The block cipher Square by Joan Daemen, Lars Knudsen and Vincent Rijmen is a precursor of Rijndael. Its round structure is very similar to the round structure of Rijndael so that this attack was improved by Joan Daemen and Vincent Rijmen to allow attacks on a round reduced Rijndael of up to six rounds.

Λ-Sets

The saturation attack is a chosen-plaintext attack, which means that we try to derive the unknown cipherkey by encrypting several properly chosen plaintexts and exploiting the particular structure of the attacked cipher. In this attack the set of chosen plaintexts is called a Λ-*set* and defined as follows.

Definition 109 A **Λ-set** is a set of 2^8 states (or intermediate results) with the following properties.

Let $x, y \in \Lambda$ be two states of the Λ-set and let $\mathcal{I} := \{0, \dots, 3\} \times \{0, \dots, N_b - 1\}$ be the index space of the individual bytes in the states within the Λ-set.

For $(i, j) \in \mathcal{I}$, x_{ij}, y_{ij} denote the byte in row i and column j of $x, y \in \Lambda$.

It holds:

$$\begin{aligned} &\exists \mathcal{I}_1, \mathcal{I}_2, \text{ with } \mathcal{I} = \mathcal{I}_1 \bigcup \mathcal{I}_2, \mathcal{I}_1 \bigcap \mathcal{I}_2 = \emptyset \text{ and} \\ &-\forall\ (i,j) \in \mathcal{I}_1 \Rightarrow x_{ij} \neq y_{ij} \quad \forall\, x, y \in \Lambda \\ &-\forall\ (i,j) \in \mathcal{I}_2 \Rightarrow x_{ij} = y_{ij} \quad \forall\, x, y \in \Lambda \end{aligned}$$

Definition 110 In a Λ-set the bytes at position $(i_1, j_1) \in \mathcal{I}_1$ are called **active bytes** and the bytes at position $(i_2, j_2) \in \mathcal{I}_2$ are called **passive bytes**.

Definition 111 $\mathcal{L} = \{\mathbf{0}, \dots, \mathbf{2^8 - 1}\}$ denotes the index space of the individual states in a Λ-set.

The reason for the choice of the plaintexts is given by the following proposition.

Proposition 27 $\bigoplus_{l \in \mathcal{L}} x_{ij}^l = 0 \ \ \forall\, (i,j) \in \mathcal{I}.$

Proof Let $(i,j) \in \mathcal{I}_1$, then all the bytes at position (i,j) of the individual states are pairwise different. Since the Λ-set contains 2^8 states, all the possible 2^8 values for the bytes are obtained and therefore sum up to zero.

Let $(i,j) \in \mathcal{I}_2$, then all the bytes at position (i,j) of the individual states are equal. Since every byte is self-inverse under $\oplus$ and the Λ-set contains 2^8 states, the bytes sum up to zero. □

Definition 112 A **Λ-maintaining** boolean transformation is a boolean transformation which maps all the 2^8 states of a Λ-set into states which form again a Λ-set.

In the saturation attack we exploit the fact that, if we choose the Λ-sets properly, all the individual steps of Rijndael are Λ-maintaining. This fact is proved by the following two propositions.

Proposition 28 *The SubBytes, the ShiftRows and the AddRoundKey steps are Λ-maintaining.*

Proof The *SubBytes* step does not change the position of the bytes of a state and it consists of one S-box which operates independently on the individual bytes of each state and is a bijection in $GF(2^8)$. There are 2^8 states in a Λ-set. If $(i,j) \in \mathcal{I}_1$, after the application of the S-box to the bytes x_{ij} the resulting bytes at this position are again pairwise different. If $(i,j) \in \mathcal{I}_2$, the resulting bytes are again all equal. It follows that the output states of the *SubBytes* step form a Λ-set.

The *AddRoundKey* step consists of the bitwise XOR of the states with a roundkey of length N_B. If we decompose this roundkey into its $\frac{1}{8}N_B$ bytes rk_l, for $l \in \{0, \dots, \frac{1}{8}N_B - 1\}$, this step equals the bitwise XOR of each byte of the state and each byte of the roundkey. It follows that if $(i,j) \in \mathcal{I}_1$ the resulting bytes are pairwise different and if $(i,j) \in \mathcal{I}_2$ the resulting bytes are all equal again. Hence, the output states of the *AddRoundKey* step form a Λ-set.

Since the *ShiftRows* step does not change the value of the individual bytes, but only changes their positions, the application of *ShiftRows* to the states of a Λ-set results in states, which again form a Λ-set. □

In general the *MixColumns* step

$$b_{ij} = MC(a_{ij}) = \text{“02”} \odot a_{ij} \oplus \text{“03”} \odot a_{i+1,j} \oplus a_{i+2,j} \oplus a_{i+3,j}$$

is not Λ-maintaining. Suppose the first two bytes a_{0j}, a_{1j} of column j of the input state of *MixColumns* are active and the last two bytes a_{2j}, a_{3j} of column j are passive.

Now we look to three different input states $a^{l_1}, a^{l_2}, a^{l_3}$ of the Λ-set with the above property, where $l_1, l_2, l_3 \in \mathcal{L}$, and assume that:

$$a_{0j}^{l_2} = (\text{“02”})^{-1} \odot \text{“03”} \odot a_{1j}^{l_1}$$
$$\text{and } a_{1j}^{l_2} = (\text{“03”})^{-1} \odot \text{“02”} \odot a_{0j}^{l_1} a_{1j}^{l_1}.$$

Applying *MixColumns* would result in the following output bytes $b_{0j}^{l_k}$:

$$b_{0j}^{l_1} = b_{0j}^{l_2} = \text{“02”} \odot a_{0j}^{l_1} \oplus \text{“03”} \odot a_{1j}^{l_1} \oplus c$$
$$\text{and} \quad b_{0j}^{l_3} = \text{“02”} \odot a_{0j}^{l_3} \oplus \text{“03”} \odot a_{1j}^{l_3} \oplus c,$$

where $c = a_{2j}^{l_k} \oplus a_{3j}^{l_k}$.

Since $b_{0j}^{l_3} \neq b_{0j}^{l_1} = b_{0j}^{l_2}$, the resulting set of states do not form a Λ-set.

Proposition 29 *If the input states of the MixColumns step have at most one active byte in each column, then the MixColumns step is Λ-maintaining.*

Proof Since the *MixColumns* step consists of one D-box operating independently on each of the columns of the input state, the condition of the proposition equals the condition that at most one byte of the input of the D-box is active.

If no byte is active, of course, the bytes of the resulting column are all passive and the states form again a Λ-set.

If one byte is active, without loss of generality we assume that this is the first byte a_{0j} of the column, we obtain the following equality for all $l \in \mathcal{L}$:

$$b_{ij}^{l} = d_i \odot a_{0j}^{l} \oplus c,$$

where:

$$d_i = \begin{cases} \text{“02”}, & \text{if } i \bmod 4 = 0 \\ \text{“01”}, & \text{if } i \bmod 4 = 1, 2 \\ \text{“03”}, & \text{if } i \bmod 4 = 3 \end{cases}$$

and:

$$c = d_{i+1} \odot a_{1j}^{l} \oplus d_{i+2} \odot a_{2j}^{l} \oplus d_{i+3} \odot a_{3j}^{l}.$$

Since the a_{0j}^{l}'s are pairwise different, so are the b_{ij}^{l}'s and the resulting states form again a Λ-set. □

Basic Four-Round Attack

This attack is a chosen-plaintext attack and we will examine it for $N_B = N_C = 128$. We choose 2^8 plaintexts so that the input states of the first round form a Λ-set with only one active byte.

Since the *AddRoundKey* (ARK), the *SubBytes* (SB) and the *ShiftRows* (SR) steps are Λ-maintaining, the input states of the first *MixColumns* (MC) step form a Λ-set with one active byte. From Proposition 29 it follows that the output states of the first MC step form a Λ-set, where all the four bytes of one column are active. This property remains until the output of the second SR step. The second SR step spreads the active bytes over all the columns so that the input states of the second MC step have one active byte per column. The output states of the second MC step form again a Λ-set with only active bytes and this remains until the input of the third MC step. After the application of the third MC step the states do not usually form a Λ-set, but we obtain the following property.

Proposition 30 *The bytes on each position* $(i, j) \in \mathcal{I}$ *of the input states of the fourth round sum up to zero.*

Proof We denote the input states of the third MC step by a^l, the output states by b^l, for $l \in \mathcal{L}$, and the individual bytes of each of them by a^l_{ij} and b^l_{ij}, where $i \in \{0, \ldots, 3\}$ and $j \in \{0, \ldots, N_b - 1\}$.

From Propositions 27, 28 and 29 it follows that all the bytes of the output states of the third MC step sum up to zero:

$$\begin{aligned}\bigoplus_{l\in\mathcal{L}} b^l_{ij} &= \bigoplus_{l\in\mathcal{L}} MC(a^l_{ij})\\ &= \bigoplus_{l\in\mathcal{L}}(\text{“02”} \odot a^l_{ij} \oplus \text{“03”} \odot a^l_{i+1,j} \oplus a^l_{i+2,j} \oplus a^l_{i+3,j})\\ &= \text{“02”} \bigoplus_{l\in\mathcal{L}} a^l_{ij} \oplus \text{“03”} \bigoplus_{l\in\mathcal{L}} a^l_{i+1,j} \oplus \bigoplus_{l\in\mathcal{L}} a^l_{i+2,j} \oplus \bigoplus_{l\in\mathcal{L}} a^l_{i+3,j}\\ &= \mathbf{0} \oplus \mathbf{0} \oplus \mathbf{0} \oplus \mathbf{0} = \mathbf{0}\end{aligned}$$

Since $ARK(b^l_{ij}, rk^{(3)}_{i+4j}) = b^l_{ij} \oplus rk^{(3)}_{i+4j}$, where $rk^{(3)}_{i+4j}$ is the $(i + 4j)$th byte of the third roundkey, and since from this it follows that:

$$\bigoplus_{l\in\mathcal{L}} ARK(b^l_{ij}, rk^{(3)}_{i+4j}) = \bigoplus_{l\in\mathcal{L}}(b^l_{ij} \oplus rk^{(3)}_{i+4j}) = \bigoplus_{l\in\mathcal{L}} b^l_{ij} \oplus \bigoplus_{l\in\mathcal{L}} rk^{(3)}_{i+4j} = \mathbf{0} \oplus \mathbf{0} = \mathbf{0},$$

this property holds until the input of the fourth round. □

Now let c_{ij}, for all $(i, j) \in \mathcal{I}$, denote the bytes of the input c of the fourth round, let d_{ij}, for all $(i, j) \in \mathcal{I}$, denote the bytes of the output of the fourth round, which is the ciphertext, and let k_{ij}, for all $(i, j) \in \mathcal{I}$, denote the bytes of the fourth roundkey. Then the following equality holds for all $(i, j) \in \mathcal{I}$:

$$d_{ij} = S_{RD}(c_{i,j+C_i}) \oplus k_{ij}.$$

It follows that each byte c_{ij} of the input state c of the fourth round can be expressed in terms of the bytes $d_{i,j-C_i}$ of the known ciphertext d and the bytes $k_{i,j-C_i}$ of the last roundkey k:

$$c_{ij} = S^{-1}_{RD}(d_{i,j-C_i} \oplus k_{i,j-C_i}) \qquad \forall\, (i, j) \in \mathcal{I}.$$

Following Proposition 30, it must hold that:

$$\bigoplus_{l \in \mathcal{L}} c_{ij}^{l} = \mathbf{0} \qquad \forall\,(i,j) \in \mathcal{I}. \tag{3.7.1}$$

The individual bytes $d_{i,j-C_i}$ of the ciphertext d are known, which means that one can now guess a value for each byte $k_{i,j-C_i}$ of the last roundkey k and check whether the following equality holds:

$$\bigoplus_{l \in \mathcal{L}} (S_{RD}^{-1}(d_{i,j-C_i}^{l} \oplus k_{i,j-C_i})) = \mathbf{0}. \tag{3.7.2}$$

One of the 2^8 possible values for each byte of the last roundkey is the right value and therefore the above equality will hold. If we assume that the 2^8 values $d_{i,j-C_i}^{l}$, $l \in \mathcal{L}$, of each byte of the ciphertext d are uniformly distributed, it follows that for each of the $2^8 - 1$ wrong values the 2^8 values c_{ij}^{l}, $l \in \mathcal{L}$, are uniformly distributed, since both the S-box S_{RD}^{-1} and the XOR operation $\oplus$ are bijective.

From this property it follows generally that:

$$Prob\left(\bigoplus_{l \in \mathcal{L}} c_{ij}^{l} = \mathbf{x}\right) = \frac{1}{2^8},\ \forall\, \mathbf{x} \in GF(2^8)$$

and in particular for $\mathbf{x} = \mathbf{0}$.

It follows that the expected number of remaining values for each byte of the last roundkey is $1 + \frac{255}{256} \approx 2$, one is the right value and one is a wrong value.

If we now do the same calculation for a second Λ-set, again approximatly two values will remain for each byte of the last roundkey, again the right value and one wrong value. Since the probablility that the two wrong remaining values are equal is $\frac{1}{255}$, we have found the right value with a probablility of $\frac{254}{255}$. The last roundkey has $\frac{1}{8}N_b = 16$ bytes and therefore we have determined the last roundkey uniquely with a probablility of $(\frac{254}{255})^{16}$, if we repeat the above calculation for the remaining 15 bytes of the last roundkey.

Retrieval of the Cipherkey

In Sect. 3.6.5 we have seen that each byte $k_{ij}^{(m)}$, $\forall\,(i,j) \in \mathcal{I}$, of the mth roundkey $k^{(m)}$, where $m \in \{1, \ldots, 4\}$, can be derived from the cipherkey $k^{(0)}$ with the following equation:

$$k_{ij}^{(m)} = \begin{cases} k_{ij}^{(m-1)} \oplus k_{i,j-1}^{(m)}, & \text{if } j = 1, 2, 3 \\ k_{i,0}^{(m-1)} \oplus S_{RD}(k_{i+1,3}^{(m-1)}) \oplus RC(m-1), & \text{if } j = 0 \end{cases}.$$

From this it follows that we can determine each byte $k_{ij}^{(m)}$, $\forall\ (i,j) \in \mathcal{I}$, of the mth roundkey $k^{(m)}$, where $m \in \{0,\dots,3\}$, uniquely from the last roundkey $k^{(4)}$ via the following equation:

$$k_{ij}^{(m)} = \begin{cases} k_{ij}^{(m+1)} \oplus k_{i,j-1}^{(m+1)}, & \text{if } j = 1,2,3 \\ k_{i,0}^{(m+1)} \oplus S_{RD}(k_{i+1,3}^{(m)}) \oplus RC(m), & \text{if } j = 0 \end{cases}.$$

Attack Complexity

In this basic attack we need two Λ-sets, which corresponds to 2^9 known plaintexts.

Checking Eq. (3.7.2) for each possible value of each byte of the last roundkey requires $16 \times 2^8 \times 2^8 = 2^{20}$ applications of the S-box S_{RD}^{-1} and the same number of XORs $\oplus$. Following Sect. 3.6.8, the complexity of a four round cipher execution where both the block length and the cipherkey length are 128 bits is:

$$80 = 2^6 + 2^4 \approx 2^6 \text{ applications of } S_{RD}$$
$$\text{and } 232 = 2^7 + 2^6 + 2^5 + 2^3 + 2 \approx 2^7 \text{ XORs } \oplus.$$

It follows that the attack complexity corresponds roughly to the number of 2^{14} 4-round cipher executions.

Extension at the End

In this extension we add a fifth round at the end. We denote the bytes of the output state e of the fifth round, which is the ciphertext, by e_{ij}, $(i,j) \in \mathcal{I}$. Following Sect. 3.6.7 we can interchange the *InvMixColumns* and the *AddRoundKey* step if we adopt the roundkey accordingly.

In order to calculate Eq. (3.7.1) we have to use the following expression for c_{ij}:

$$\begin{aligned} c_{i,j+C_i} = \; & S_{RD}^{-1}((\text{"0E"} \odot (S_{RD}^{-1}(e_{i,j-C_i} \oplus k_{i,j-C_i}^{(5)}) \\ & \oplus(\text{"0B"} \odot (S_{RD}^{-1}(e_{i+1,j-C_i} \oplus k_{i+1,j-C_{i+1}}^{(5)}) \\ & \oplus(\text{"0D"} \odot (S_{RD}^{-1}(e_{i+2,j-C_i} \oplus k_{i+2,j-C_{i+2}}^{(5)}) \\ & \oplus(\text{"09"} \odot (S_{RD}^{-1}(e_{i+3,j-C_i} \oplus k_{i+3,j-C_{i+3}}^{(5)}) \oplus k_{ij}^{*(4)}), \end{aligned} \tag{3.7.3}$$

where $k_{ij}^{*(4)} = MC^{-1}(k_{ij}^{(4)})$.

It follows that we have in addition to the one byte $k_{ij}^{*(4)}$ of the fourth roundkey the four additional bytes $k_{i+q,j-C_{i+q}}^{(5)}$, for $q \in \{0,\dots,3\}$, of the fifth roundkey $k^{(5)}$ to be guessed, in order to check whether Eq. (3.7.1) holds or not.

This means that we have $(2^8)^5 = 2^{40}$ combinations of 2^8 values of the five bytes. If we guess the right combination Eq. (3.7.1) will hold and again if we assume that the bytes e_{ij} of the ciphertext e are unifromly distributed then Eq. (3.7.1) will hold for every wrong combination with probability $\frac{1}{256}$. It follows that the amount of the $(2^{40} - 1)$ wrong combinations is reduced to $\frac{2^{40}-1}{2^8}$ after the checking of (3.7.1) with the first Λ-set so that the amount of the remaining possible combinations is $1+\frac{2^{40}-1}{2^8}$.

If we repeat the whole calculation with another different Λ-set the amount of the remaining wrong combinations will be $\frac{2^{40}-1}{2^{16}}$. Again the right combination will sum up to zero so that the amount of the remaining possible combinations is $1 + \frac{2^{40}-1}{2^{16}}$.

In general the amount of the remaining possible combinations after the calculation of Eq. (3.7.1) with k different Λ-sets is $1 + \frac{2^{40}-1}{2^{8k}}$.

After the calculation of (3.7.1) with five Λ-sets we will obtain two remaining possible combinations so that the calculation with the sixth Λ-set will determine the right combination with probability $\frac{254}{255}$.

We have to repeat the whole attack four times in order to obtain all of the sixteen bytes of the last roundkey.

Attack Complexity

This extension needs six different Λ-sets which corresponds to $6 \cdot 2^8$ chosen plaintexts.

The calculation of (3.7.3) requires four multiplications $\odot$.

As shown in subsection "The Finite Field $GF(2^8)$" of Sect. 3.3.3 the multiplication can be done efficiently via the application of *xtime*. The multiplication by "0E", "0B" and "0D" requires three applications of *xtime* and two XORs $\oplus$ and the multiplication by "09" requires three applications of *xtime* and one XOR $\oplus$.

If we follow Sect. 3.6.8 and simplify the *xtime* operation to equal one XOR operation we obtain that the calculation of (3.7.3) requires five applications of the S-box S_{RD}^{-1} and 27 XORs $\oplus$.

We have to check (3.7.1) 2^8 times for every of the 2^{40} possible combinations. And we have to do this six times for every needed Λ-set.

After that we have uniquely determined four of the sixteen bytes of the last roundkey so that we have to repeat the whole calculation three more times.

This leads to a complexity of:

$$4 \times 6 \times 2^{40} \times 2^8 \times 5 \approx 2^{54}\, {S^{-1}}_{RD}\text{'s}$$
$$\text{and } 4 \times 6 \times 2^{40} \times 2^8 \times 27 \approx 2^{58} \text{ XORs.}$$

Since the complexity of a five-round cipher equals:

$$100 = 2^6 + 2^5 + 2^2 \approx 2^6\, {S^{-1}}_{RD}\text{'s}$$
$$\text{and } 272 = 2^8 + 2^4 \approx 2^8 \text{ XORs,}$$

the complexity of this attack corresponds roughly to 2^{49} five-round cipher executions.

Extension at the Beginning

Consider a set of 2^{32} chosen plaintexts so that the input states of the first *MixColumns* step contain one column C which ranges over all 2^{32} possible values and three columns which are constant.

Since the *MixColumn* step and the *AddRoundKey* step do not change the positions of the bytes, this property remains until the input of the second round.

Now we consider the 2^{32} plaintexts as a set of 2^{24} Λ-sets where each Λ-set has one active byte in column C and the other bytes are passive.

We cannot separate the plaintext and calculate (3.7.1) for each of the 2^{24} Λ-sets independently. But since (3.7.1) must hold for every Λ-set, it still must hold if we calculate it for all of the 2^{32} plaintexts.

It follows that we can determine the last roundkey uniquely with 2^{25} different Λ-sets.

Attack Complexity

This extension needs 2^{25} Λ-sets which corresponds to 2^{33} chosen plaintexts.

To find one byte of the last roundkey we have to calculate (3.7.1) for each possible value of this byte and for each of the 2^{25} Λ-sets and this must be repeated sixteen times to obtain all bytes of the last roundkey. This leads to a complexity of:

$$\begin{gathered} 16 \times 2^8 \times 2^{25} \times 2^8 = 2^{45}\, S_{RD}^{-1}\text{'s} \\ \text{and } 16 \times 2^8 \times 2^{25} \times 2^8 = 2^{45}\, \text{XORs} \end{gathered}$$

which corresponds roughly to 2^{38} five-round cipher executions.

Six-Round Attack

If we apply both extensions we need $5 \times 2^{25} \approx 2^{27}$ Λ-sets which corresponds to 2^{35} chosen plaintexts.

To obtain a 4-byte column of the last roundkey we have to calculate (3.7.1) over equation (3.7.3) for each possible value of the five key bytes and for each of the 2^{25} Λ-sets and this must be repeated four times to get all bytes of the last roundkey. This leads to a complexity of:

$$\begin{gathered} 4 \times 2^{40} \times 2^{25} \times 2^8 \times 5 \approx 2^{77}\, S_{RD}^{-1}\text{'s} \\ 4 \times 2^{40} \times 2^{25} \times 2^8 \times 27 \approx 2^{79}\, \text{XORs.} \end{gathered}$$

Since the complexity of a six-round cipher equals:

$$\begin{gathered} 120 = 2^6 + 2^5 + 2^4 + 2^3 \approx 2^6\, S_{RD}^{-1}\text{'s} \\ \text{and } 328 = 2^8 + 2^6 + 2^3 \approx 2^8\, \text{XORs,} \end{gathered}$$

the complexity of this attack corresponds roughly to 2^{71} six-round cipher executions.

3.7.2 *Further Cryptanaylsis*

There are many actual approaches to cryptanalize Rijndael and we will now give a short view on three of these. All three approaches take advantage of the specific structure of Rijndael but do not yield to an actual attack to cryptanalize it.

In [8] Niels Ferguson, John Kelsey, Stefan Lucks, Bruce Schneier, Mike Stay, David Wagner and Doug Whiting introduce the *partial sum technique*, which can be used to lower the attack requirements of the six-round saturation attack to 2^{46} cipher executions and to launch an attack over seven and eight rounds of Rijndael. The seven-round attack requires $2^{128} - 2^{119}$ chosen plaintexts and 2^{120} cipher executions. The eight-round attack requires the same amount of chosen plaintexts and 2^{204} cipher executions. Of course, these attacks are faster than the exhaustive key search attack, which requires 2^{N_C} cipher executions, but the number of cipher executions is still to great to be feasible.

In [9] Niels Ferguson, Richard Schroeppel and Doug Whiting show that the full fourteen-round Rijndael cipher, with $N_C = 256$, can be expressed as a single algebraic formula, which consists of 2^{70} terms. This is an interesting result but the authors are not aware of any technique to exploit this fact and launch an attack.

In [17] Josef Pieprzyk and Nicolas T. Courtois introduce the *XSL attack*, which can applied on all block ciphers with a XSL round structure. The round transformation of a XSL block cipher consists of a XOR with the roundkey (X), a non-linear substitution layer (S) and a linear diffusion layer (L). As we have seen, Rijndael fulfills this requirement. Firstly, the authors of [17] show that the cryptanalysis of Rijndael can be reduced to the problem of solving *multivariate quadratic equations*, called *MQ-equations*. In general this problem is NP-hard but its workload decreases if the number of equations exceeds the number of variables, see [21]. The authors show, that this can be achieved but T.T. Moh [15] and Dan Coppersmith [4] state that this fact cannot be used to launch an actual attack on Rijndael.

3.8 The Extended Euclidean Algorithm

We will now define the *Extended Euclidean Algorithm* in mainly same way as it is done in [14]. We will show that given two polynomials $m(x), a(x) \in F[x]$ the Extended Euclidean Algorithm determines uniquely $b(x), c(x) \in F[x]$, with $(m(x) \otimes c(x)) \oplus (a(x) \otimes b(x)) = gcd(m(x), a(x))$.

In Sect. 3.3.2 we have shown that $a(x) \odot b(x) = \mathbf{1}$.

In the last section we will show that $deg(b(x)) < deg(m(x)) = d$.

Hence, $b(x)$ is the multiplicative inverse of $a(x)$ under field multiplication $\odot$, defined in Sect. 3.3.2.

3.8.1 The Euclidean Algorithm

We will start with the definition of the *Euclidean domain*.

Definition 113 Let D be a set.
An **integral domain** is a triple $(D, +, \cdot)$ with the following properties:

- $(D, +, \cdot)$ is a Ring with 0 as the additive neutral element and 1 as the multiplicative neutral element
- If $ab = ac$ and $a \neq 0$, then $b = c$, for all $a, b, c \in D$

Definition 114 Let S be an ordered set.
An **Euclidean domain** $\mathcal{E}$ is an integral domain $(D, +, \cdot)$ together with a function $g : D \to S$, with the following properties:

- $g(a) \leq g(ab)$, if $b \neq 0$
- $\forall\, a, b \in D \backslash \{0\}$, there exist unique $q, r \in D$, such that $a = qb+r$, with $g(r) < g(b)$

Remark 31 $(F[x], +, \otimes)$ together with $g : F[x] \to \mathbb{N}$ is an Euclidean domain, where $F[x]$ is the set of polynomials over a field F, $+$ is the polynomial addition, $\otimes$ is the polynomial multiplication and $g(f(x)) := deg(f(x))$.

The following proposition is the inductive basis for the Euclidean Algorithm for polynomials.

Proposition 31 *For any elements* $m(x), a(x), q(x) \in F[x]$, *it holds that:*

$$gcd(m(x), a(x)) = gcd(a(x), m(x) - q(x) \otimes a(x)).$$

Proof Let $\mathcal{D}[x] \subset F[x]$ denote the set of common divisors of $m(x)$ and $a(x)$ and let $\mathcal{D}'[x] \subset F[x]$ denote the set of common divisors of $a(x)$ and $(m(x) - q(x) \otimes a(x))$.

If $d(x) \in \mathcal{D}[x]$
$\Rightarrow d(x)|(m(x) - q(x) \otimes a(x)) \Rightarrow d(x) \in \mathcal{D}'[x]$.

If $d(x) \in \mathcal{D}'[x]$
$\Rightarrow d|(m(x) = (m(x) - q(x) \otimes a(x)) + q(x) \otimes a(x)) \Rightarrow d(x) \in \mathcal{D}[x]$.

It follows that $\mathcal{D}[x] = \mathcal{D}'[x]$ and from this follows the proposition. □

The input of the Euclidean Algorithm are two polynomials $m(x), a(x) \in F[x]$, with $deg(m(x)) \geq deg(a(x))$, and its output is $gcd(m(x), a(x)) \in F[x]$, which is **1** if $m(x)$ is irreducible.

For a given input $m(x), a(x) \in F[x]$, it follows from Definition 114 that there exist unique $q_1(x), r_1(x) \in F[x]$, with:

$$m(x) = q_1(x) \otimes a(x) + r_1(x) \text{ and } deg(r_1(x)) < deg(a(x)).$$

And from Proposition 31 it follows that:

$$gcd(m(x), a(x)) = gcd(a(x), r_1(x)).$$

Again there exist unique $q_2(x), r_2(x) \in F[x]$ with:

$$a(x) = q_2(x) \otimes r_1(x) + r_2(x) \text{ and } deg(r_2(x)) < deg(r_1(x)),$$

and it follows that:

$$gcd(m(x), a(x)) = gcd(a(x), r_1(x)) = gcd(r_1(x), r_2(x)).$$

Since the sequence $(deg(r_k(x)))_k$ is strictly decreasing and $deg(r_k(x)) \in \mathbb{N}$, it follows that after a finite number $n \in \mathbb{N}$ of steps $deg(r_n(x)) = 0$.

A polynomial of degree zero divides any other polynomial.

If $r_n(x) \neq 0$, it follows that

$$gcd(m(x), a(x)) = \cdots = gcd(r_{n-1}(x), r_n(x)) = r_n(x)$$

and if $r_n(x) = 0$, it follows that

$$gcd(m(x), a(x)) = \cdots = gcd(r_{n-2}(x), r_{n-1}(x)) = r_{n-1}(x).$$

The Euclidean Algorithm for Polynomials

- input: $m(x), a(x) \in F[x]$
- set: $r_{-1}(x) := m(x)$, $r_0(x) := a(x)$ and $k := 1$
- while $r_k(x) \neq 0$ do:
  - $r_k(x) := r_{k-2}(x) - q_k(x) \otimes r_{k-1}(x)$
  - increase k by 1
- output: $r_{k-1}(x) = gcd(m(x), a(x)) \in F[x]$

3.8.2 The Extended Euclidean Algorithm

The extended version of the Euclidean Algorithm has the same input as the Euclidean Algorithm, two polynomials $m(x), a(x) \in F[x]$, but in addition to $gcd(m(x), a(x)) \in F[x]$, the output also includes two polynomials $b(x), c(x) \in F[x]$ with $m(x) \otimes c(x) + a(x) \otimes b(x) = gcd(m(x), a(x))$.

In the previous section we have seen that if we define $r_{-1}(x) := m(x)$ and $r_0(x) := a(x)$, there exist unique sequences $(q_k(x))_k$, $(r_k(x))_k$,
for $k \in \{1, 2, \ldots, n\}$, with:

$$r_{k-2} = q_k(x) \otimes r_{k-1}(x) + r_k(x) \text{ and } deg(r_k(x)) < deg(r_{k-1}(x)).$$

Definition 115 Let $c_k(x) \in F[x]$.
The sequence $(c_k(x))_k$ is defined via:

$$c_k(x) := \begin{cases} 1, & \text{if } k = -1 \\ 0, & \text{if } k = 0 \\ c_{k-2}(x) - q_k(x) \otimes c_{k-1}(x), & \text{if } k \geq 1. \end{cases}$$

Definition 116 Let $b_k(x) \in F[x]$.
The sequence $(b_k(x))_k$ is defined via:

$$b_k(x) := \begin{cases} 0, & \text{if } k = -1 \\ 1, & \text{if } k = 0 \\ b_{k-2}(x) - q_k(x) \otimes b_{k-1}(x), & \text{if } k \geq 1. \end{cases}$$

With these definitions we are able to prove the following proposition, which proves the correctness of the Extended Euclidean Algorithm.

Proposition 32 *The following property holds for all* $k \in \{-1, 0, 1, 2, \ldots, n\}$:

$$r_k(x) = c_k(x) \otimes m(x) + b_k(x) \otimes a(x)$$

Proof For $k = -1$, we have:

$$r_{-1}(x) = m(x), c_{-1}(x) = 1 \text{ and } b_{-1}(x) = 0 \Rightarrow m(x) = 1 \otimes m(x) + 0 \otimes a(x).$$

For $k = 0$, we have:

$$r_0(x) = a(x), c_0(x) = 0 \text{ and } b_0(x) = 1 \Rightarrow a(x) = 0 \otimes m(x) + 1 \otimes a(x).$$

If we now assume that the proposition is proved for $k - 2$ and $k - 1$, we have the following equations:

$$r_{k-2}(x) = c_{k-2}(x) \otimes m(x) + b_{k-2}(x) \otimes a(x) \tag{3.8.1}$$

$$r_{k-1}(x) = c_{k-1}(x) \otimes m(x) + b_{k-1}(x) \otimes a(x) \tag{3.8.2}$$

$$r_k(x) = r_{k-2}(x) - q_k(x) \otimes r_{k-1}(x) \tag{3.8.3}$$

If we insert the Eqs. (3.8.1) and (3.8.2) in Eq. (3.8.3), we obtain:

$$r_k(x) = c_k(x) \otimes m(x) + b_k(x) \otimes a(x).$$ □

The Extended Euclidean Algorithm for Polynomials

- input: $a(x), b(x) \in F[x]$
- set:

- $r_{-1}(x) := m(x)$, $r_0(x) := a(x)$
- $c_{-1}(x) := 1$, $s_0(x) := 0$
- $b_{-1}(x) := 0$, $t_0(x) := 1$
- $k := 1$

• while $r_k(x) \neq 0$ do:

- $r_k(x) := r_{k-2}(x) - q_k(x) \otimes r_{k-1}(x)$
- $c_k(x) := c_{k-2}(x) - q_k(x) \otimes c_{k-1}(x)$
- $b_k(x) := b_{k-2}(x) - q_k(x) \otimes b_{k-1}(x)$
- increase k by 1

• output: $r_{k-1}(x), s_{k-1}(x), t_{k-1}(x) \in F[x]$, where:

- $c_{k-1}(x) = c(x)$
- $b_{k-1}(x) = b(x)$
- $r_{k-1}(x) = gcd(m(x), a(x)) = c(x) \otimes m(x) + b(x) \otimes a(x)$

3.8.3 *Results*

For all the results in this section we assume that $deg(m(x)) > deg(a(x)) > 0$.

Lemma 22 *For all $k \in \{1, 2, \dots, n\}$ it holds that:*

$$deg(q_k(x)) = deg(r_{k-2}(x)) - deg(r_{k-1}(x)) > 0.$$

Proof From Definition 114 and the construction of the Euclidean Algorithm it follows for all $k \in \{1, 2, \dots, n\}$ that:

$$deg(r_{k-2}(x)) > deg(r_{k-1}(x)) > deg(r_k(x)) \geq 0$$

and:

$$\begin{aligned} deg(r_{k-2}(x)) &= deg(q_k(x) \otimes r_{k-1}(x) + r_k(x)) \\ &= deg(q_k(x) \otimes r_{k-1}(x)) \\ &= deg(q_k(x)) + deg(r_{k-1}(x)). \end{aligned}$$

It follows that:

$$deg(q_k(x)) = deg(r_{k-2}(x)) - deg(r_{k-1}(x)) > 0.$$

□

Lemma 23 *For all $k \in \{0, 1, \dots, n\}$ it holds that:*

$$deg(b_k(x)) \geq deg(b_{k-1}(x)).$$

Proof For $k = 0$ it follows that:

$$deg(b_0(x)) = deg(1) = 0 = deg(0) = deg(b_{-1}(x))$$

For $k = 1$ it follows that:

$$deg(b_1(x)) \geq 0 = deg(1) = deg(b_0(x))$$

We now assume that the lemma is proved for $k - 2$ and $k - 1$.

It follows from Definition 116 that we have the following equality:

$$deg(b_k(x)) = deg(b_{k-2}(x) - q_k(x) \otimes b_{k-1}(x))$$

Since $deg(b_{k-1}(x)) \geq deg(b_{k-2}(x))$ and $deg(q_k(x)) > 0$ (Lemma 22), it follows that:

$$\begin{aligned} deg(b_k(x)) &= deg(b_{k-2}(x) - q_k(x) \otimes b_{k-1}(x)) \\ &= deg(q_k(x) \otimes b_{k-1}(x)) \\ &= deg(q_k(x)) + deg(b_{k-1}(x)) \\ &> deg(b_{k-1}(x)). \end{aligned}$$ □

Proposition 33 *For all $k \in \{1, 2, \ldots, n\}$ it holds that:*

$$deg(b_k(x)) < deg(m(x))$$

Proof We will show that $deg(b_k(x)) = deg(m(x)) - deg(r_{k-1}(x))$.

Since $deg(r_{k-1}(x)) > deg(r_k(x)) \geq 0$, for all $k \in \{1, 2, \ldots, n\}$, this yields to $deg(b_k(x)) < deg(m(x))$.

For $k = 1$ it follows from $b_1(x) = -q_1(x)$ that:

$$deg(b_1(x)) = deg(q_1(x)) = deg(m(x)) - deg(a(x)) = deg(m(x)) - deg(r_0(x)).$$

We now assume that the proposition is proved for $k - 1$.

$$\begin{aligned} deg(b_k(x)) &= deg(b_{k-2}(x) - q_k(x) \otimes b_{k-1}(x)) \\ &= deg(q_k(x) \otimes b_{k-1}(x)) \\ &= deg(q_k(x)) + deg(b_{k-1}(x)) \\ &= deg(q_k(x)) + deg(m(x)) - deg(r_{k-2}(x)) \\ &= deg(r_{k-2}(x)) - deg(r_{k-1}(x)) + deg(m(x)) - deg(r_{k-2}(x)) \\ &= deg(m(x)) - deg(r_{k-1}(x)). \end{aligned}$$ □

References

1. R. Anderson, E. Biham, L. Knudsen, Serpent: a proposal for the advanced encryption standard, in *1st AES Conference* (1999)
2. E. Biham, A. Shamir, *Differential Cryptanalysis of the Data Encryption Standard* (Springer, New York, 1993)
3. C. Burwick, D. Coppersmith, E. D'Avignon, R. Gennaro, S. Halevi, C. Jutla, S.M. Matyas, L. O'Connor, M. Peyravian, D. Safford, N. Zunic, MARS a candidate cipher for AES, in *1st AES Conference* (1999)
4. D. Coppersmith, Re: impact of Courtois and Pieprzyk results, Entry at the AES discussion forum (2002). http://aes.nist.gov/aes/
5. J. Daemen, Cipher and hash function design strategies based on linear and differential cryptanalysis, Doctoral dissertation K.U. Leuven (1995)
6. J. Daemen, V. Rijmen, AES proposal: Rijndael, in *1st AES Conference* (1999)
7. J. Daemen, L. Knudsen, V. Rijmen, *The Block Cipher SQUARE, Fast Software Encryption'97* (Springer, New York, 1997)
8. N. Ferguson, J. Kelsey, S. Lucks, B. Schneier, M. Stay, D. Wagner, D. Whiting, *Improved Cryptanalysis of Rijndael, Fast Software Encryption 2000* (Springer, New York, 2001), pp. 213–231
9. N. Ferguson, R. Schroeppel, D. Whiting, *A Simple Algebraic Representation of Rijndael*, Lecture Notes in Computer Science (Springer, New York, 2001)
10. S.W. Golomb, *Shift Register Sequences* (Holden-Day Inc., San Francisco, 1967)
11. T. Jakobsen, L.R. Knudsen, *The Interpolation Attack on Block Ciphers, Fast Software Encryption'97* (Springer, New York, 1997), pp. 28–40
12. R. Lidl, H. Niederreiter, *Introduction to Finite Fields and Their Applications* (Cambridge University Press, Cambridge, 1986)
13. M. Matsui, Linear cryptanalysis method for DES cipher, *Advances in Cryptology, Proceedings of Eurocrypt'93* (Springer, New York, 1994), pp. 386–397
14. R.J. McEliece, *Finite Fields for Computer Scientists and Engineers* (Kluwer Academic Publishers, Boston, 1987), pp. 3–9
15. T.T. Moh, On the Courtois-Pieprzyk's attack on Rijndael (2002). http://www.usdsi.com/aes.html
16. K. Nyberg, Differentially uniform mappings for cryptography, *Advances in Cryptology, Proceedings of Eurocrypt'93* (Springer, New York, 1994), pp. 55–64
17. J. Pieprzyk, N.T. Courtois, Cryptanalysis of block ciphers with overdefined systems of equations, *Advances in Cryptology - ASIACRYPT 2002*, vol. 2501, Lecture Notes in Computer Science (Springer, New York, 2002), pp. 267–287
18. B. Preneel, Analysis and design of cryptographic hash functions, Doctoral dissertation K.U. Leuven (1993)
19. R.L. Rivest, M.J.B. Robshaw, R. Sidney, Y.L. Yin, The RC6 block cipher, *1st AES Conference* (1999)
20. B. Schneier, J. Kelsey, D. Whiting, D. Wagner, C. Hall, N. Ferguson, Twofish: a 128-bit block cipher, *1st AES Conference* (1999)
21. A. Shamir, A. Kipnis, Cryptanalysis of the HFE public key cryptosystem, in *Proceedings of Crypto'99* (Springer, New York, 1999)

Chapter 4
Elliptic Curve Cryptosystems

In the last 15 years much research has been done concerning practical applications of elliptic curves like integer factorization [46], primality proving [3], algebraic geometry codes [89] and public-key cryptosystems [36, 58].

In this section we shall discuss the mathematical background of elliptic curve public-key schemes up to the first implementation ideas. We will restrict ourselves to public-key cryptosystems and digital signature schemes since almost all of these schemes can be extended to other areas of public-key cryptology.

Starting with a short introduction into the history of public-key cryptology and the presentation of the RSA and ElGamal cryptosystems we give in Sect. 4.1 a short survey how to solve the underlying problems of integer factorization and finding the discrete logarithm in a cyclic group. In the next chapter we shall discuss the theory of elliptic curves giving necessary definitions and theorems for the rest of this paper. The main interest will be taken into the additive (pseudo-) group of rational points of an elliptic curve defined over the finite field $\mathbb{F}_q$ (or the ring $\mathbb{Z}_n$). In Sect. 4.3 some algorithms and techniques are developed for efficient m-fold addition of rational points and even finding points on a given curve. Afterwards we will be able to present two rather different types of elliptic curve public-key cryptosystems.

At first we present several cryptoschemes based on integer factorization in Sect. 4.4. Beside discussing possible attacks referring to the recent research, we present the elliptic curve method for integer factorization.

Secondly we shall discuss elliptic curve cryptosystems based on the discrete logarithm problem in the group of rational points in Sect. 4.5. Again we shall present several possible attacks and elaborate necessary conditions for cryptographically good elliptic curves, which are curves where the discrete logarithm gets computational infeasible. Since it will be shown that these cryptosystems have a great advantage over other publicly known public-key schemes nowadays, we will spend much time in the discussion of the mentioned discrete logarithm. The question how to construct such curves will also be answered afterwards.

A. Ahlswede et al. (eds.), *Hiding Data – Selected Topics*,
Foundations in Signal Processing, Communications
and Networking 12, DOI 10.1007/978-3-319-31515-7_4

For a short summary of the connection between the related areas we refer to the diagram on the next page. Although necessary and further references to literature are given the author tried to write a self-containing paper as far as possible.

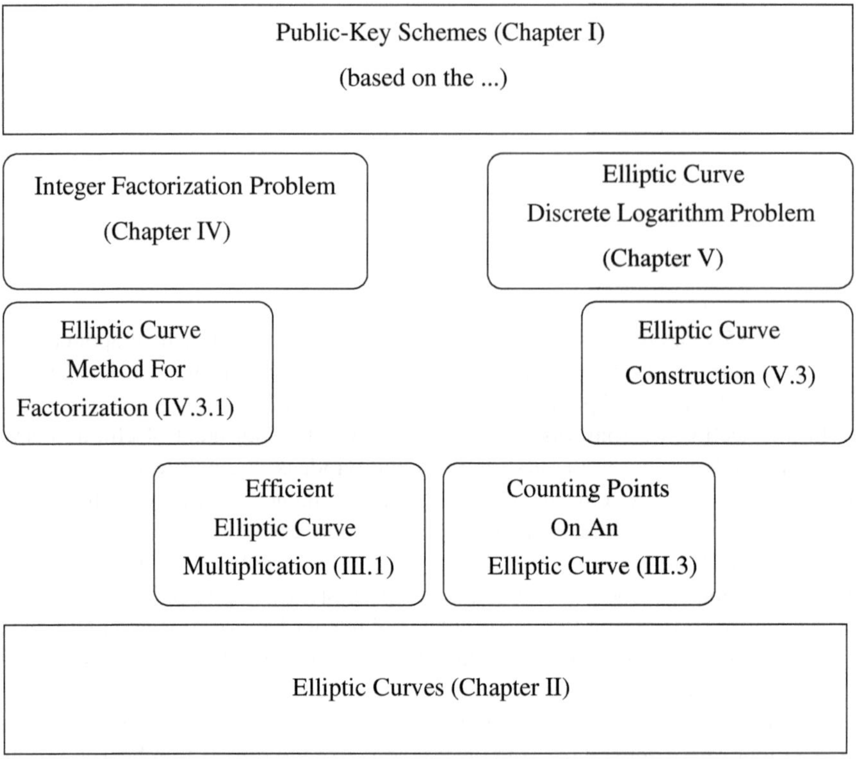

4.1 Cryptography

4.1.1 Secret-Key Cryptography

The first purpose of cryptography is to achieve privacy, i.e. to assure that two persons Alice and Bob, denoted A and B respectively, are able to transmit a message over an insecure channel, such that only the recipient is able to read this message. This was generally done by secret-key cryptography.

We shall denote by $\mathcal{M}$ the set of all possible plaintext messages, by $\mathcal{C}$ the set of all possible ciphertext messages and by $\mathcal{K}$ the set of all possible keys.

Then a secret-key cryptosystem consists of a family of pairs of functions

$$c_j : \mathcal{M} \longrightarrow \mathcal{C}, \qquad d_j : \mathcal{C} \longrightarrow \mathcal{M}, \quad j \in \mathcal{K}$$

such that

$$d_j(c_j(m)) = m, \quad \text{for all } m \in \mathcal{M}, j \in \mathcal{K}.$$

The first step to use a secret-key system is the agreement upon a secret-key $j \in \mathcal{K}$ for both persons A and B. This has to be done over a secure channel, e.g. by a personal meeting or a believed courier. Later A can send the message $m \in \mathcal{M}$ by using the encryption method $\tilde{m} = c_j(m)$ and sending $\tilde{m}$ to B. B afterwards can decrypt $m = d_j(\tilde{m})$. It's easy to see that the properties of the functions c_j and d_j are very important and that the cryptosystem fails, if an eavesdropper, denoted E, is able to get m or j given $\tilde{m}$ and all about the cryptosystem.

Although messages have been encrypted with secret keys already in ancient times, the mathematical foundations of cryptology and especially secret-key cryptography are due to Shannon (1949) [81]. For a survey on the history of cryptography until 1945 see Kahn [34]. Shannon demonstrated that the one-time pad, i.e. a cryptosystem, where keys are random binary strings which are exclusive-ored by the message to obtain the encrypted message, is perfect, i.e. the random variables of the plaintext and the cryptogram are independent. It follows that E is not able to gain knowledge about the plaintext, even with infinite computer resource.

The Data Encryption Standard (DES) is the most widely used secret-key cryptosystem today, although the keylength of 56 bits is to short to obtain secure encryption. In June 1998, the distributed.net team won the RSA Lab's DES-III 56-bit challenge by the brute force method, i.e. testing every key $j \in \mathcal{K} = \{1, \ldots, 2^{56}\}$, in less than 24 h. So further improvements were made to achieve TripleDES with a keylength of 128 bits.

For further secret-key cryptosystems like RC4 and IDEA (which are often used in Internet applications like SSL) see [73] and even for a good mathematical background and reference section [56].

Although secret-key cryptography has the advantage to be extremely fast (over 1 GBit/s), it has the following deficiencies, which make it unsuitable for use in certain applications:

(i) **Key Distribution Problem**: Two users have to select a key j before they can communicate over an insecure channel. This is a real problem if a secure channel for selecting a key may not be available like in the Internet (all transmitted data can be observed by E).

(ii) **Key Management Problem**: When n users want to communicate in a network every pair of users must share a secret-key for a total of $n(n-1)/2 = O(n^2)$ keys. In the Internet for instance, n is about 1.47×10^8 in September 1998. Thus there are about 10^{16} keypairs needed.

(iii) **No Digital Signature**: As a digital analogy of a hand-written signature a digital signature is needed to do for example banking or merchandising. An important property of a digital signature would be the ability to convince any third party that the message in fact originated from the sender. In a secret-key cryptosystem B cannot convince a third party that a message received from A in fact originated from A, since A and B have the same capabilities for encryption and decryption.

Especially for military purposes, where many secret communication is used, the disadvantages of secret-key cryptography above was a great problem. Already in 1944 an unknown author at Bell Labs [22] had the genius idea for secure telephone speech without distributing a secret-key. He suggested that the recipient should mask the senders speech by adding noise to the line. Afterwards the recipient could subtract the noise and would get the original message. Although the system was not used in practice there is a new idea of encryption: no common secret-key is needed for both parties. But the recipient has to take part in the encipherment now.

In 1997 a cryptographer employed at Bell Labs got a copy of a memorandum [65] from the desk of John F. Kennedy about the problem of securing nuclear weapons with launch codes. Steve Bellowin [65] claims that after asking the question if authentication is possible already before 1970 the NSA was able to produce digital signatures. Since all reports are classified by now it is not possible to verify that the US military used public-key cryptography before 1976 as follows in the rest of this chapter.

4.1.2 Public-Key Cryptography

In their paper "New directions in cryptography" Diffie and Hellman (1976) [17] introduced the first publicly known public-key protocol based on the discrete logarithm in a finite field. In public-key cryptography communication over a secure channel is no longer necessary (cf. Sect. 4.1.1(i)). Two persons A and B can calculate one common secret key $j \in \mathcal{K}$ from private and publicly known informations. This common secret key can then be used for a secret-key cryptosystem such as DES.

In order to obtain a public-key cryptosystem it is important to assume that the eavesdropper has got unlimited computational power. It was already pointed out by Shannon in his pioneering paper [81] that complexity of encoding and decoding might be considered. However, Diffie and Hellman introduced the concept of a one-way function, i.e. a function, which is easy to evaluate but hard to invert, defined in Sect. 4.1.3. So using a one-way function $c : \mathcal{M} \longrightarrow \mathcal{C}$ as a key j, the encoding, i.e. the evaluation of $\tilde{m} = c(m)$, can be done rather fast, but in order to decrypt the transmitted message the eavesdropper has to apply the inverse function c^{-1} to recover the original message $m = c^{-1}(\tilde{m}) = c^{-1}(c(m))$, which is a task of much higher complexity and can not be done in reasonable time.

Notice that already Ellis, a mathematician at the British Government Communications Headquarter, gave an existence theorem [20] for public-key encryption in 1970. In 1974 Williamson, a colleague of Ellis, published a practical implementation using finite rings [93]. In 1976 he proposed an easier scheme [94]. The Diffie and Hellman scheme differs from this scheme only in the fact that Diffie and Hellman used a finite field and not only a ring. This papers were classified up to Dec. 1997 by the British government and refering to the GCHQ webside more documents concerning the contribution of government research to public-key cryptography are on the way for publishing.

We shall now present the key-exchange protocol of Diffie and Hellman already in such a form that it is clear how it will work in a multiuser system, e.g. the Internet (cf. Sect. 4.1.1(ii)).

Diffie–Hellman Key Exchange Scheme

(i) **(Setup)** Select a finite group $GF(p)$, p a large prime, and a primitive element $\alpha \in GF(p)$. The order of α is known to be $p-1$. Every person i chooses a random private key $a_i \in \{1, 2, \dots, p-1\}$, computes $b_i = \alpha^{a_i}$ and stores b_i in a public directory.

(ii) **(Communication)** If persons i and j want to communicate, they calculate their common key

$$c = k_{ij} = b_i^{a_j} = (\alpha^{a_i})^{a_j} = (\alpha^{a_j})^{a_i} = b_j^{a_i} = k_{ji} = c$$

and encrypt/decrypt their messages using this common key.

(iii) **(Cryptanalysis)** In order to break the key c one has to know one of the numbers

$$a_i = \log_\alpha b_i, \quad a_j = \log_\alpha b_j, \tag{4.1.1}$$

where $\log_\alpha$ is the discrete logarithm to the base α in $GF(p)$, to obtain

$$c = k_{ij} = \alpha^{a_i a_j}.$$

For a definition of the discrete logarithm see Definition 120.

Observe in the communication section (ii) that there is only one key c between persons i and j. Public-key cryptography thus overcomes the key distribution and management problems inherent with secret-key cryptosystems (cf. Sect. 4.1.1(i), (ii)).

For encryption, the communication partners could for instance split their messages into blocks of length $\lceil \log_2 p \rceil$ and add $c = k_{ij} = k_{ji}$ to each of these blocks. If p is large enough, a third person will not be able to decipher the text. Additionally, every user in the network has the necessary information to calculate k_{ij}. Since p and α are publicly known, every user can deduce a_i and a_j from b_i and b_j, since $a_i \longrightarrow \alpha^{a_i}$ is bijective. However, in order to obtain a_i or a_j, a third person has to apply the discrete logarithm $\log_\alpha b_i$ or $\log_\alpha b_j$, which is a computationally hard task. Algorithms to solve this problem are presented in Sect. 4.1.5. The best known algorithm for arbitrary groups takes $O(\sqrt{p})$ steps (cf. subsection "Square Root Methods" of Sect. 4.1.5). In contrast, persons i and j have to exponentiate in order to obtain $c = k_{ij}$. This can be done in $O(\log p)$ steps using the so-called repeated squaring method proposed in Sect. 4.1.3.

By now it is not known if there is even another way of finding $c = k_{ij}$ given $\alpha, \alpha^{a_i}, \alpha^{a_j}$, denoted as the Diffie–Hellman problem.

4.1.3 Trapdoor One-Way Functions

To understand public-key cryptography we need to define a trapdoor one-way function.

Definition 117 A **one-way function** $f : \mathcal{M} \longrightarrow \mathcal{C}$ is an invertible function such that for all $m \in \mathcal{M}$ it is possible to evaluate $f(m)$ in polynomial time, while for most $\tilde{m} \in \mathcal{C}$ one requires an exponential time calculation to obtain $f^{-1}(\tilde{m})$, even when probabilistic algorithms are allowed.

In practice we can translate exponential time into "computationally infeasible", which means infeasible using the best known algorithms and best available computational power. For a survey over polynomial and exponential time see [38].

By now it is not known whether one-way functions exist, although there are several candidates like $f(x) = \alpha^x$ in $GF(p)$. This one was conjectured by Diffie and Hellman and made precise by Hellman and Pohlig [30], who found, that additionally $p - 1$ must have a large prime factor.

Definition 118 A one-way function $f : \mathcal{M} \longrightarrow \mathcal{C}$ is said to be a **trapdoor one-way function**, denoted **TOF**, if there is some extra information t, called **trapdoor**, with which f can be efficiently inverted, i.e. inverted in polynomial time.

To construct a public-key cryptosystem, we need a family

$$f_j : \mathcal{M} \longrightarrow \mathcal{C}, \quad j \in \mathcal{K}$$

of TOFs with the following properties:

(i) For each $j \in \mathcal{K}$ the trapdoor $t(j)$ is easy to obtain.
(ii) It is possible to describe a fast algorithm for computing f_j, such that it is infeasible to recover j (and further $t(j)$) from this description.

When such a family of trapdoor one-way functions exist, we can set up a public-key cryptosystem.

Let G be a multiplicatively written finite group of order n. Assume that the group operation is easy to compute, i.e. an efficient (polynomial time) algorithm is known for computing $\alpha \cdot \beta$ for all $\alpha, \beta \in G$.

We will first present the repeated squaring method (also called square-and-multiply method), a method which computes the nth power of a given number $\alpha \in G$ in $O(\log_2 n)$ steps.

Let

$$n = \sum_{i=0}^{t} a_i\, 2^i, \quad a_i \in \{0,\ 1\},\ t = \lfloor \log_2 n \rfloor$$

be the binary representation of n. Then

$$x^n = x^{a_0 + a_1 2 + \cdots + a_t 2^t} = x^{a_0} \cdot (x^2)^{a_1} \cdot (x^4)^{a_2} \cdot \cdots \cdot (x^{2^t})^{a_t}.$$

With this product representation we get the following algorithm:

Repeated squaring method
Require: $x \in G$ and $n = \sum_{i=0}^{t} a_i 2^i, \quad a_i \in \{0, 1\}$
  $s \Leftarrow 1$ {identity element}
  for $i = 0$ to t **do**
    if $a_i = 1$ **then**
      $s \Leftarrow s \cdot x$
    end if
    $x \Leftarrow x \cdot x$ {squaring}
  end for
Ensure: $s = x^n$

We compute $x, x^2, x^4, \ldots, x^{2^t}$ by repeated squaring (totally $t = \lfloor \log_2 n \rfloor$ multiplications). Further, after each squaring, we look if the coefficient a_i is 0 or 1. If $a_i = 0$ then x^{2^i} does not contribute to the product, if $a_i = 1$ then x^{2^i} occurs as a factor in the product $x^n = \prod_{i=1,a_i=1}^{t} x^{2^i}$. So, to obtain x^n as a product of the squares $(x^{2^i})_{i=1}^{t}$ we need at most $t = \lfloor \log_2 n \rfloor$ multiplications, so that the number of group operations is smaller than $2\lfloor \log_2 n \rfloor$.

Now we want to explore two different public-key systems, RSA and El Gamal, using different trapdoor one-way functions based on finite groups.

RSA Cryptosystem

The Rivest–Shamir–Adleman (RSA) cryptosystem was invented 1977 [70] as the first realization of the Diffie and Hellman public-key model. The RSA cryptosystem is the most widely used public-key cryptosystem today. However C. Cocks, a further colleague of Ellis, proposed already in 1973 [12] a public-key cryptoscheme, which is nearly the same than the RSA scheme. He directly followed the existence prove of Ellis for construction. But this paper was also classified up to Dec. 1997.

Let p and q be two "big" primes and $n = pq$.

We know that the group $G = \mathbb{Z}_n^* := \{x \in \mathbb{Z}_n : \exists y \in \mathbb{Z}_n \text{ such that } x \cdot y = 1\}$ has these two properties:

(i) **Efficiency**: It exists an efficient algorithm for multiplying group elements $\alpha, \beta \in G$.
(ii) **Security**: Evaluating the order $\Phi(n) = (p-1)(q-1)$ of the group is infeasible without a specific trapdoor information, e.g. a prime p or q.

Thus the group order $\Phi(n)$ seems to be a TOF. (Φ denotes the Euler phi-function).

RSA-Cryptosystem

(i) **(Setup)** Each person i selects two "large" prime numbers p and q and forms the product $n = pq$.
Further, each person selects at random a "large" number d, such that $\gcd(d, (p-1)(q-1)) = 1$, and then computes its multiplicative inverse e, hence $e \cdot d \equiv 1$

(mod $(p-1)(q-1)$). Then each person i stores (e, n) in a public and d in a private directory.

(ii) **(Communication)** If j wants to submit a message m to person i, he encrypts it using the encoding function

$$E_i(m) = m^e \mod n =: c.$$

Person i can easily decrypt c by application of the decoding function

$$D_j(c) = c^d \mod n = (m^e)^d \mod n = m^{ed} \mod n = m \mod n.$$

(iii) **(Cryptanalysis)** The security of the RSA cryptosystem is based on factorization, but there are also attacks, e.g. the Hastad attack below, which exploits the scheme and recovers at least parts of the plaintext without factoring n.

The problem of computing $\Phi(n)$, given only n, is computationally equivalent to the problem of factoring n. Even no efficient algorithm is known for taking the eth root in $\mathbb{Z}_n^*$ without the knowledge of p and q. By now it has not been shown that factoring is really a "hard" problem. The only thing we know is, that up to now the fastest known algorithms for integer factorization are much slower than the best prime number tests, respectively.

For an introduction into integer factoring algorithms, see Sect. 4.1.6, and especially for elliptic curve factoring, which was initially proposed by Lenstra [46], see subsection "Elliptic Curve Method" of Sect. 4.4.3.

We will remark that even if the factorization of n is unknown there is a possibility to get information about the plaintext only using the scheme itself. We will give a short example for an attack on RSA.

Hastad Attack

Assume now that $(n_i, e)_i$ are k different RSA public keys, $\{n_i\}_i$ relatively prime. Then j could encrypt a message m with the k public keys and send $E_i(m)$, $i = 1, \dots, k$. Note that $E_i(m)$ is a polynomial of degree e in m.

Theorem 63 (Hastad, [29]) *Let $n = \min\{n_i\}$. Given a set of equations*

$$\sum_{j=0}^{h} a_{ij} m^j = 0 \bmod n_i,\ i = 1, \dots, k,$$

where $m < n$ and $\gcd((a_{ij})_{j=0}^h, n_i) = 1$ for all i. Then the message m can be recovered in polynomial time in e, k and $\log n_i$ if

$$\prod_{i=1}^{k} n_i > n^{h(h+1)/2}(k+h+1)^{(k+h+1)/2} 2^{(k+h+1)^2/2}(h+1)^{(h+1)}.$$

Hastad proved that this theorem holds for small e in RSA.

El Gamal Cryptosystem

Let G be a finite group of order n and assume that the discrete logarithm problem in G defined in Sect. 4.1.5 is intractable. The following public-key scheme based on discrete exponentiation, which exploits the properties of a TOF, was proposed 1985 by T. El Gamal [19].

El Gamal Cryptosystem

(i) **(Setup)** Select a finite group G and an element $\alpha \in G$.
Each user i chooses a random integer l_i as his private key and α^{l_i} as his public key.

(ii) **(Communication)** i wishes to send to user j a message $m \in G$:

(enc) i generates a random integer k and evaluates α^k.
i gets j's public key α^{l_j} and computes $(\alpha^{l_j})^k$ and $m\alpha^{l_j k}$.
i sends j the pair $(\alpha^k, m\alpha^{l_j k})$.

(dec) j computes $(\alpha^k)^{l_j}$, evaluates the inverse $(\alpha^{kl_j})^{-1}$ and gets
$m = (m\alpha^{l_j k})(\alpha^{kl_j})^{-1}$.

(iii) **(Cryptanalysis)** The security of the El Gamal cryptosystem and the Diffie–Hellman key exchange as in Sect. 4.1.2 are equivalent, this means that the security of the El Gamal protocol is also based on the discrete logarithm problem.

It is understood that for a secure and efficient implementation two conditions should hold:

(i) **Efficiency**: the group operation in G should be easy/fast to apply.
(ii) **Security**: the discrete logarithm problem (see Sect. 4.1.5) in the cyclic subgroup of G generated by α should be hard.

El Gamal used in his original paper the multiplicative group of a finite field $\mathbb{Z}_p$.

Beside this, there have been other finite groups considered to be used in the El Gamal cryptosystem like the multiplicative group of a finite field $\mathbb{F}_{2^k}$ or the Jacobian of an hyperelliptic curve defined over a finite field (introduced by N. Koblitz, 1989 [37]).

In this section we will especially mention the use of the group of points on an elliptic curve over a finite field, which was introduced independently by V. Miller [58] and N. Koblitz [36] in 1985.

El Gamal even designed a digital signature scheme which makes use of the group G. In spite of presenting this scheme, which can briefly be found in [51], we will introduce the NIST Digital Signature Standard in the next section.

4.1.4 Digital Signature Standard (DSS)

In 1991 the U.S. government's National Institute of Standards and Technology (NIST) proposed a Digital Signature Algorithm (DSA).

The role of DSA is expected to be analogous to the Data Encryption Standard (DES): it is supposed to provide a standard digital signature method for use by government and commercial organizations. But we already know while DES is a classical ("secret-key") cryptosystem, in order to construct digital signatures it is necessary to use public-key cryptosystems.

The DSA is posed on the discrete logarithm problem (Definition 120) in a prime finite field $\mathbb{F}_p$. The DSA is very similar to a signature scheme that was originally proposed in [74] by Schnorr (1990) and also to the El Gamal signature scheme [19].

At first we have to define a special function H:

Definition 119 The function $H : \mathcal{M} \longrightarrow \mathbb{Z}$ defines a **hash function**, if $H(x)$ is easy to compute for any x, but no one can feasibly find two different values of x that give the same $H(y)$ (so-called "collision resistant") and given $y \in H(\mathcal{M})$, no one can feasibly find an x such that $H(x) = y$ (so-called "preimage resistant").

DSA Algorithm

(i) **(Setup)** Each user chooses

(a) a prime q of about 160 bit using a random number generator and a primality test,
(b) a prime p such that $p \equiv 1 \pmod q$ of about 500 bits,
(c) a generator g of the unique cyclic subgroup of $\mathbb{F}_p^*$ of order q,
(d) a random integer x, $0 \leq x \leq q$, as a private key and
(e) $y = g^x \mod q$ as a public key

(ii) **(Signing)** i wants to sign a message m:

(a) i applies a hash function H to m to obtain $H(m), 0 < H(m) < q$.
(b) i picks a random integer $k, 0 < k < q$.
(c) i computes $r = (g^k \mod p) \mod q$.
(d) i finds an integer s such that

$$sk \equiv H(m) + xr \pmod q. \tag{4.1.2}$$

(e) i's signature of m is $(r, s) \pmod q$.

(iii) **(Verifying)** j wishes to verify the signature (r, s) of a message m from i:

(a) j computes $u_1 = s^{-1}H(m) \mod q, \quad u_2 = s^{-1}r \mod q$.
(b) j evaluates $v = g^{u_1}y^{u_2} \mod p$.
(c) j verifies if $v = r \mod q$.

To prove the correctness of j's verification observe that by (4.1.2)

$$\begin{aligned} k &\equiv s^{-1}H(m) + s^{-1}xr \quad &(\text{mod } q) \\ &\equiv u_1 + u_2 x \quad &(\text{mod } q) \end{aligned}$$

and raising g to the powers $u_1 + xu_2$ gives

$$\begin{aligned} &(g^k \equiv g^{u_1} g^{xu_2} \mod p) \quad (\text{mod } q) \\ \Longleftrightarrow \\ &r \equiv (g^{u_1} y^{u_2} \mod p) \quad (\text{mod } q) \end{aligned}$$

This signature scheme has the advantage that signatures are fairly short, consisting of two numbers of 160 bits (the magnitude of q). On the other hand, the security of the system seems to depend upon intractability of the discrete logarithm problem in the multiplicative group of the rather large field $\mathbb{F}_p$ ($p \sim 2^{500}$). Although to break the system it would suffice to find discrete logarithms in the smaller subfield generated by g, in practice this does not seem to be easier than finding arbitrary discrete logarithms in $\mathbb{F}_p^*$. Thus the DSA seems to have attained a fairly high level of security without sacrificing small signature storage and implementation time.

There are further important topics in public-key cryptography we will not discuss here. For more information on the following items see for instance [38]:

Coin-flip: needed, if, for example, two game players in different cities want to determine by e-mail who starts.

Secret sharing: needed, if some secret information must be available to k subordinates working together but not to $k-1$ of them.

Zero knowledge proof: needed, if we want to convince someone that we have successfully solved a problem, e.g. factoring a 1000-bit number, without conveying any knowledge of the solution.

4.1.5 Discrete Logarithms in Finite Groups

This section is based partly on Chap. 6 of [54]. At first we will give a precise definition for the discrete logarithm problem.

Definition 120 (*DLP*) The **discrete logarithm problem** in a cyclic finite group G to the base $\alpha \in G$ is the following problem:
If $\beta \in G$ is given, find an integer l such that $\beta = \alpha^l$, provided that such l exists. Hence, $l = \log_\alpha \beta$ is the discrete logarithm of β to the base α.

Remark 32 If the group operation in G is written additively, then find l such that $\beta = l\alpha$. This will be the case for a kind of elliptic curve cryptosystems, cf. Definition 148.

For the development of cryptosystems it is interesting to measure the time spend to solve the DLP. So we will give a definition to measure the complexity of the presented algorithms.

Definition 121 Let

$$L_n(\gamma, c) := O(e^{c((\ln n)^\gamma (\ln \ln n)^{1-\gamma})}),$$

where n is the size of the input space, $0 \leq \gamma \leq 1$ and c is a constant.

We get a polynomial algorithm in the input size $\ln n$, if $\gamma = 0$, while we get an fully exponential algorithm in $\ln n$, if $\gamma = 1$. If $0 < \gamma < 1$, then $L_n(\gamma, c)$ is said to be subexponential. For further information on this topic, see [38, Chap. 2].

In order to solve the DLP there are mainly four methods, which we present in the next four subsections.

Square Root Methods

In this subsection we present methods for computing logarithms in arbitrary cyclic groups. Let $m = \lceil \sqrt{\#G} \rceil$.

Baby-Step Giant-Step Method

```
1: for i = 0 to m − 1 do
2:     compute (i, α^i)          % i = log_α α^i
3: end for
4: sort (i, α^i) by the second component
5: for j = 0 to m − 1 do
6:     γ ⇐ βα^(−jm)
7:     binary search i such that (i, γ) = (i, α^i)
8:     if i exists then
9:         print(jm); STOP {DLP solved}
10:    end if
11: end for
```

In step 1–3 we compute a table with m entries consisting of an integer i and the corresponding value α^i. Note that $i = \log_\alpha \alpha^i$. In step 4 we sort this table (i, α^i) by the second component. Now we search i such that $\alpha^i = \beta\alpha^{-jm}$ for $k \in \mathbb{Z}_m$. If the search succeeds we get $\beta = \alpha^{i+jm}$ and thus $\log_\alpha \beta = i + jm$. To sort the list of $O(m)$ entries and search it for each value of j requires $O(m \log m)$ operations.

Pollard ρ-Method

In [68] Pollard proposed a method, so-called Pollard ρ-method, to find logarithms in probabilistic polynomial time with expected running time $O(m)$, removing the necessity of precomputing a list of logarithms. This is done by dividing G into three sets and defining a sequence of group elements, which implicitly defines two further sequences. Using some facts about these three sequences, one can minimize the number of logarithms to test. This method can be easily implemented for massive parallel computing, which can decrease the total running time.

All square root methods get infeasible, if the order of G is large enough.

The Silver–Pohlig–Hellman Method

This type of method also works in arbitrary groups, but exploits the subgroup structure.

Let G be an additive group of order $N = \#G = \prod_{i=1}^{t} {p_i}^{\lambda_i}$, p_i prime, $\lambda_i \in \mathbb{N}$, for each i, $1 \leq i \leq t$. Let $\mathcal{O}$ denote the identity element of the additive group (cf. ECDLP Definition 148). Then we can apply the method of Silver–Pohlig–Hellman [30, 80] as follows:

At first we find the exact order n of α, i.e. we find the smallest $n = \prod_{i=1}^{t} {p_i}^{r_i}, r_i \in \mathbb{N}$, such that $n\alpha = \mathcal{O}$. This can be done by computing $(N/p_i)\alpha$ for all i, $1 \leq i \leq t$, and then $(N/p_i^2)\alpha$, whenever $(N/p_i)\alpha = \mathcal{O}$, and so one, until n is found.

If a positive integer $l \leq n-1$ exists such that $l\alpha = \beta$ then we can find it by determine $l \pmod{{p_i}^{r_i}}$ for each i, $1 \leq i \leq t$, and then using the Chinese Remainder Theorem to compute $l \pmod{n}$.

So suppose p is a fixed prime divisor of n, and

$$l \equiv \sum_{i=0}^{r-1} l_i p^i \pmod{p^r},$$

where $0 \leq l_i \leq p-1$. In order to find l_0 let $\gamma = \alpha(n/p)$. Then

$$\beta(n/p) = \alpha l(n/p) = \gamma l = \gamma l_0.$$

Now we can determine the logarithm of γl_0 to the base γ in the cyclic group of order p in G (note the additive structure of G, and that $p\gamma = n\alpha = \mathcal{O}$) using one of the methods in subsection "Square Root Methods" of Sect. 4.1.5.

Once we know l_0, we find l_1 considering the equalities

$$\begin{aligned} &\tfrac{n}{p^2}\beta = (l_0 + l_1 p)\tfrac{n}{p^2}\alpha \\ \Longleftrightarrow\quad & l_1 \tfrac{N}{p}\alpha = \tfrac{N}{p^2}(\beta - l_0\alpha) \\ \Longleftrightarrow\quad & \gamma l_1 = \tfrac{N}{p^2}(\beta - l_0\alpha). \end{aligned}$$

Hence, we can again use one of the methods of subsection "Square Root Methods" of Sect. 4.1.5 in order to obtain l_1. We continue this process inductively finding $l_2, l_3, \ldots, l_{r-1}$. The running time is given by $O(\sum_{i=1}^{t} \lambda_i(\log_2 n + \sqrt{p_i}\log_2 p_i))$ group operations.

Instead of an additive group we could also apply this method to a multiplicative group, e.g. the group $\mathbb{F}_p^*$.

Hence, whenever the security of a cryptosystem is based on the DLP in a group G, it is important to select a group with the property that $\#G$ is divisible by some suitable large prime factor.

The Index-Calculus Method

The most powerful method for computing logarithms in a group is commonly refered to the index-calculus method. Basic ideas are from [92]. This method can not be applied to arbitrary groups. For a generic description of the method, see [54, 6.6]. Adleman [1] described the method for the group $\mathbb{F}_p^*$ and analysed the complexity of the algorithm. Most algorithms have running times about $L_p(1/2, c)$, c a constant. Recently, D. Gordon [25, 26] used the number field sieve (cf. subsection "The Pollard $p-1$ Method" of Sect. 4.1.6) to obtain the heuristic asymptotic running time $L_p(1/3, 3^{2/3})$ (at least in the case where p is a prime). For references on the implementation over fields $\mathbb{F}_q$ with special q, see [54, Sect. 6.7].

Isomorphisms Between Groups

Even though any two groups of order N are isomorphic, an efficient algorithm to compute logarithms on one does not necessarily imply an efficient algorithm for the other. Let G be a cyclic group. Hence $G \cong (\mathbb{Z}_n, +)$ and logarithms in $(\mathbb{Z}_n, +)$ can easily be computed by the extended Euclidean algorithm. So Definition 120 can be restated as:
Find a (computationally efficient) algorithm for computing an isomorphism from a cyclic group of order n to $(\mathbb{Z}_n, +)$.

This technique will be used in Sect. 4.5.2 in order to reduce the elliptic curve discrete logarithm problem Definition 148 to a less difficult discrete logarithm problem or even to solve the problem completely.

4.1.6 Factorization of Composite Numbers

As mentioned in the subsection "RSA Cryptosystem" of Sect. 4.1.3 the security of the RSA cryptosystem is based on the difficulty to factor a composite number $n = pq$, p and q large primes. Since there exist even elliptic curve cryptosystems with the security based on factorization in Sect. 4.4, we will present several factorization methods. One method would be trial division, which is computationally hard for great primes p and q, because one has to check all primes less than $\sqrt{n}$ in the worst case $(L_n(1, c))$. There are better factorization methods, we will discuss next.

The Pollard ρ-Method

In 1975 Pollard [67] proposed the Pollard ρ-method (even called Monte Carlo method): First we choose an easily evaluatable map $f : \mathbb{Z}_n \to \mathbb{Z}_n$, and some particular value $x = x_0$, e.g. $x_0 = 1$ or x_0 a random integer. Next we compute the successive iterates of f:

$$x_{j+1} = f(x_j), \quad j = 0, 1, 2, \dots, l.$$

Then we make comparisons between different x_j's, hoping to find two which are in different residue classes modulo n but in the same residue class modulo some divisor of n. Once we find such x_j, x_k, we have found $\gcd(x_j - x_k, n) | n$.

Example 9 Let us factor 91 by selecting $f(x) = x^2 + 1$, $x_0 = 1$. Hence, $x_1 = 2$, $x_2 = 5$, $x_3 = 26$, etc. Computing $\gcd(x_3 - x_2, n) = \gcd(21, 91)$ yields the nontrivial factor 7.

The Pollard $p - 1$ Method

Assume n is a composite number and p a divisor of n. Then we can use the following algorithm proposed by Pollard [67]. Let $B_1 \in \mathbb{N}$ be an upper bound for the factor of n to search for.

Pollard $p - 1$ method
Require: $n > 1$ and $1 < B_1 < n$

```
1: Let k = lcd(2, 3, ..., B1)
2: for l = 1 to maxLoops do
3:     Choose a random a ∈ {2, 3, ..., n − 2}.
4:     Compute b = a^k mod n using repeated squaring Section (4.1.5)
5:     Evaluate d = gcd(b − 1, n) using the Euclidean algorithm.   Ensure: d
6:     if d ≠ 1 and d ≠ n then
7:         Output d; STOP {nontrivial divisor found}
8:     end if
9: end for
```

may be a nontrivial divisor of n

In step 1 we could even choose a $k \in \mathbb{N}$ such that k is a multiple of at most all integers less than B_1. And instead of vary a in step 3 we could also vary k or the bound B_1 after each loop.

To understand how the algorithm works, suppose that $c|k$ for all $c \in \{1, \ldots, B_1\}$ and further suppose that p is a prime divisor of n such that

$$p - 1 = \prod_{i=1}^{s} {p_i}^{h_i},$$

where p_i is a prime such that ${p_i}^{h_i} \leq B_1$ for all i, $1 \leq i \leq s$. Then it follows that $p - 1$ divides k, since k is a multiple of the prime powers ${p_i}^{h_i}$, $1 \leq i \leq d$. Hence, $a^k \equiv 1 \pmod{p}$ by Fermat's Little Theorem. Then $p | \gcd(a^k - 1, n)$. Only if $a^k \equiv 1 \pmod{n}$ the algorithm fails and yields a nontrivial factor of n.

The Pollard $p - 1$ method becomes infeasible if all of the prime divisors p of n have $p - 1$ divisible by a relatively large prime.

Example 10 (i) Let $n = 540143$, $B_1 = 8$, $k = \text{lcd}(2, 3, \ldots, 8) = 840$ and $a = 2$. Then $b = 2^{840} \mod n = 53047$ and $\gcd(53047 - 1, n) = 421$, which yields $540143 = 421 \cdot 1283$.

(ii) Let $n = 491389 = 383 \cdot 1283$. If we would choose $B_1 < 191$ there would be no chance to get a factorization of n. Let $p = 383$, so $p - 1 = 383 - 1 = 2 \cdot 191$, where 191 is a prime. The same yields for $1283 - 1 = 2 \cdot 641$, where 641 is a prime. Since $191 \nmid k$ and $641 \nmid k$, $\gcd(a^k - 1, n)$ always yields 1.

Because for a fixed n the groups $\mathbb{F}_p^*$, p a prime divisor of n, are also fixed. So if all such finite groups $\mathbb{F}_p^*$ have an order divisible by a large prime, we can not succeed with a small bound B_1, which is necessary for an efficient algorithm. Using elliptic curves in subsection "Elliptic Curve Method" of Sect. 4.4.3 this problem can be solved.

For further speed up's and a method to find higher divisors using a second step, we refer to [60].

Sieve Based Methods

A sieve based integer factoring method tries to construct a solution to the congruence

$$a^2 \equiv b^2 \pmod{n}.$$

Hence $\gcd(a - b, n)$ is a divisor of n. In oder to achieve this, sieve based methods try to factor many congruences of the form

$$c \equiv d \pmod{n},$$

with some special relations between c and d. Two factor bases B_c and B_d consisting of a fixed set of prime numbers are used to factor each c and d, respectively. This yields congruences of the form

$$\prod p_i^{l_i} \equiv \prod q_i^{l_i} \pmod{n}, \qquad (4.1.3)$$

where $q_i \in B_c$, the factor base of c, and $p_i \in B_d$, the factor base associated with d. The main idea now is to collect $\#B_c + \#B_d$ congruences of the form (4.1.3) in order to find a set of these congruences which when multiplied together yields squares on both sides. This set is found by solving a set of linear equations $\pmod{2}$. Hence a sieve based factoring method consists of two essential steps:

(i) Collecting a set of equations by 'sieving'.
(ii) Solving this set of equations (i.e. using a matrix).

Notice that the factor bases can be precomputed and used for further integer factorizations.

There are two main sieving methods using this idea known today. The quadratic sieve method proposed by Pomerance [69] and improved by R.D. Silverman [83] has a running time of $L_n(1/2, 1)$ in $\ln n$. The (general) number field sieve (GNFS) proposed by A.K. Lenstra et al. in 1991 finds its successfully factored congruences by sieving over the norms of two sets of integers. These norms are represented by polynomials. The NFS may factor integers of the form $n = r^e - s$, where r and $|s|$ are small positive integers, $r > 1$ and e is large. Whereas the GNFS may factor any integer. The running time is conjectured to be $L_n(1/3, c)$, where $c \approx 1.5$ for the NFS and $c \approx 1.9$ for the GNFS. For the collected papers dealing with the development of the (G)NFS see [47].

4.2 Elliptic Curves

We shall now introduce some definitions and basic properties about elliptic curves. We will refer partly on the book of Menezes [51]. Proofs not given here can be found in the book of Silverman [82] or even Husemöller [31], unless stated otherwise.

At first we will define elliptic curves as the solution of a smooth Weierstrass equation. Then we state the chord-and-triangle addition law, which is a group operation on the set of points of an elliptic curve. Further we shall point out a few theorems about the group structure and the number of points on elliptic curves over finite fields. After a short introduction into divisor theory in subsection "Divisor Theory" of Sect. 4.2.3 we will state some properties of the Weil-Pairing in subsection "The Weil Pairing" of Sect. 4.2.3, which is needed in Sect. 4.5.

4.2.1 Definitions

In the further sections we will denote a (perfect) field by K and its algebraic closure by $\overline{K}$.

Definition 122 The homogeneous equation

$$C: \; Y^2Z + a_1XYZ + a_3YZ^2 = X^3 + a_2X^2Z + a_4XZ^2 + a_6Z^3, \tag{4.2.1}$$

where $a_1, a_2, a_3, a_4, a_6 \in \overline{K}$, is called **Weierstrass equation.**

Definition 123 The **projective plane** (over K), denoted $\mathbb{P}^2$ or $\mathbb{P}^2(\overline{K})$, is the set of all triples $(x, y, z) \in K^3$, such that at least one coordinate is non-zero, modulo the equivalence relation given by $(x_1, y_1, z_1) \sim (x_2, y_2, z_2)$. (x_1, y_1, z_1) is equivalent to (x_2, y_2, z_2) if there exists a $\lambda \in \overline{K}^*$ with $x_1 = \lambda x_2, y_1 = \lambda y_2$ *and* $z_1 = \lambda z_2$. An equivalence class $\{(\lambda x, \lambda y, \lambda z)\}$ is denoted by $(x : y : z)$.

We will use the set of K-rational points in $\mathbb{P}^2$, defined by

$$\mathbb{P}^2(K) = \{(x : y : z) \in \mathbb{P}^2 : x, y, z \in K\}.$$

Definition 124 A projective point $P \in \mathbb{P}^2(\overline{K})$ of the function C in (4.2.1) is called **singular**, if all three partial derivatives $\frac{\delta F}{\delta x}, \frac{\delta F}{\delta y}, \frac{\delta F}{\delta z}$ vanish at $P = (x : y : z)$, where F is defined by

$$F(X, Y, Z) = Y^2X + a_1XYZ + a_3YZ^2 - X^3 - a_2X^2Z - a_4XZ^2 - a_6Z^3 = 0.$$

C is said to be **smooth** (or non-singular) if C is not singular for all points $P \in \mathbb{P}^2(\overline{K})$.

Now we will make the important definition of an elliptic curve:

Definition 125 An **elliptic curve** is a pair $(E, \mathcal{O})$, consisting of the set of all solutions in $\mathbb{P}^2(\overline{K})$ of a smooth Weierstrass equation. $\mathcal{O} = (0 : 1 : 0) \in \mathbb{P}^2$ is called the point at infinity.

Remark 33 (i) An elliptic curve is usually only denoted by E instead of $(E, \mathcal{O})$.

(ii) To ease notation, we will usually write the Weierstrass equation using affine coordinates $x = X/Z$ and $y = Y/Z$,

$$E : y^2 + a_1xy + a_3y = x^3 + a_2x^2 + a_4x + a_6 \tag{4.2.2}$$

with the point $\mathcal{O} = (0 : 1 : 0)$ at infinity and $a_1, a_2, a_3, a_4, a_6 \in K$. Then E is said to be defined over K, denoted E/K.

(iii) The set of **K-rational points** is defined by

$$E(K) := \{(x, y) \in K^2 : (x, y) \in E\} \cup \mathcal{O}.$$

More precisely the K-rational points are exactly the points on E which are invariant under the Galois-group $G_{\overline{K}/K}$.

(iv) In the algebraic literature elliptic curves are defined as algebraic curves of genus 1. Silverman [82] shows, using the Riemann–Roch theorem, that every elliptic curve can be written as a cubic plane, and conversely, every smooth Weierstrass cubic plane is an elliptic curve.

The *function field* $K(E)$ of E over K is the field of fractions of the coordinate ring $K[E] = K[x, y]/(f)$ of E over K, where $f \in K[x, y]$ is given by the rewritten elliptic curve

$$E : f(x, y) = y^2 + a_1xy + a_3y - x^3 - a_2x^2 - a_4x - a_6 = 0. \tag{4.2.3}$$

$K[E]$ is an integral domain if its field of functions $K(E)$ is the set of equivalence classes of quotients $\frac{g}{h}$, $g, h \in K[E]$, $h \neq 0$, where $\frac{g_1}{h_1} \sim \frac{g_2}{h_2}$ if $g_1h_2 = g_2h_1$. In the same way we can define $\overline{K}(E)$, the function field of E over K, where the elements of $\overline{K}(E)$ are rational functions. Let $\overline{K}(E)^*$ denote the invertible elements of $\overline{K}(E)$.

If $f \in \overline{K}(E)^*$ and $P \in E \setminus \{\mathcal{O}\}$ then f is *regular* at P, if there exists $g, h \in \overline{K}[E]$ with $h(P) \neq 0$ such that $f = g/h$. Hence if f is regular, we can evaluate $f(P) = g(P)/h(P)$, where $f(P)$ does not depend on the choice of g and h. $f(\mathcal{O})$ can also be defined, cf. [51].

Definition 126 (i) A projective plane V is called a **projective variety** if its homogeneous ideal $\{f \in \overline{K}[X] : f$ is homogeneous and $f(P) = 0 \,\forall P \in V\}$ is a prime ideal in $\overline{K}[X]$.

(ii) Let V_1 and V_2 be projective varieties. We say V_1/K and V_2/K are **isomorphic** over K, denoted $V_1/K \simeq_p V_2/K$, if there are morphisms $\Phi : V_1/K \to V_2/K$ and $\Psi : V_2/K \to V_1/K$ such that $\Psi \circ \Phi = id_{V_1}$ and $\Phi \circ \Psi = id_{V_2}$, where

id_{V_1}, id_{V_2} are the identity maps on V_1/K and V_2/K respectively, and Φ, Ψ can be defined over K.

(iii) Let E_1 and E_2 be elliptic curves. Then E_1 is said to be **isomorphic** to E_2 over K, denoted $E_1/K \simeq E_2/K$, if $E_1/K \simeq_p E_2/K$. This is a equivalence relation.

Assume Φ to be a non-constant rational map. Then composition with Φ induces an injection of function fields fixing K,

$$\Phi^* : K(E_2) \to K(E_1)$$

$$\Phi^* f = f \circ \Phi.$$

Definition 127 Let E_1 and E_2 be elliptic curves. Let $\Phi : E_1/K \to E_2/K$ be a map of elliptic curves defined over K.

(i) Define the **degree of** Φ by

$$\deg \Phi = \begin{cases} 0, & \Phi \text{ constant} \\ [K(E_1) : \Phi^* K(E_2)], & \text{otherwise.} \end{cases}$$

(ii) Φ is denoted to be **separable**, if the extension $K(E_1)/\Phi^* K(E_2)$ is separable.

An *automorphism* of an elliptic curve E is an isomorphism $E \to E$. The set of automorphisms is denoted by Aut E or $\mathrm{Aut}_K E$.

Theorem 64 ([82], III.3.1(b)) *Two elliptic curves E_1/K and E_2/K given by*

$$E_1 : y^2 + a_1 xy + a_3 y = x^3 + a_2 x^2 + a_4 x + a_6 \tag{4.2.4}$$
$$E_2 : y^2 + \tilde{a}_1 xy + \tilde{a}_3 y = x^3 + \tilde{a}_2 x^2 + \tilde{a}_4 x + \tilde{a}_6 \tag{4.2.5}$$

are isomorphic over K, if and only if there exists $u, r, s, t \in K$, $u \neq 0$, such that

$$(x, y) \to (u^2 x + r, u^3 y + u^2 s x + t), \tag{4.2.6}$$

*which is denoted a **good change of variables**, transforms E_1 to E_2.*

Remark 34 The transformation from equation E_2 to E_1 is done by the following good change of variables $(x, y) \to (u^{-1}(x - z),\ u^{-3}(y - sx - t + rs))$.

So the only change of variables fixing $\mathcal{O} = (0 : 1 : 0)$ and preserving the Weierstrass form of the equation is (4.2.6) with $u, r, s, t \in K$, $u \neq 0$. Then assuming $E_1 \simeq E_2$ over K and using the notation of Theorem 64, equation E_1 (4.2.4) is transformed to

$$(u^3 y + u^2 s x + t)^2 + (a_1(u^2 x + r) + a_3)(u^3 y + u^2 s x + t)$$
$$= (u^2 x + r)^3 + a_2(u^2 x + r)^2 + a_4(u^2 x + r) + a_6,$$

which is equivalent to

$$\begin{aligned} & u^6y^2 + u^5(2s + a_1)xy + u^3(a_3 + a_1r + 2t)y \\ = \quad & u^6x^3 + u^4(3r - s^2 - a_1s + a_2)x^2 \\ & + u^2(2a_2r - a_3s - a_1rs + a_4 - a_1t - 2st + 3r^2)x \\ & + a_6 + a_2r^2 + a_4r + r^3 - a_3t - a_1rt - t^2. \end{aligned}$$

Assume $u \neq 0$. So dividing by u^6 and comparing with E_2 (4.2.5), we will get the following dependences:

$$\begin{aligned} u\tilde{a}_1 &= a_1 + 2s \\ u^2\tilde{a}_2 &= a_2 - a_1s - s^2 + 3r \\ u^3\tilde{a}_3 &= a_1r + a_3 + 2t \\ u^4\tilde{a}_4 &= a_4 - a_1(rs + t) - a_3s + 2a_2r + 3r^2 - 2st \\ u^6\tilde{a}_6 &= a_b + a_2r^2 + a_4r + r^3 - a_3t - a_1rt - t^2 \end{aligned} \tag{4.2.7}$$

This yields the following corollary:

Corollary 6 *Two elliptic curves E_1/K and E_2/K are isomorphic over K, if and only if there exists $u, r, s, t \in K,\ u \neq 0$, such that (4.2.7) is satisfied.*

Let E be an affine Weierstrass equation (4.2.2). Define

$$\begin{aligned} b_2 &:= a_1^2 + 4a_2, \\ b_4 &:= 2a_4 + a_1a_3, \\ b_6 &:= a_3^2 + 4a_6, \\ b_8 &:= a_1^2a_6 + 4a_2a_6 - a_1a_3a_4 + a_2a_3^2 - a_4^2, \\ c_4 &:= b_2^2 - 24b_4. \end{aligned} \tag{4.2.8}$$

Definition 128 (*discriminant and j-invariant*)

(i) The quantity

$$\Delta(E) := \ -b_2^2b_8 - 8b_4^3 - 27b_6^2 + 9b_2b_4b_6 \tag{4.2.9}$$

is called the **discriminant** of the Weierstrass equation E.

(ii) If $\Delta(E) \neq 0$, the quantity

$$j(E) := \ {c_4}^3/\Delta \tag{4.2.10}$$

is said to be the **j-invariant** of E.

Let $P = (x_0,\ y_0)$ be a point satisfying a Weierstrass equation (4.2.3). Assume that P is a singular point on the curve $f(x, y) = 0$ (i.e. $\frac{\delta f}{\delta y}(P) = 0$), then the Taylor expansion of $f(x, y)$ at P has the form

$$f(x, y) - f(x_0, y_0) = [(y - y_0) - \alpha(x - x_0)][(y - y_0) - \beta(x - x_0)] - (x - x_0)^3$$

for some $\alpha, \beta \in \overline{K}$.

Definition 129 (*node/cusp*) With the above notation the singular point P is a node if $\alpha \neq \beta$, and P is a cusp if $\alpha = \beta$.

Theorem 65 (classification of Weierstrass equations)

(i) Let C be a Weierstrass equation (4.2.1). Then

(a) C is smooth (i.e. an elliptic curve), if and only if $\Delta \neq 0$.
(b) C has a node if and only if $\Delta = 0$ *and* $c_4 \neq 0$.
(c) C has a cusp if and only if $\Delta = c_4 = 0$.

(ii) If two elliptic curves E_1/K *and* E_2/K *are isomorphic over K, then they have the same j-invariant (4.2.10). The converse is true if* $K = \overline{K}$.

Example 11 In Figs. 4.1, 4.2 and 4.3 we have plotted some affine Weierstrass equations (4.2.2) with real values for x and y.

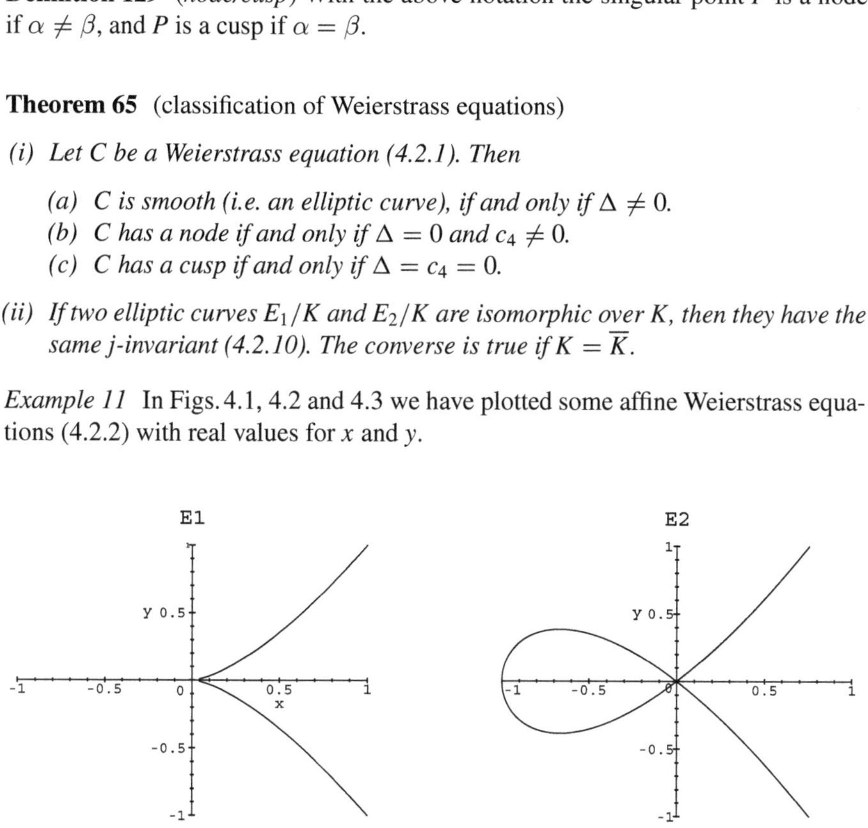

Fig. 4.1 Curves with $\Delta = 0$ and a singular point at $(0, 0)$: $E_1 : y^2 = x^3$ and $E_2 : y^2 = x^3 + x^2$

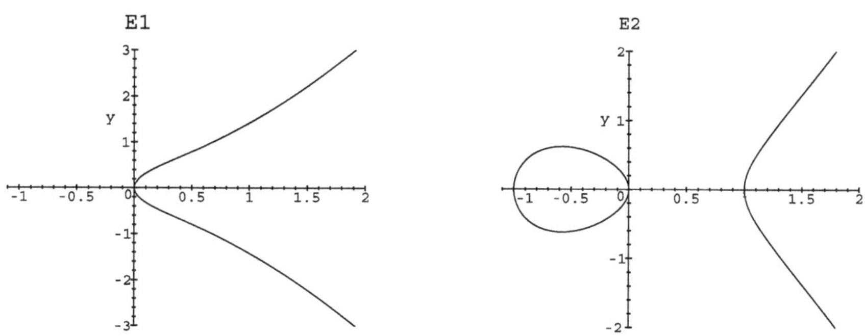

Fig. 4.2 Curves with $j = 1728$: $E_1 : y^2 = x^3 + x,\ \Delta(E_1) = -512$ and $E_2 : y^2 = x^3 - x,\ \Delta(E_2) = 64$

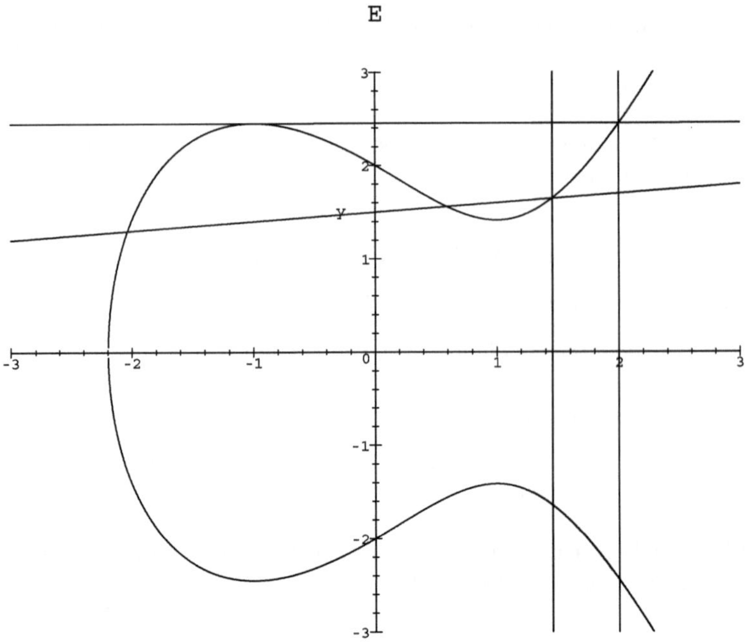

Fig. 4.3 The chord-and-triangle law ($E : y^2 = x^3 - 3x + 4$)

Observe that in Fig. 4.1 $\Delta = 0$, and so there are two possibilities for the singular point (i.e. either a node or a cusp). Even interesting are the two graphs for the same j in Fig. 4.2.

4.2.2 Group Law

In this section let E be an elliptic curve given by a Weierstrass equation (4.2.2).

It is well known that the points on an elliptic curve form an abelian group, which is very important for elliptic curve cryptography. For this reason we will study the abelian group very exactly.

Definition 130 (*chord-and-triangle law*) Let $P, \ Q \in$E$\subset \mathbb{P}^2, \ L \subset \mathbb{P}^2$ the line $\overline{PQ}$ (tangent line to E, if $P = Q$), and R the third point of intersection of L with E. Let L' be the line connecting R and $\mathcal{O}$. Then $P \oplus Q$ is the point such that L' intersects E at R, $\mathcal{O}$ and $P \oplus Q$.

Remark 35 The existence of the intersection point is given by the fact that the points of E satisfy an equation of degree 3 in $\mathbb{P}^2$ and L intersects E at exactly 3 points (if L is tangent to E, then count with multiplicities).

In Fig. 4.3 the composition law is illustrated. In the next lemma the additive structure of the chord-and-triangle law is determined:

Lemma 24 *The chord-and-triangle law (Definition 130) has the following properties:*

(i) If a line L intersects E at the (not necessarily distinct) points P, Q, R, then

$$(P \oplus Q) \oplus R = \mathcal{O}.$$

(ii) $P \oplus \mathcal{O} = P$ for all $P \in E$. (identity element)
(iii) $P \oplus Q = Q \oplus P$ for all P, $Q \in E$. (commutativity)
(iv) Let $P \in E$. There exists a point $\ominus P \in E$, such that $P \oplus (\ominus P) = \mathcal{O}$. (inverse element)
(v) Let $P, Q, R \in E$. Then

$$(P \oplus Q) \oplus R = P \oplus (Q \oplus R). \text{ (associativity)}$$

Proof Note that we always work with multiplicities if a line is a tangent line.

(i) Trivial from Definition 130.
(ii) Let $Q = \mathcal{O}$. Then the lines L and L' are the same in Definition 130. We get

$$L \cap E = \{P, \mathcal{O}, R\} \text{ and } L' \cap E = \{R, \mathcal{O}, P \oplus \mathcal{O}\}.$$

Hence $P \oplus \mathcal{O} = P$.
(iii) Definition 130 is symmetric in P and Q.
(iv) Let R be the third point of intersection of $L = \overline{P\mathcal{O}}$ with E. Then

$$\mathcal{O} = (P \oplus \mathcal{O}) \oplus R = P \oplus R.$$

by (i) and (ii).
(v) See [82]. □

Theorem 66 *(i) $(E, \oplus)$ is an abelian group with identity $\mathcal{O}$.*
(ii) Let E be defined over K. Then

$$E(K) := \{(x, y) \in K^2 : y^2 + a_1xy + a_3y = x^3 + a_2x^2 + a_4x + a_6\} \cup \{\mathcal{O}\}$$

is a subgroup of E.

Proof (i) Clear from Lemma 24.
(ii) If P and Q have coordinates in K, then the equation of the line connecting them has coefficients in K. Assuming E is defined over K, the third point of intersection will have coordinates given by a rational combination of the coefficients of the line and of E, so will be in K, since K is a field.

Definition 131 In the further sections we will only use "+" and "−" for "⊕" and "⊖", respectively. For $m \in \mathbb{Z}$ and $P \in E$ we write

$$mP = \underbrace{P + \cdots + P}_{\text{m terms}}$$

for all $m > 0$, $0P = \mathcal{O}$ and $mP = (-m)(-P)$ for all $m < 0$.

In the following theorem we shall summarize a few basic facts about arithmetic on the abelian group of points on an elliptic curve given by a Weierstrass equation:

Theorem 67 (arithmetic algorithms) *Let E be an elliptic curve given by a Weierstrass equation (4.2.2). For all $P, Q, R \in E$ (label the coordinates of the points by the point symbol):*

(i) If $P = (x_P, y_P) \neq \mathcal{O}$, then

$$-P = (x_P, -y_P - a_1 x_P - a_3).$$

(Observe that P and $-P$ are the only points on E with x-coordinate equal to x_P)

(ii) Let $P \neq \pm Q$, then $R = P + Q$ can be computed in polynomial time by

$$\begin{aligned} x_R &= \lambda^2 + a_1\lambda - a_2 - x_P - x_Q \\ y_R &= -(\lambda + a_1)x_R - \gamma - a_3, \end{aligned} \tag{4.2.11}$$

where $\gamma = y_P - \lambda x_P$ and λ is defined by

$$\lambda = \begin{cases} \frac{y_Q - y_P}{x_Q - x_P}, & \text{if } P \neq Q \\ \\ \frac{3x_P^2 + 2a_2 x_P + a_4 - a_1 y_P}{2y_P + a_1 x_P + a_3}, & \text{if } P = Q. \end{cases} \tag{4.2.12}$$

(iii) Duplication formula for $R = 2P = P + P$

$$x_R = \frac{x_P^4 - b_4 x_P^2 - 2b_6 x_P - b_8}{4x_P^3 + b_2 x_P^2 + 2b_4 x_P + b_6} \tag{4.2.13}$$

with y_R as in (iii) and b_2, b_4, b_6, b_8 as in (4.2.8).

Proof Following the ideas of [82] we rewrite the Weierstrass equation as

$$f(x, y) := y^2 + (a_1 x + a_3)y - x^3 - a_2 x^2 - a_4 x - a_6 = 0. \tag{4.2.14}$$

(i) Let the line L through P and $\mathcal{O}$ also intersect E at R. The line L is given by $L: x - x_P = 0$. Inserting this into the equation of E yields a quadratic polynomial

$f(x_P, y)$ in y. We get two roots y_P and $\tilde{y}_P$ for $f(x_P, y)$, where $-P = (x_P, \tilde{y}_P)$. So we can factor

$$f(x_P, y) = c(y - y_P)(y - \tilde{y}_P) = cy^2 + c(-\tilde{y}_P - y_P)y + cy_P\tilde{y}_P,$$

which yields after coefficient comparing with (4.2.14) $c = 1$ and $-\tilde{y}_P - y_P = a_1x_P + a_3$, which proves (i).

(ii) Let $P, Q \in E \setminus \{\mathcal{O}\}$ and $P \neq -Q$
Observe that if $P = -Q$ then $x_P = x_Q$ and $y_P = -x_Q - a_1x_Q - a_3$ from (i) and this gives $P + Q = \mathcal{O}$.
Let L be the line passing through P and Q if $P \neq Q$ or the tangent line to the curve E at P if $P = Q$, respectively. Then L has the form

$$L:\ y = \lambda x + \gamma. \tag{4.2.15}$$

To calculate the slope γ of L for $P \neq Q$ is easy, since it is only the secant of P and Q. For $P = Q$, as usual, the tangent line to the curve (4.2.14) at $P = (x_P,\ y_P)$ is the line

$$\frac{\delta f}{\delta x}(P)(x - x_P) + \frac{\delta f}{\delta y}(P)(y - y_P) = 0$$

which yields

$$(a_1y_P - 3x_P{}^2 - 2a_2x_P - a_4)(x - x_P) + (2y_P + a_1x_P + a_3)(y - y_P) = 0.$$

This is equivalent to (4.2.15).
To find the third point $S \in E$ of intersection of L with the curve, we substitute $y = \lambda x + \gamma$ into (4.2.14) to get a cubic polynomial

$$\begin{aligned} f(x, \lambda x + \gamma) &= -x^3 - a_2x^2 - a_4x - a_6 + (\lambda x + \gamma)^2 \\ &\quad + a_1x(\lambda x + \gamma) + a_3(\lambda x + \gamma) \\ &= -x^3 - (a_2 - \lambda^2 - a_1\lambda)x^2 - (a_4 - 2\lambda\gamma - a_1\gamma - a_3\lambda)x \\ &\quad - a_6 + \gamma^2 + a_3\gamma \\ &= 0, \end{aligned}$$

which can be factored into the equation

$$\begin{aligned} c(x - x_P)(x - x_Q)(x - x_S) &= cx^3 - c(x_P + x_Q + x_S)x^2 \\ &\quad + c(x_Px_Q + x_Px_S + x_Qx_S)x - cx_Px_Qx_S \\ &= 0, \end{aligned}$$

since it has three roots x_P, x_Q and the unknown x_S. So coefficient comparing at x^3 and x^2 yields $c = -1$ and $(x_P + x_Q + x_S) = -a_2 + \lambda^2 + a_1\lambda$.

This gives the formula for x_S, and substitution of x_S in L (4.2.15) gives $y_S = \lambda x_S + \gamma$. So we know $P + Q + S = \mathcal{O}$ by Lemma 24(i), but we want to derive $P + Q = R$. Hence we can use (i) to calculate the coordinates for $S = -R$: $x_R = x_S$ and $y_R = -y_S - a_1 x_S - a_3 = -(a_1 + \lambda)x_R - \gamma - a_3$.

(iii) See [82]. □

Note that if two elliptic curves are isomorphic then they are also isomorphic as abelian groups. The converse statement is not true in general as seen in Example 14.

Curves over K, char(K) = 2

In this paragraph we shall reduce the Weierstrass equations in order to ease notation and to speed up the computational evaluation of the Weierstrass equation.

Let E be the Weierstrass equation (4.2.2)

$$E: \; y^2 + a_1 xy + a_3 y = x^3 + a_2 x^2 + a_4 x + a_6$$

defined over K, $\mathrm{char}(K) = 2$.

So we will search for a good change of variables, which satisfies Theorem 64, in order to obtain an isomorphism. At first we compute $j(E) = \frac{c_4^3}{\Delta} = \frac{b_2^6}{\Delta} = \frac{a_1^{12}}{\Delta}$ in K. Then there are two possibilities:

(I) $j(E) = 0$: So $a_1 = 0$ and we can rewrite (4.2.2) to

$$\begin{aligned} y^2 + a_3 y &= x^3 + a_2 x^2 + a_4 x + a_6 \\ &= x^3 + a_2 x^2 + a_2^2 x + a_2^3 + a_4 x + a_4 a_2 + a_6 + a_2^2 (x + a_2) \\ &= (x^3 + a_2)^3 + (a_4 + a_2^2)(x + a_2) + a_6 + a_4 a_2. \end{aligned}$$

Now a possible good change of variables is $(x, y) \to (x + a_2, y)$, with $t = s = 0$, $u = 1$ and $r = a_2$. So the elliptic curve E is isomorphic to

$$E_1/K: \; y^2 + \tilde{a}_3 y = x^3 + \tilde{a}_4 x + \tilde{a}_6. \tag{4.2.16}$$

Observe that the formula (4.2.7) yields $\tilde{a}_6 = a_6 + a_2 a_4$, $\tilde{a}_4 = a_4 - a_2^2$. Further for E_1 $\Delta = \tilde{a}_3^4$ and $j(E_1) = 0$.

Addition Formula

From (4.2.11) and (4.2.12) with $\tilde{a}_1 = \tilde{a}_2 = 0$ we get

(i) $P \in E_1 \Rightarrow -P = (x_P, y_P + \tilde{a}_3)$

(ii) $P, \; Q \in E_1, \; Q \neq -P$, then $R = (x_R, y_R) = P + Q$ is given by

$$R = \begin{cases} \left(\left(\frac{y_P+y_Q}{x_P+x_Q}\right)^2 + x_P + x_Q,\ \frac{y_P+y_Q}{x_P+x_Q}(x_P + x_R) + y_P + \tilde{a}_3\right), & P \neq Q \\ \left(\frac{x_P^4+\tilde{a}_4^2}{\tilde{a}_3^2},\ \frac{x_P^2+\tilde{a}_4}{\tilde{a}_3}(x_P + x_R) + y_P + \tilde{a}_3\right), & P = Q. \end{cases} \tag{4.2.17}$$

(II) $j(E) \neq 0$: So $a_1 \neq 0$ and we can find as in (I) a good change of variables

$$(x, y) \to \left(a_1^2 x + \frac{a_3}{a_1},\ {a_1}^3 y + \frac{a_1^2 a_4 + a_3^2}{a_1^3}\right),$$

which transforms E to

$$E_2/K:\ y^2 + xy = x^3 + \tilde{a}_2 x^2 + \tilde{a}_6.$$

Further for E_2 $\Delta = \tilde{a}_6$ and $j(E_2) = 1/\tilde{a}_6$.

Addition Formula

From (4.2.11) and (4.2.12) with $\tilde{a}_1 = 1,\ \tilde{a}_3 = \tilde{a}_4 = 0$ we get

(i) $P \in E_2 \Rightarrow -P = (x_P, y_P + x_P)$
(ii) $P,\ Q \in E_2,\ Q \neq -P$, then $R = (x_R, y_R) = P + Q$ is given by

$$R = \begin{cases} \left(\lambda^2 + \lambda + x_P + x_Q + \tilde{a}_2,\ \lambda(x_P + x_R) + x_R + y_P\right), & \text{if } P \neq Q \\ \left(x_P^2 + \frac{\tilde{a}_P}{x_P^2},\ x_P^2 + (x_P + \frac{y_P}{x_P} + 1)x_R\right), & \text{if } P = Q, \end{cases} \tag{4.2.18}$$

where $\lambda = \left(\frac{y_P+y_Q}{x_P+x_Q}\right)$

Curves over K, char(K) $\neq$ 2,3

Like in the last paragraph elliptic curves which are not defined over fields of char$(K) = 2$ or 3 can be dramatically simplified in order to ease computation. Let again E/K be an elliptic curve given by (4.2.2). If char$(K) \neq 2$, then we can simplify the equation by completing the square. Replacing y by $y - \frac{1}{2}(a_1 x + a_3)$ gives an equation

$$E_3/K:\ y^2 = x^3 + \frac{1}{4}b_2 x^2 + \frac{1}{2}b_4 x + \frac{1}{4}b_6.$$

See again Theorem 64, formula (4.2.7) which yields $b_2,\ b_4$ and b_6 as defined in (4.2.8).

If further char$(K) \neq 2,\ 3$ then a good change of variables

$$(x, y) \to \left(\frac{x - 3b_2}{36},\ \frac{y}{216}\right)$$

eliminates the x^2-term, yielding

$$E_4/K: \; y^2 = x^3 \underbrace{-\,27c_4}_{a} x + \underbrace{54b_2^3 - 1944b_2b_4 + 11664b_6}_{b}$$

Observe that $E/K \simeq E_3/K \simeq E_4/K$.

Theorem 68 *Let K be a field, char(K) $\neq 2, \; 3$, E an elliptic curve defined over K given by a Weierstrass equation (4.2.2). Then E can be written in the standard form*

$$E_{a,b}: \; y^2 = x^3 + ax + b, \qquad a, \; b \in K. \tag{4.2.19}$$

Further $\Delta(E_{a,b}) = -16(4a^3 + 27b^2) \neq 0$ *and* $j(E_{a,b}) = -1728(4a)^3/\Delta$.

The proof is clear from the above and a straightforward calculation yields $\Delta(E_{a,b})$ from (4.2.9) and $j(E_{a,b})$ from (4.2.10).

Remember taking a Weierstrass equation $E_{a,b}, \; a, b \in K$, we get an elliptic curve, if $\Delta \neq 0$.

Addition Formula

From (4.2.11) and (4.2.12) with $a_1 = a_2 = a_4 = 0$ we get

(i) $P \in E_{a,b} \; \Rightarrow \; -P = (x_P, -y_P)$
(ii) $P, \; Q \in E_{a,b}, \; Q \neq -P$, then $R = (x_R, y_R) = P + Q$ is given by

$$R = \begin{cases} \left(\left(\frac{y_Q - y_P}{x_Q - x_P}\right)^2 - x_P - x_Q, \; \frac{y_Q - y_P}{x_Q - x_P}(x_P - x_R) - y_P \right), & \text{if } P \neq Q \\ \left(\left(\frac{3x_P{}^2 + a}{2y^P}\right)^2 - 2x_P, \; \frac{3x_P{}^2 + a}{2y_P}(x_P - x_R) - y_P \right), & \text{if } P = Q. \end{cases} \tag{4.2.20}$$

Example 12 Let $E = E_{1,6}: y^2 = x^3 + x + 6$ be defined over $K = \mathbb{Z}_{13}$, the finite field of 13 elements. E is an elliptic curve, since $\Delta = 10 \neq 0$. Notice that $E_{1,3}$ isn't an elliptic curve over $\mathbb{Z}_{13}$ since $\Delta = 0$.

Let us first determine the points on E. This can be done by looking at each possible $x \in \mathbb{Z}_{13}$ computing

$$f(x) = x^3 + x + 6 \mod 13$$

and then trying to solve the equation

$$y^2 \equiv x^3 + x + 6 \pmod{13}.$$

For a given x we can test if $f(x)$ is a quadratic residue by applying Euler's criterion. So one obtains Table 4.1.

Thus, $E(\mathbb{Z}_{13})$ has 13 points. Hence, $E(\mathbb{Z}_{13})$ has a prime order and it follows $E(\mathbb{Z}_{13}) \simeq \mathbb{Z}_{13}$ and any point $P \neq \mathcal{O}$ is a generator of $E(\mathbb{Z}_{13})$. Suppose taking the generator $P = (2, 4)$ we can quickly derive by the above addition formula

Table 4.1 Searching for a quadratic residue (QR) in $\mathbb{Z}_{13}$

x	f(x)	QR(13)?	y
0	6	N	
1	8	N	
2	3	Y	4, 9
3	10	Y	6, 7
4	9	Y	3, 10
5	6	N	
6	7	N	
7	5	N	
8	6	N	
9	3	Y	4, 9
10	2	N	
11	9	Y	3, 10
12	4	Y	2, 11

$$E(\mathbb{Z}_{13}) = \{\mathcal{O}, (2,4), (9,9), (11,10), (12,11), (3,7), (4,3), (4,10),$$
$$(3,6), (12,2), (11,3), (9,4), (2,9)\}.$$

Some examples for additions: $(2,4)+(2,4) = (9,9)$, $10(2,4) = (11,3)$, $(2,4)+(9,9) = (9,9)+(2,4) = (11,10)$ and $(3,7)+(3,6) = \mathcal{O}$, since $(3,7) = -(3,6) = (3,-6) = (3,7)$.

Corollary 7 *Let $E_{a,b}$, $E_{\tilde{a},\tilde{b}}$ be elliptic curves given by (4.2.19). Then $E_{a,b} \simeq E_{\tilde{a},\tilde{b}}$ over K if and only if there exists $u \in K^*$ such that $u^4\tilde{a} = a$, $u^6\tilde{b} = b$.*

4.2.3 Elliptic Curves over the Finite Field $\mathbb{F}_q$

By now let E be an elliptic curve over the finite field $K = \mathbb{F}_q, q = p^m$, where $\mathbb{F}_q$ denotes the finite field containing q elements, $m \in \mathbb{N}$ and p a prime. So $\overline{\mathbb{F}}_q = \bigcup_{m \geq 1} \mathbb{F}_{q^m}$. E can be given by a smooth Weierstrass equation (4.2.2), where the coefficients are also defined over $\mathbb{F}_q$.

Number of Points

In this paragraph we shall explore $\#E(\mathbb{F}_q)$, the number of points on an elliptic curve.

Lemma 25 *Let E be an elliptic curve defined over $\mathbb{F}_q$, $char(\mathbb{F}_q) = p$, Φ an endomorphism on E.*

(i) *If Φ is a non-zero separable isogeny, i.e. a morphism satisfying $\Phi(\mathcal{O}) = \mathcal{O}$, then* $\# \ker \Phi = \deg \Phi$.
(ii) *End(E)* $:= \{$*isogeny* $\Phi : E \to E : \Phi(\mathcal{O}) \neq \{\mathcal{O}\}\}$ *is the* **endomorphism ring of E** *with the addition and multiplication*

$$(\phi + \psi)(P) = \phi(P) + \psi(P)$$

$$(\phi\psi)(P) = \phi(\psi(P)),$$

where $\phi, \psi \in End(E)$.
(iii) *If Φ is the qth-power Frobenius endomorphism $\Phi_E : (x, y) \to (x^q, y^q)$ then $1 - \Phi_E$ is separable and* $\deg \Phi_E = q$.
(iv) *The degree map* $\deg : End(E) \to \mathbb{Z}$ *is a positive definite quadratic form.*

Proofs can be found in Silverman [82] ((i) III.4.10c, (ii)II.4, (iii) III.5.5, II.2.11c, (iv) III.6.3).

In the following theorem, which was originally proved 1937 by Hasse, $\#E(\mathbb{F}_q)$ is shown to be restricted for any $q = p^m$, p prime.

Theorem 69 (Hasse inequality) *Let $E/\mathbb{F}_q$ be an elliptic curve with $\#E(\mathbb{F}_q) = q + 1 - t$. Then $|t| \leq 2\sqrt{q}$.*

Proof Following [82] we choose a Weierstrass equation E defined by (4.2.2) over $\mathbb{F}_q$. Define the Frobenius endomorphism

$$\begin{aligned}\Phi_E : E &\to E \\ (x, y) &\mapsto (x^q, y^q).\end{aligned}$$

Since the Galois group $G_{\overline{\mathbb{F}}_q/\mathbb{F}_q}$ is generated by the qth-power map and by Remark 33(iii), we get for all $P \in E(\overline{\mathbb{F}}_q)$

$$P \in E(\mathbb{F}_q) \iff \Phi_E(P) = P.$$

Hence

$$E(\mathbb{F}_q) = \ker(1 - \Phi_E) = \{P \in E(\overline{\mathbb{F}_q}) : \Phi_E(P) = P\},$$

so

$$\#E(\mathbb{F}_q) = \# \ker(1 - \Phi_E) = \deg(1 - \Phi_E),$$

by Lemma 25(i), (iii). Since the degree map on End(E) is a positive definite quadratic form we obtain for all $m, n \in \mathbb{Z}$:

$$\begin{aligned}0 &\leq \deg(m - n\Phi_E) \\ &= m^2 + mn(\deg(1 - \Phi_E) - \deg \Phi_E - \deg 1) + n^2 \deg \Phi_E \\ &= m^2 + mn(\#E(\mathbb{F}_q) - q - 1) + n^2 q\end{aligned}$$

With $m = -(\#E(\mathbb{F}_q) - q - 1)$ and $n = 2$ we get

$$\begin{aligned} & 0 \leq -(\#E(\mathbb{F}_q) - q - 1)^2 + 4q \\ \iff & \\ & |\#E(\mathbb{F}_q) - q - 1| \leq 2\sqrt{q}. \end{aligned}$$

□

Definition 132 Let p be a prime and $w \in \mathbb{F}_p$. Then define the **(extended) Legendre symbol** by

$$\left(\frac{w}{p}\right) = \begin{cases} +1, & \text{if } w \text{ is a quadratic residue in } \mathbb{F}_p^*, \\ 0, & \text{if w=0}, \\ -1, & \text{else, i.e. if } w \text{ is a quadratic non-residue in } \mathbb{F}_p^*. \end{cases}$$

Theorem 70 *Let $E/\mathbb{F}_p$, $p > 3$ prime, be an elliptic curve. Then*

$$\#E(\mathbb{F}_p) = 1 + \sum_{x \in \mathbb{F}_p} \left[\left(\frac{x^3 + ax + b}{p} \right) + 1 \right].$$

Proof Let $f(x) = x^3 + ax + b$. For all $x \in \mathbb{F}_p$ there are

$$\begin{Bmatrix} 2 \\ 1 \\ 0 \end{Bmatrix} \text{ values of } y \text{ corresponding to } x, \text{ if } \begin{cases} f(x) \text{ is a quadratic residue in } \mathbb{F}_p^* \\ \text{f(x)=0} \\ \text{else} \end{cases}$$

Finally add 1 for $\mathcal{O}$. □

By the proof of the last theorem we can easily count the number of rational points on E over $\mathbb{F}_p$. We have used this already in the example of the last paragraph. But since the running time is $O(p^{1+\epsilon})$ this gets infeasible for great primes p.

Let Φ_E be the Frobenius endomorphism. From the general theory of separable endomorphisms we know that

$$\deg(1 - \Phi_E) = \deg \Phi_E - \text{tr}(\Phi_E) + 1,$$

where $\text{tr}(\Phi_E) = t$ denotes the Frobenius trace of Φ_E. Hence

$$\#E(\mathbb{F}_q) = q + 1 - \text{tr}(\Phi_E).$$

Definition 133 Let $E/\mathbb{F}_q$ be an elliptic curve. The **characteristic polynomial of Frobenius** is

$$f_E(T) = \det(1 - \Phi_E T) = 1 - \text{tr}(\Phi_E)T + qT^2 \in \mathbb{Z}[T], \tag{4.2.21}$$

where Φ_E is the Frobenius endomorphism.

Schoof [75] described an algorithm to compute $\#E(\mathbb{F}_q)$ for a given q in polynomial time, see Sect. 4.3.3. In this discussion we will need the following theorem.

Theorem 71 *Let f_E be the characteristic polynomial of Frobenius of an elliptic curve $E/\mathbb{F}_q$. Let $\#E(\mathbb{F}_q) = q + 1 - t$. Then*

$$f_E(\Phi_E) = 0,$$

where Φ_E is the Frobenius endomorphism.

If an elliptic curve is defined over $\mathbb{F}_q,\ q = p^m,\ p$ prime, then it is also defined over an extension field $\mathbb{F}_{q^k},\ k \in \mathbb{N}$.

Definition 134 Defining $N_k = \#E(\mathbb{F}_{q^k}),\ k = 1, 2, \ldots$ and the formal power series

$$Z(E/\mathbb{F}_q; T) = e^{\sum_k N_k T^k / k}, \tag{4.2.22}$$

we obtain the **zeta-function** of the elliptic curve, where T is again an indeterminate.

The following Weil Theorem (which was proved by Hasse in 1934) enables us to compute $N_k = \#E(\mathbb{F}_{q^k}),\ k \geq 2$, knowing $N_1 = \#E(\mathbb{F}_q)$.

Theorem 72 (Weil Theorem, [82], V.2) *The zeta-function of an elliptic curve E defined over $\mathbb{F}_q$ is a rational function of T having the form*

$$Z(E/\mathbb{F}_q; T) = \frac{f_E(T)}{(1-T)(1-qT)} = \frac{1 - tT + qT^2}{(1-T)(1-qT)}, \tag{4.2.23}$$

where t depends on the elliptic curve E itself, i.e.

$$t = q + 1 - N_1 = q + 1 - \#E(\mathbb{F}_q). \tag{4.2.24}$$

Since by Theorem 69 $t^2 \leq 4q$ there is a factorization

$$1 - tT + qT^2 = (1 - \alpha T)(1 - \bar{\alpha} T), \tag{4.2.25}$$

where $\alpha, \bar{\alpha}$ are complex conjugate roots and $|\alpha| = |\bar{\alpha}| = \sqrt{q}$.

To calculate $\#E(\mathbb{F}_{q^k}) = N_k,\ k \geq 2$, observe that

$$Z(E/\mathbb{F}_q; T) = e^{\sum_k N_k T^k / k} = \frac{1 - tT + qT^2}{(1-T)(1-qT)} = \frac{(1-\alpha T)(1-\bar{\alpha} T)}{(1-T)(1-qT)}.$$

After applying the logarithm (using $\ln(1-T) = -\sum_r T^r/r$) we get

$$\begin{aligned}\sum_k N_k T^k/k &= \ln(1-\alpha T) + \ln(1-\bar{\alpha}T) - \ln(1-T) - \ln(1-qT) \\ &= -\sum_r \alpha^r T^r/r - \sum_r \bar{\alpha}^r T^r/r + \sum_r T^r/r + \sum_r q^r T^r/r \\ &= \sum_r (q^r + 1 - \alpha^r - \bar{\alpha}^r) T^r/r\end{aligned}$$

which yields

$$N_k = q^k + 1 - \alpha^k - \bar{\alpha}^k \qquad \forall k \geq 1.$$

Hence knowing $\#E(\mathbb{F}_q) = N_1$ we get

$$N_2 = q^2 + 1 - \alpha^2 - \bar{\alpha}^2$$

by calculating $t = q + 1 - N_1$ and finding the roots $\alpha, \bar{\alpha}$ of $f_E(T) = qT^2 - tT + 1$.

Example 13 Let $E/\mathbb{Z}_{13} : y^2 = x^3 + x + 6$. We want to calculate $\#E(\mathbb{Z}_{13^{19}})$.

From Example 12 we know $\#E(\mathbb{Z}_{13}) = 13$. Hence $t = 1$. Therefore we have the following condition for α and $\bar{\alpha}$:

$$\begin{aligned}(1-\alpha T)(1-\bar{\alpha}T) &= 1 - (\alpha + \bar{\alpha})T + \alpha\bar{\alpha}T^2 \\ &= 1 - 1 \cdot T + 13 \cdot T^2.\end{aligned}$$

Hence $\alpha + \bar{\alpha} = 1$ and $\alpha\bar{\alpha} = 13$ which yields $\alpha = \frac{1}{2} + i\frac{\sqrt{51}}{2}$. We get

$$\#E(\mathbb{Z}_{13^k}) = 13^k + 1 - \alpha^k - \overline{\alpha}^k.$$

Computing this for $k = 19$ yields

$$E(\mathbb{Z}_{13^{19}}) = 13 \cdot 112455406954768477177 = 13 \cdot P21,$$

where $P21$ is a 21 digit prime.

If we want to construct elliptic curves without counting the number of points explicitly we can use the following helpful lemma.

Lemma 26 *Let p be an odd prime such that $p \equiv 2 \pmod 3$. Then*

$$\#E_{0,b}(\mathbb{F}_p) = p + 1,$$

independently of $b \in \mathbb{F}_p^*$.

Proof Observe that the map $x \mapsto x^3 + b$ is a permutation on $\mathbb{F}_p$, since $p \equiv 2 \pmod 3$. Hence, there are $(p-1)/2$ elements $x \in \mathbb{F}_p$, such that $x^3 + b$ is a non-zero quadratic residue in $\mathbb{F}_p$. These x serve as the first coordinate in order to get the points $(x, \pm\sqrt{x^3+b})$. Knowing further that $\mathcal{O} \in E_{0,b}(\mathbb{F}_p)$ and calculating the x-coordinate of $(x, 0)$, which yields $((-b)^{1/3}, 0) \in E_{0,b}(\mathbb{F}_p)$, gives $\#E_{0,b}(\mathbb{F}_p) = 2 \cdot (p-1)/2 + 1 + 1 = p + 1$. □

Beside this $\#E_{0,b}(\mathbb{F}_p)$ is a cyclic group.

Lemma 27 *Let p be a prime satisfying $p \equiv 3 \pmod 4$. Then for $a \in \mathbb{F}_p^*$ we have*

$$\#E_{a,0}(\mathbb{F}_p) = p + 1.$$

Proof Let $f(x) = x^3 + ax$. $f(x)$ is an odd function, i.e. $f(-x) = -f(x)$. Since $p \equiv 3 \pmod 4$ $(p-1)/2$ is odd and -1 is a quadratic residue modulo p. Hence, for every $w \in \mathbb{F}_p^*$ either w or $-w$ is a quadratic residue modulo p. Consider now the $(p-1)/2$ pairs $(x, -x)$, $0 < x \le (p-1)/2$. For each pair, either $f(x) = f(-x) = 0$ or $f(x)$ is a quadratic residue or $f(-x)$ is a quadratic residue. In each of this three cases there exist 2 points on $E_{a,0}(\mathbb{F}_p)$ associated to the pair $(x, -x) : (\pm x, 0), (x, \pm\sqrt{f(x)})$ or $(-x, \pm\sqrt{-f(x)})$, respectively. Together with $(0, 0)$ and $\mathcal{O}$ we get $p + 1$ points on $E_{a,0}(\mathbb{F}_p)$. □

Lemma 28 *The number of isomorphism classes of elliptic curves E over $\mathbb{F}_p$, $p > 3$ prime, is given by*

$$\#'\{E : E/\mathbb{F}_p \text{ elliptic curve}\}/\cong \mathbb{F}_p = p,$$

where $\#'$ denotes the weighted cardinality, the isomorphism classes of E being counted with weight $(\mathrm{Aut}\ E)^{-1}$.

Menezes showed for the case of Lemma 28 in [51] that, if $p = q$ is a prime, there exists an elliptic curve $E/\mathbb{F}_p$ with $\#E(\mathbb{F}_p) = p + 1 - t$ for all $|t| \le 2\sqrt{q}$.

Theorem 73 ([46]) *There exist effectively computable positive constants c_1, c_2 such that for each prime $p > 3$ the following is valid:*

(i) If S is any set of integers in $I_1 = [p + 1 - \sqrt{p},\ p + 1 + \sqrt{p}]$ then

$$\#'\{E : E/\mathbb{F}_p \text{ ell. curve}, \#E(\mathbb{F}_p) \in S\}/\cong \mathbb{F}_p \le c_1 \#S\sqrt{p}(\log p)(\log\log p)^2.$$

(ii) If S is any set of integers in $I_2 = [p + 1 - 2\sqrt{p},\ p + 1 + 2\sqrt{p}]$ then

$$\#'\{E : E/\mathbb{F}_p \text{ ell. curve}, \#E(\mathbb{F}_p) \in S\}/\cong \mathbb{F}_p \ge c_2(\#S - 2)\frac{\sqrt{p}}{(\log p)}.$$

This theorem, proved by Lenstra using Lemma 28, states that if E varies over all elliptic curves over $\mathbb{F}_p$, then the values of $\#E(\mathbb{F}_p)$ are nearly uniformal distributed in I_1.

Group Structure of Several Curves

Theorem 74 *Let E be an elliptic curve defined over $\mathbb{F}_q$. Then $E(\mathbb{F}_q)$ is an abelian group of rank 1 or 2. The type of this group is $(n_1,\ n_2)$, i.e.*

$$E(\mathbb{F}_q) \simeq \mathbb{Z}_{n_1} \oplus \mathbb{Z}_{n_2}, \quad \textit{where } n_2|n_1 \textit{ and } n_2|q-1.$$

Note that if two elliptic curves are isomorphic over $\mathbb{F}_q$ then the abelian groups are also isomorphic over $\mathbb{F}_q$. But the converse is not true:

Example 14 From table 3.1 in [51] we see that the two elliptic curves $E_1 : y^2 = x^3 + 2$ and $E_2 : y^2 = x^3 + 4$ defined over $\mathbb{F}_5$ have order 6. Hence $E_1(\mathbb{F}_5) \cong \mathbb{Z}_6 \cong E_2(\mathbb{F}_5)$. But there is no $u \in \mathbb{F}_5^*$ such that $4u^6 = 2$, hence by Corollary 7 $E_1 \ncong E_2$ over $\mathbb{F}_5$. Perhaps the two curves are isomorphic over another field.

Lemma 29 ([76], (4.2)) *There exists an elliptic curve $E/\mathbb{F}_q$, q a prime power, such that $\#E(\mathbb{F}_q) = q + 1 - t$, if and only if one of the following conditions holds:*

(i) $t \not\equiv 0 \pmod p$ *and* $t^2 \leq 4q$,
(ii) (a) m *is odd, and* $t = 0$,
(b) m *is odd,* $t^2 = 2q$ *and* $p = 2$,
(c) m *is odd,* $t^2 = 3q$ *and* $p = 3$,
(iii) (a) m *is even,* $t^2 = 4q$,
(b) m *is even,* $t^2 = q$ *and* $p \not\equiv 1 \pmod 3$,
(c) m *is even,* $t = 0$*and* $p \not\equiv 1 \pmod 4$.

Supersingular Curves

Definition 135 Let E be an elliptic curve defined over $\mathbb{F}_q$, q a prime power, $\#E(\mathbb{F}_q) = q + 1 - t$. E is said to be **supersingular** if $p|t$. Otherwise E is called **non-supersingular**.

Corollary 8 *Let E be an elliptic curve over $\mathbb{F}_q$, $q = p^m$ a prime power. Then E is supersingular if and only if one of the following assumptions holds.*

(i) The qth-*Frobenius trace* $tr(\Phi_E) \equiv 0 \pmod p$ *or equivalently* $\#E(\mathbb{F}_q) \equiv 1 \pmod p$.
(ii) $j(E) = 0$, *assuming that* $p = 2$ *or* $p = 3$ *(cf. subsection "Curves over K, char(K) = 2" of Sect. 4.2.2(I)).*
(iii) $t^2 = 0,\ q,\ 2q,\ 3q$ *or* $4q$.

Proof (i) trivial, (ii) cf. [82]

(iii) If E is supersingular, i.e. $p|t$, we know that $t \equiv 0 \pmod p$. Thus $t^2 = 0, q, 2q, 2q$ or $4q$ by Lemma 29. Conversely, apply this lemma on these cases: $t^2 =$

0: $t = 0$. Hence $\#E(\mathbb{F}_q) \equiv q + 1 \equiv 1 \pmod p$, then use (i).
q: $t = \pm p^{\frac{m}{2}}$, $m \geq 2$ even. Thus $p|t$.

$2q$: $t = \pm\sqrt{2} \cdot 2^{\frac{m}{2}}$, $m \geq 1$ odd. Thus $p = 2|2^{\frac{m+1}{2}} = t$.
$3q$: $t = \pm\sqrt{3} \cdot 3^{\frac{m}{2}}$, $m \geq 1$ odd. Thus $p = 3|3^{\frac{m+1}{m}} = t$.
$4q$: $t = \pm 2p^{\frac{m}{2}}$, $m \geq 2$ even. Thus $p|t$. □

The following result gives the group structure of supersingular curves

Lemma 30 ([76], (4.8)) *Let* $\#E(\mathbb{F}_q) = q + 1 - t$.

(i) If $t^2 = q, 2q, 3q$ *or* ($t = 0$ *and* $q \not\equiv 3 \pmod 4$) *then* $E(\mathbb{F}_q)$ *is cyclic.*
(ii) If $t^2 = 4q$ *then* $E(\mathbb{F}_q) \cong \mathbb{Z}_{\sqrt{q}\mp 1} \oplus \mathbb{Z}_{\sqrt{q}\mp 1}$ *if* $t = \pm 2\sqrt{q}$.
(iii) If $t = 0$ *and* $q \equiv 3 \pmod 4$ *then* $E(\mathbb{F}_q)$ *is either cyclic or isomorphic to* $\mathbb{Z}_{(q+1)/2} \oplus \mathbb{Z}_2$.

Theorem 75 ([82], Proposition V.3.1) *Let E be a non-supersingular elliptic curve defined over* $\mathbb{F}_q$. *Then the endomorphism ring End(E) of E is an order in an imaginary quadratic field.*

Thus a non-supersingular elliptic curve E has complex-multiplication since End(E) is strictly larger than $\mathbb{Z}$.

n-Torsion Group E[n]

Definition 136 Let $E/\mathbb{F}_q$ be an elliptic curve and $n \in \mathbb{N} \setminus \{0\}$. The n-torsion subgroup of E, denoted $E[n]$, is the set of points of order n in $E(\overline{\mathbb{F}}_q)$, i.e.

$$E[n] = \{P \in E(\overline{\mathbb{F}}_q) : nP = \mathcal{O}\}.$$

Furthermore $E[n](\mathbb{F}_q) = \{P \in E(\mathbb{F}_q) : nP = \mathcal{O}\}$.

Theorem 76 ([82], III.6.4 and V.3.1) *Let E be an elliptic curve defined over* $\mathbb{F}_q$, *char*$(\mathbb{F}_q) = p$, $n \in \mathbb{N}$, $n \neq 0$.

(i) If n is prime to q then $E[n] \simeq \mathbb{Z}_n \oplus \mathbb{Z}_n$.
(ii) If $n = p^e$, $e \in \mathbb{N}$, *then*

$$E[p^e] \simeq \begin{cases} \{\mathcal{O}\}, & \text{if } E \text{ is supersingular} \\ \mathbb{Z}_{p^e}, & \text{if } E \text{ is non-supersingular} \end{cases}$$

Example 15 Let $E/\mathbb{F}_q : y^2 = x^3 + ax + b$, $q = p^m$, where p is a prime greater 3. Then we know that $P = (x_P, y_P) \in E(\mathbb{F}_q)$ has order 2 if and only if $P = -P = (x_P, -y_P)$, i.e. $y_P = 0$. Let x_1, x_2, x_3 be the roots of $x^3 + ax + b$. Since the discriminant $\Delta \neq 0$, all roots x_i are distinct. Hence

$$E[2] = \{\mathcal{O}, (x_1, 0), (x_2, 0), (x_3, 0)\}.$$

Anomalous Curves

Definition 137 Let $E/\mathbb{F}_q$ be an elliptic curve with $q = p^m$ a prime power.

(i) E is denoted **anomalous** if $E(F_q)$ contains a (rational) point $P \in E[p] \setminus \{\mathcal{O}\}$.
(ii) E is denoted **totally anomalous** if $\#E(\mathbb{F}_q) = q$.

Lemma 31 *Let $E/\mathbb{F}_q$ be an elliptic curve. Then E is anomalous if and only if one of the following conditions holds:*

(i) The qth*-Frobenius trace $tr(\Phi_E) \equiv 1 \pmod{p}$ or equivalently $\#E(\mathbb{F}_q) \equiv 0 \pmod{p}$.*
(ii) E is totally anomalous, provided $q = p \geq 7$ is prime.

By McKee [50] the density of (totally) anomalous curves over $\mathbb{F}_p$ is at most $O\left(\frac{1}{\sqrt{p}} \log p \log \log p\right)$.

Complementary Group

Definition 138 Let $p > 3$ be a prime. Let $E_{a,b}/\mathbb{F}_p : y^2 = x^3 + ax + b$ be an elliptic curve. Then define the **complementary group**

$$\overline{E_{a,b}(\mathbb{F}_p)} = \{(x, y) \in \mathbb{F}_p \times \mathbb{F}_p : y^2 v = x^3 + ax + b\} \cup \{\mathcal{O}\},$$

where v is a fixed quadratic non-residue in $\mathbb{F}_p^*$. The **twist** $\hat{E}/\mathbb{F}_p$ of $E/\mathbb{F}_p$ is the curve $E : y^2 = x^3 + v^2 ax + v^3 b$, where $v \in \mathbb{F}_p^*$ is any non-square.

Lemma 32 *Let $p > 3$ be a prime.*

(i) $\overline{E_{a,b}(\mathbb{F}_p)}$ forms an abelian group with the identity element $\mathcal{O}$ and the addition law identical to (4.2.20).
(ii) For all $x \in \mathbb{F}_p$ there exists an $y \in \mathbb{F}_p$ such that $(x, y) \in E_{a,b}(\mathbb{F}_p)$ or $(x, y) \in \overline{E_{a,b}(\mathbb{F}_p)}$.
(iii) If $\#E_{a,b}(\mathbb{F}_p) = p + 1 - t$ then $\#\overline{E_{a,b}(\mathbb{F}_p)} = p + 1 + t$.

The same is valid for twist $\hat{E}$ of E.

For more information on the complementary group, see [82], Chap. X: Twisted group.

Isomorphism Classes

Theorem 77 ([51]) *Let $q = p^m$, p a prime. Then the number of isomorphism classes of elliptic curves over the finite field $\mathbb{F}_q$ are given as follows:*

(i) If $p = 2$, then

$$\#\{E : E/\mathbb{F}_{2^m} \text{ ell.curve}\}/\cong\mathbb{F}_{2^m} = \begin{cases} 2q-2, & E/\mathbb{F}_{2^m} \text{ non-supersingular} \\ 3, & E/\mathbb{F}_{2^m} \text{ supersing.}, m \text{ odd} \\ 7, & E/\mathbb{F}_{2^m} \text{ supersing.}, m \text{ even} \end{cases}$$

(ii) If $p > 3$, then

$$\#\{E : E/\mathbb{F}_q \text{ ell. curve}\}/\cong\mathbb{F}_q = \begin{cases} 2q+6, & q \equiv 1 \mod 12 \\ 2q+2, & q \equiv 5 \mod 12 \\ 2q+4, & q \equiv 7 \mod 12 \\ 2q, & q \equiv 11 \mod 12 \end{cases}$$

Notice that in Theorem 77(ii) these are the only possibilities for $q \mod 12$, since $\gcd(q, 6) = 1$.

For more details on isomorphism classes over $\mathbb{F}_{2^m}$, especially for supersingular curves, cf. [51], Chap. 3.

Divisor Theory

We will only give a short introduction into the divisor theory in order to do calculus. For a deeper treatment of this topic see [82] or for arbitrary genus [44], Chap. 2. Let $E/\mathbb{F}_q$ be an elliptic curve, q a prime power. For convenience we define $K = \mathbb{F}_q$.

Definition 139 (i) A **divisor** D of an elliptic curve E/K is a formal sum of $\overline{K}$ points

$$D = \sum_{P \in E} n_P(P) \tag{4.2.26}$$

with $n_P \in \mathbb{Z}$ and $n_P = 0$ for all but finitely many $P \in E$.

(ii) $\text{supp}(D) = \{P \in E : n_P \neq 0\}$ is called **support** of a divisor D.

(iii) Let $D = \sum_{P \in E} n_P(P)$ be a divisor. If $n_P \geq 0$ for all $P \in \text{supp}\{D\}$, then D is denoted **positive**.

(iv) The degree of D is defined by

$$\deg D = \sum_{P \in E} n_P \in \mathbb{Z}.$$

Let $\text{Div}(E)$ denote the set of all divisors and $\text{Div}^0(E) = \{D \in \text{Div}(E) : \deg D = 0\}$ the divisors of degree 0. Then $\text{Div}(E)$ is the free abelian group generated by the points of E under the addition

$$\sum_{P \in E} n_P(P) + \sum_{P \in E} m_P(E) = \sum_{P \in E} (n_P + m_P)(P).$$

Hence $\text{Div}^0(E)$ is a subgroup of $\text{Div}(E)$.

Let $f \in \overline{K}(E)^*$. For each $P \in E$ there exists functions $s, t_P \in \overline{K}(E)$, $t_P(P) = 0$, $s(P) \neq 0, \infty$ and a $d \in \mathbb{Z}$ independent of t_P, such that $f = t_P^d s$. Then $\text{ord}_P(f) = d$ is the **order** of f at P. A **local parameter** at P is a function t_P such that $\text{ord}_P(t_P) = 1$. With the following lemma we can easily derive the order of a function at a point P.

Lemma 33 *Let $P \in E$, $f \in \overline{K}(E)$.*

(i) P is a zero of f, i.e. $f(P) = 0 \iff \text{ord}_P(f) > 0$
In this case $\text{ord}_P(f)$ is the multiplicity of the zero.
(ii) P is a pole of f, i.e. $f(P) = \infty \iff \text{ord}_P(f) < 0$
In this case $\text{ord}_P(f)$ is the multiplicity of the pole.

Example 16 Let $P \in E(\mathbb{F}_q) \setminus \{\mathcal{O}\}$ where $E_{a,b}/\mathbb{F}_q$ is an elliptic curve.

(i) If $P \notin E[2]$ then $t_P = x - x_P$, since t_P has a zero with multiplicity 1 at P, i.e. $\text{ord}_P(x - x_p) = 1$.
(ii) If $P \in E[2]$ then $t_P = y$, since $P = (x_P, 0)$, i.e. $\text{ord}(y) = 1$. Note that if P has order 2, $P = (x_P, y_P) = (x_P, -y_P) = -P$, since $2P = \mathcal{O}$.
(iii) Using projective coordinates we have $\text{ord}_{(0:1:0)}(X/Y) = 1$. Since $x = X/Z$ and $y = Y/Z$ in affine coordinates we get $\text{ord}_{\mathcal{O}}(x/y) = 1$. Thus $t_{\mathcal{O}} = x/y$ or $t_{\mathcal{O}} = -x/y$.

Defining the *divisors of a function* $f \in \overline{K}(E)$ as

$$\div(f) = \sum_{P \in E} \text{ord}_P(f)(P)$$

is possible, since f has only a finite number of zeros and poles.

Theorem 78 ([82], Proposition II.3.1) *Let E be an elliptic curve, $f \in \overline{K}(E)^*$.*

(i) $\div(f) = 0 \iff f \in \overline{K}^*$.
(ii) $\deg(\div(f)) = 0$, i.e. $\div(f) \in Div^0(E)$.

Definition 140 A divisor $D \in \text{Div}^0(E)$ is **principal** if $D = \div(f)$ for some $f \in \overline{K}(E)^*$.

Example 17 Let $E/\mathbb{F}_q : y^2 = x^3 + ax + b$, $\text{char}(K) \neq 2, 3$.

(i) If $P = (c, d) \notin E[2]$ then $\div(x - c) = (P) + (-P) - 2(\mathcal{O})$.
(ii) If $P = (c, 0) \in E[2]$ then $\div(x - c) = 2(P) - 2(\mathcal{O})$.
(iii) If $P_1, P_2, P_3 \in E[2]$ then $\div(y) = (P_1) + (P_2) + (P_3) - 3(\mathcal{O})$.

Let $\text{Div}_p(E) \subseteq \text{Div}^0(E)$ be the set of all principal divisors. If $f_1, f_2 \in \overline{K}(E)$ then

$$\begin{aligned}\div(f_1 f_2) &= \sum_{P \in E} \text{ord}_P(f_1 f_2)(P) = \sum_{P \in E} \text{ord}_P(f_1)(P) + \sum_{P \in E} \text{ord}_P(f_2)(P) \\ &= \div(f_1) + \div(f_2).\end{aligned}$$

Hence $\text{Div}_p(E)$ is a subgroup of $\text{Div}^0(E)$. The 0-part of the divisor class group (or the Picard group) of E is the quotient group $\text{Pic}^0(E) = \text{Div}^0(E)/\text{Div}_p(E)$.

Two divisors $D_1, D_2 \in \text{Div}^0(E)$ are said to be *linearly equivalent*, denoted $D_1 \sim D_2$, if $D_1 - D_2 \in \text{Div}_p(E)$.

Theorem 79 ([82], Proposition III.3.4) *Let $E/\mathbb{F}_q$ be an elliptic curve.*

(i) For each divisor $D \in Div^0(E)$ there exists a unique point $Q \in E$ such that $D \sim (Q) - (\mathcal{O})$. Let $\sigma : Div^0(E) \to E$ be the map given by this association.
(ii) σ induces a bijection of sets

$$\Phi : Pic^0(E) \to E$$

with the inverse map $\chi : E \to Pic^0(E)$, $P \mapsto \overline{(P) - (\mathcal{O})}$, i.e. class of $((P) - (\mathcal{O}))$.
(iii) If E is given by a Weierstrass equation then the "chord-and-triangle law" (4.2.11) on E and the group law induced from $Pic^0(E)$ by using σ are the same, i.e. if $P, Q \in E$ then $\chi(P \oplus Q) = \chi(P) + \chi(Q)$, where "+" is the addition of divisor classes in $Pic^0(E)$ and $\oplus$ is the addition on E.

It can be shown that σ is given by $\sigma(n_P(P)) = n_P P$. Hence we get a useful corollary to characterize principal divisors:

Corollary 9 *Let $D = \sum n_P(P)$ be an divisor. Then D is principal if and only if $\sum n_P = 0$ and $\sum n_P P = \mathcal{O}$*

Proof From Definition 140 every principal divisor has $\deg D = \sum n_P = 0$. Since $D \in \text{Div}^0(E)$ Theorem 79(i), (ii) implies

$$D \sim 0 \Longleftrightarrow \sigma(D) = \mathcal{O} \Longleftrightarrow \sum n_P \sigma((P) - (\mathcal{O})) = \sum n_P P - n_P \mathcal{O} = \mathcal{O},$$

where $0 = \sum 0 \cdot (P)$. □

Remark 36 Any divisor $D \in \text{Div}^0(E)$, E an elliptic curve, can be written as $D = (Q) - (\mathcal{O}) + \div(f)$ for an unique $Q \in E$ and some $f \in \overline{K}(E)$, which is determined up to multiplication by a nonzero element of $\overline{K}$, since $D \sim (Q) - (\mathcal{O})$ for a unique $Q \in E$ and so $D - (Q) - (\mathcal{O}) \in \text{Div}_p(E)$.

Lemma 34 *Let $D_1, D_2 \in Div^0(E)$ such that $D_1 = (P_1) - (\mathcal{O}) + \div(f_1)$ and $D_2 = (P_2) - (\mathcal{O}) + \div(f_2)$, with $P_1, P_2 \in E(K) \setminus \{\mathcal{O}\}$ and $f_1, f_2 \in K(E)$. Then*

$$D_1 + D_2 = (P_3) - (\mathcal{O}) + \div(f_1 f_2 f_3),$$

where $P_3 = P_1 + P_2$ and $f_3 = \frac{l}{v}$, with $l = \overline{P_1 P_2}$ and $v = \overline{P_3 \mathcal{O}}$ if $P_3 \neq \mathcal{O}$, else $v = 1$.

Proof Since $P_1, P_2 \neq 0$, we have

$$\div (l) = (P_1) + (P_2) + (-P_3) - 3(\mathcal{O}), \tag{4.2.27}$$

noting that $-P_3$ is the third point of intersection on l, and

$$\div (v) = \begin{cases} (P_3) + (-P_3) - 2(\mathcal{O}), & \text{if } P_3 \neq \mathcal{O} \\ 0, & \text{if } P_3 = \mathcal{O} \end{cases} \tag{4.2.28}$$

Hence,

$$\begin{aligned} D_1 + D_2 &= (P_1) + (P_2) - 2(\mathcal{O}) + \div(f_1) + \div(f_2) \\ &= \div(l) + (P_3) - \div(v) - (\mathcal{O}) + \div(f_1 f_2) \\ &= (P_3) - (\mathcal{O}) + \div(l) - \div(v) + \div(f_1 f_2) \\ &= (P_3) - (\mathcal{O}) + \div(f_1 f_2 f_3), \end{aligned}$$

since $\div(f_3) = \div(l) - \div(v)$. □

Observe that all the computations take place in the field K. For an algorithm how to compute f_3 we refer to [51].

Corollary 10 *Let $A_1 = (P_1)$ and $A_2 = (P_2)$ be positive divisors of degree 1. Let P_3 and $h = f_3$ as in Lemma 34. Then*

$$\div(h) = A_1 + A_2 - A_3 - (\mathcal{O}), \ \textit{where } A_3 = (P_3)$$

Proof trivial.

Example 18 Let $E_{a,b}/\mathbb{F}_q$, $\text{char}(\mathbb{F}_q) \neq 2, 3$ be an elliptic curve. We want to evaluate $h = l/v$ in $K(x, y)$ for $P_1 \neq -P_2$. Since $\div(v) = (P_3) + (-P_3) - 2(\mathcal{O})$, we get $v(x, y) : x - x_{P_3} = 0$. Note that (P_3) and $(-P_3)$ are the zeros of v. Since $\div(l) = (P_1) + (P_2) + (-P_3) - 3(\mathcal{O})$, we can get the defining equations $y_{P_1} = \delta x_{P_1} + \alpha$ and $y_{P_2} = \delta x_{P_2} + \alpha$ for the straight line $l(x, y) : \ y = \delta x + \alpha$. It is easy to see that δ is the slope of $\overline{P_1 P_2}$. (As usual take the tangent line to E, if $P_1 = P_2$.) If $P_1 = -P_2$ then $\div(l) = (P_1) + (-P_1) - 2(\mathcal{O})$ and $\div(v) = 0$. Then we can take $h(x, y) : \ x - x_{P_1} = 0$.

Let $\Omega(E)$ ($\Omega^1(E)$) denote the $\mathbb{F}_q$-vector-space of *(holomorphic) differential forms* on an elliptic curve E.

Theorem 80 ([82], Proposition II.4.3) *Let $P \in E$ and $t \in \overline{K}(E)$ be a local parameter at P.*

(i) For every $\omega \in \Omega(E)$ there exists a unique function $g \in \overline{K}(E)$, depending on ω and t, such that $\omega = g\dot{t}$.
(ii) Let $f \in \overline{K}(E)$ be regular at P. Then $f/\dot{t}$ is also regular at P.
(iii) Let $\omega \in \Omega(E)$. $ord_P(\omega/\dot{t})$ depends only on ω and P.

Definition 141 Let $\omega \in \Omega(E)$.

(i) The **divisor of** ω is given by

$$\div(\omega) = \sum_{P \in E} \mathrm{ord}_P(\omega) \in \mathrm{Div}(E).$$

(ii) ω is denoted **holomorphic**, if $\mathrm{ord}_P(\omega) \geq 0$ for all $P \in E$. If $\mathrm{ord}_P(\omega) \leq 0$ for all $P \in E$ ω is **non-vanishing.**

Example 19 Let $E > y^2 = x^3 + ax + b$ be an elliptic curve. Let P_1, P_2, P_3 be the points of order 2. Then $\div(\dot{x}) = (P_1) + (P_2) + (P_3) - 3(\mathcal{O})$, since $\dot{x} = (\dot{x} - x_{P_i}) = -\dot{x}^2(1/x)$. Hence $\div(\dot{x}/y) = 0$ and $\dot{x}/y$ is as well holomorphic as non-vanishing.

Corollary 11 *Let E be an elliptic curve. Then* $\Omega^1(E) \simeq \overline{K}$.

For a proof see [82] Sect. II.5. This result is a direct corollary of the Riemann–Roch theorem, since the genus of an elliptic curve is 1.

The Weil Pairing

Let n be a positive integer relatively prime to q. Let $P, Q \in E[n]$ and $D_P \in \overline{(P) - (\mathcal{O})}$, $D_Q \in \overline{(Q) - (\mathcal{O})}$ divisors with different support. Then the Weil-Pairing e_n is a function

$$\begin{aligned} e_n : E[n] \times E[n] &\rightarrow \overline{\mathbb{F}}_q^*. \\ (P, Q) &\mapsto f_P(D_Q)/f_Q(D_P), \end{aligned}$$

where f_P, f_Q are functions on E such that $f_P = nD_P$ and $f_Q = nD_Q$.

The Weil e_n-pairing has for all $P,\ Q,\ R \in E[n]$ these important properties (cf. [82] III.8):

(i) Identity: $e_n(P,\ P) = 1$.
(ii) Bilinearity: $e_n(P + Q,\ R) = e_n(P,\ R)\, e_n(Q,\ R)$ and $e_n(P,\ Q + R) = e_n(P,\ Q)\, e_n(P,\ R)$.
(iii) Alternation: $e_n(P,\ Q) = e_n(Q,\ P)^{-1}$.
(iv) Non-degeneracy: If $S \in E[n]$ then $e_n(S, \mathcal{O}) = 1$. If $e_n(S,\ T) = 1$ for all $S \in E[n]$, then $T = 0$.
(v) Galois compatibility: If $E[n] \subseteq E(F_{q^k})$, then $e_n(P,\ Q) \in \mathbb{F}_{q^k}$.

Remark 37 Miller has developed an efficient probabilistic polynomial-time algorithm for computing the Weil-Pairing. For a summarized explanation and example computations, see [51, Chap. 5]. For a short implementation. see [32], Appendix A.12.2.

Lemma 35 ([51]) *Let* $E(\mathbb{F}_q)$ *be an elliptic curve with*

(i) group type (n_1, n_2), *and* $P \in E(\mathbb{F}_q)$ *such that* $ord(P)|n_1$. *Then for all* $P_1, P_2 \in E(\mathbb{F}_q)$ P_1 *and* P_2 *are in the same coset of* $< P >$ *if and only if* $e_{n_1}(P, P_1) = e_{n_1}(P, P_2)$.

(ii) $E[n] \subseteq E(\mathbb{F}_q)$, *where* $n \in \mathbb{N}$ *is coprime to* q *and* $P \in E[n]$ *such that* $ord(P) = n$. *Then for all* $P_1, P_2 \in E[n]$, P_1 *and* P_2 *are in the same coset of* $< P >$ *within* $E[n]$ *if and only if* $e_{n_1}(P, P_1) = e_{n_1}(P, P_2)$.

4.2.4 Elliptic Curves over the Ring $\mathbb{Z}_n$

Definition 142 Let $n \in \mathbb{N}$ with $\gcd(n, 6) = 1$. The equation

$$E_{a,b} : y^2 = x^3 + ax + b \tag{4.2.29}$$

defines an **elliptic curve over the ring $\mathbb{Z}_n$**, where $a, b \in \mathbb{Z}_n$ and $\gcd(4a^3 + 27b^2,\ n) = 1$.

$$E_{a,b}(\mathbb{Z}_n) := \{(x, y) \in \mathbb{Z}_n \times \mathbb{Z}_n : y^2 = x^3 + ax + b\} \cup \mathcal{O}_n \tag{4.2.30}$$

denotes the points on $E_{a,b}$, where $\mathcal{O}_n$ is a point at infinity.

A pseudo-addition can be defined on $E_{a,b}(\mathbb{Z}_n)$ in the same way addition on $E_{\bar{a},\bar{b}}(\mathbb{F}_p)$, p prime, $\bar{a}$, $\bar{b}$ the congruence class containing a or $b \mod p$, is defined by simply replacing all operations in $\mathbb{F}_p$ with operations in $\mathbb{Z}_n$. Therefore the definition

$$P_p := \begin{cases} (\bar{x}, \bar{y}), & \text{if } P = (x, y) \\ \mathcal{O}_p, & \text{if } P = \mathcal{O}_n, \end{cases}$$

$\mathcal{O}_p$ being the point at infinity in $E_{\bar{a},\bar{b}}(\mathbb{F}_p)$, is very useful, since $P_p \in E_{\bar{a},\bar{b}}(\mathbb{F}_p)$.

But since division is not always possible modulo n, an elliptic curve over $\mathbb{Z}_n$ does not form a group.

Example 20 Let $n = 493$ and $P = (97, 319)$, $Q = (12, 124)$ points in $E_{0,480}(\mathbb{Z}_n)$. Using the addition-formula (4.2.20) in order to calculate the x-coordinate of $P + Q$ we get

$$\begin{aligned} x_{P+Q} &= \left(\frac{y_P - y_Q}{x_P - x_Q}\right) - x_P - x_Q \mod n \\ &= \left(\frac{195}{85}\right) - 109 \mod 493 \end{aligned}$$

Since $\gcd(85, 493) = 17 > 1$ there is no inverse of 85 in $\mathbb{Z}_n$. Hence, we can not evaluate the formula!

Note that we found the idea of the elliptic curve factorization method in the example above: When the pseudo-addition in $E(\mathbb{Z}_n)$, n a composite number, is not defined over $\mathbb{Z}_n$, then we automatically find an nontrivial divisor of n. This nontrivial divisor is given by $\gcd(x_P - x_Q, n)$ if $P \neq Q$ or $\gcd(2y_P, n)$ if $P = Q$, where $P, Q \in E_{a,b}(\mathbb{Z}_n)$ such that (4.2.20) is undefined.

If $n = pq$, p and q prime, then using the Chinese Remainder Theorem any $c \in \mathbb{Z}_n$ can be uniquely represented by a pair $(c_p, c_q) \in \mathbb{Z}_p \times \mathbb{Z}_q$. Hence, every point $P = (x, y) \in E_{a,b}(\mathbb{Z}_n)$ can be uniquely represented by a pair $(P_p, P_q) = ((x_p, y_p), (x_q, y_q))$ such that $P_p \in E_{\bar{a}_p,\bar{b}_p}(\mathbb{F}_p)$ and $P_q \in E_{\bar{a}_q,\bar{b}_q}(\mathbb{F}_q)$ with $\mathcal{O}_n$ represented by $(\mathcal{O}_p, \mathcal{O}_q)$, consisting of points at infinity on $E_{\bar{a}_p,\bar{b}_p}(\mathbb{F}_p)$ and $E_{\bar{a}_q,\bar{b}_q}(\mathbb{F}_q)$, respectively.

Now if the pseudo-addition on $E_{a,b}(\mathbb{Z}_n)$ is well-defined, then this addition is equivalent to the componentwise addition on $E_{\bar{a}_p,\bar{b}_p}(\mathbb{F}_p) \times E_{\bar{a}_q,\bar{b}_q}(\mathbb{F}_q)$. Notice that $(P_p, \mathcal{O}_q)$ for $P_p \in E_{\bar{a}_p,\bar{b}_p}(\mathbb{F}_p) \setminus \{\mathcal{O}_p\}$ has no representation in $E(\mathbb{Z}_n)$. The same yields for $(\mathcal{O}_p, P_q)$ for $P_q \in E_{\bar{a}_q,\bar{b}_q}(\mathbb{F}_q) \setminus \{\mathcal{O}_q\}$.

More generally let n be a arbitrary composite number and $P, Q \in E_{a,b}(\mathbb{Z}_n)$. If $P + Q$ is undefined, we already saw that the addition law (4.2.20) may yield a non-trivial divisor of n (more precise the addition law must yield a non-trivial divisor of n, see Lemma 39). If $P + Q$ is well-defined then $(P + Q)_p = P_p + Q_p$ for all prime divisors p of n. By repeated application of the pseudo-addition we obtain a pseudo-multiplication kP for $k \in \mathbb{Z}$. Thus again if kP is undefined, we get a non-trivial divisor of n, else $(kP)_p = kP_p$ for all prime divisors p of n.

Since factoring a "big" composite number $n = pq$ is hard the probability that the pseudo-addition/multiplication fails, i.e. yields a non-trivial factor of n, is very small. In practice only factors of up to 30 decimal digits let the pseudo-addition fail with a reasonable probability. See ECM in subsection "Elliptic Curve Method" of Sect. 4.4.3 for more details.

For ease of notation we skip "pseudo-" if the situation is clear.

The following lemma is crucial for elliptic curve cryptosystems over the ring $\mathbb{Z}_n$:

Lemma 36 *Let $E_{a,b}/\mathbb{Z}_n$ be an elliptic curve such that n is the product of two primes p and q. Let $N_n = \mathrm{lcm}(\#E_{\bar{a}_p,\bar{b}_p}(\mathbb{F}_p), \#E_{\bar{a}_q,\bar{b}_q}(\mathbb{F}_q))$. Then for any $P \in E_{a,b}(\mathbb{Z}_n)$ and any $k \in \mathbb{Z}$*

$$(kN_n + 1)P = P \text{over} E_{a,b}(\mathbb{Z}_n).$$

The proof can be obtained using the Chinese Remainder Theorem.

4.2.5 *Elliptic Curves over $\mathbb{Q}$*

Definition 143 The equation

$$E_{a,b} : y^2 = x^3 + ax + b \tag{4.2.31}$$

defines an **elliptic curve over the rational numbers** $\mathbb{Q}$, if $a, b \in \mathbb{Q}$ and $\Delta(E_{a,b}) = -16(4a^3 + 27b^2) \neq 0$.

Definition 144 Let the **torsion subgroup $\mathbf{E_{tors}}$** of an elliptic curve E be the set of points of finite order, i.e.

$$E_{tors} = \bigcup_{n=1}^{\infty} E[n],$$

where $E[n]$ denotes the set of points of order n in E.

Mordell proved that the abelian group $E(\mathbb{Q})$ is finitely generated, i.e. it consists of a finite torsion subgroup E_{tors} and the subgroup generated by a finite number of points of infinite order:

$$E(\mathbb{Q}) \simeq E_{tors} \oplus \mathbb{Z}^r,$$

where r, the number of generators needed for the infinite part, is denoted by *rank*.

Lemma 37 *Let $E_{a,b}$ be an elliptic curve over $\mathbb{Q}$ and p be a prime such that p does not divide either the denominator of a or b or the discriminant $\Delta(E_{a,b})$. Then $E : y^2 = x^3 + \hat{a}x + \hat{b}$ with $\hat{b} = b \pmod{p}$ and $\hat{a} = a \pmod{p}$ is an elliptic curve over $\mathbb{F}_p$, denoted* $\mathbf{E}$ (*mod* $\mathbf{p}$).

4.3 Elliptic Curves: Algorithms

4.3.1 Efficient m-fold Addition in $E(\mathbb{F}_p)$

For this section let

$$E = E_{a,b} : y^2 = x^3 + ax + b$$

be an elliptic curve defined over $\mathbb{F}_q$, where $q = p^m$ and p is a prime greater 3. Since (scalar) multiplication, i.e. k-fold addition, is the most time consuming part in the communication part of the public-key schemes following in the next two chapters, we want to give several methods to evaluate kP faster than calculating $\underbrace{P + P + \cdots + P}_{k}$, where $P \in E_{a,b}(\mathbb{F}_q)$ and k an integer. Notice that this method would take $O(k)$ multiplications in $\mathbb{F}_q$.

Let $P = (x_1, y_1)$ and $kP = (x_k, y_k)$. By $x(P)$ we will denote the x-coordinate of P, i.e. $x(P) = x_1$. Similarly $y(P) = y_1$.

Addition–Subtraction Method

Let d denote the time to double the same point and a (=s) the time to add (subtract) two distinct points in $E(\mathbb{F}_q)$. Than we get the following table using the addition formula (4.2.20) of Sect. 4.2:

Operation	Multiplications in $\mathbb{F}_q$	Inversions in $\mathbb{F}_q$
Doubling d	3	1
Addition a	4	1

Remark 38 If $q = 2^m$, we can decrease the evaluation of the addition formulas in subsection "Curves over K, $\text{char}(K) = 2$" of Sect. 4.2.2 easily to 2 multiplications, one squaring and one inversion in $\mathbb{F}_{2^m}$, see [32], Appendix A 10.2.

The principal question now is: What is the smallest number of additions necessary to compute kP, provided we may only sum two already-computed summations? Or equivalently: What is the shortest addition chain for k? To use a more mathematical point of view we will first introduce a few definitions referring to [27].

Definition 145 An **addition chain** for k is a list of positive integers

$$a_1 = 1, \quad a_2, \dots, a_t = k, \qquad 1 \leq i \leq t, \; a_i \in \mathbb{N},$$

such that, for every $i > 1$, there is some j and k with $1 \leq j \leq k < i$ and

$$a_i = a_j + a_k.$$

Let $l(k)$ denote the length of the shortest addition chain for k.

If we find a short addition chain we immediately get a fast algorithm to compute kP. Hence it would be interesting to know $l(k)$, the length of the shortest addition chain, but $l(k)$ is only known exactly for small values k. For k large we have

$$l(k) = \log_2 k + (1 + o(1))\frac{\log_2 k}{\log_2 \log_2 k}. \tag{4.3.1}$$

The upper bound is given by the m-ary method below, the lower bound was shown by Erdös in [21]. The problem of finding the shortest addition chain was shown by Downey et al. [18] to be NP-complete.

Beside this we will give the first algorithm using additions chains, but not necessarily the shortest. We assume that the binary representation of k,

$$k = \sum_{i=0}^{t} a_i 2^i, \quad a_i \in \{0, 1\},$$

is given. Note that $t = \lfloor \log_2 k \rfloor$.

Repeated Doubling method
Require: $P \in E(\mathbb{F}_q)$ and $k = \sum_{i=0}^{t} a_i 2^i,\ a_i \in \{0, 1\}, t = \lfloor \log_2 k \rfloor$
    Set $R = P$
    for $i = t$ downto 0 **do**
        evaluate $R = R + R$
        if $a_i = 1$ **then**
            compute $R = R + P$
        end if
    end for
Ensure: $R = kP$

This method takes $\lfloor \log_2 k \rfloor$ doublings d and $\lfloor \log_2 k \rfloor$ additions a in the worst case and $\frac{1}{2}\lfloor \log_2 k \rfloor$ additions a in the average case. It is the direct analogue of the repeated squaring method in subsection "Square Root Methods" of Sect. 4.1.5. Since $\lfloor \log_2 k \rfloor$ doublings are the lower bound needed, we will try to improve this bound next.

Therefore notice that subtracting the point $P = (x_1, y_1)$ we only need to add the point $-P = (x_1, -y_1)$, i.e. inversion in $E(\mathbb{F}_q)$ is not expensive or, to be more precise, inversion in $E(\mathbb{F}_q)$ is done in almost the same time as addition in $E(\mathbb{F}_q)$. Even in the binary case $q = 2^m$ subtraction is cheap, see subsection "Curves over K, char$(K) = 2$" of Sect. 4.2.2.

Definition 146 An **addition–subtraction chain** for k is a list of integers

$$a_1 = 1, \quad a_2, \dots, a_t = k, \qquad 1 \le i \le t,\ a_i \in \mathbb{Z}, \tag{4.3.2}$$

such that, for every $i > 1$, there is some j and k with $1 \le j \le k < i$ and

$$a_i = a_j + a_k.$$

Let $l(k)$ again denote the length of the shortest addition–subtraction chain for k.

It's easy to see that an addition chain is always an addition–subtraction chain and the next example shows that addition–subtraction chains may be shorter.

Example 21 Let $k = 63$. The shortest addition chain is

$$1, 2, 3, 5, 10, 15, 30, 60, 63$$

and an immediately shorter addition–subtraction chain is given by

$$1, 2, 4, 8, 16, 32, 64, 63.$$

This gives the following algorithm:

Addition and Subtraction method
Require: $P \in E(\mathbb{F}_q)$ and $k = \sum_{i=0}^{t} a_i 2^i, \quad a_i \in \{-1, 0, 1\}$
Set $R = P$
for $i = t$ downto 0 **do**
Evaluate $R = 2R$
Compute $R = R + a_i P$, where addition/subtraction is possible
end for
Ensure: $R = kP$

Note that now $t \leq \lfloor \log_2 k \rfloor$, thus this algorithm will work faster in general. For a better implementation, see [32].

How to find a representation (4.3.2) such that the addition/subtractions are minimal (since doubling is even less expensive than addition/subtraction)? We first give some useful definitions:

Definition 147 Let the **weight** of a representation (4.3.2) be given by

$$w(k) = \sum_{i=0}^{t} |a_i|.$$

A **nonadjacent form NAF(k)** of an integer k is a representation (4.3.2) with $a_i a_{i+1} = 0$ for all $i \geq 0$.

Example 22 Let $k = 29$. Then

$$NAF(29) = 1 \cdot 2^5 + 0 \cdot 2^4 + 0 \cdot 2^3 + (-1) \cdot 2^2 + 0 \cdot 2^1 + 1 \cdot 2^0 = [1, 0, 0, -1, 0, 1],$$

since $29 = 32 - 4 + 1$.

Theorem 81 ([27]) *Every integer k has exactly one $NAF(k)$. The weight $w(NAF(k))$ is the minimum for all $w(k_i)$, k_i a representation of k as in (4.3.2).*

Morain and Olivos showed the following theorem:

Theorem 82 ([61]) *The length of NAF(k) is at most one bit longer than the binary representation of k. The expected number of nonzeros in a NAF of length t is $t/3$.*

In 1989 Jedwad and Mitchell proposed an algorithm to find a NAF for any k.

NAF Find Algorithm

Require: $k = \sum_{i=0}^{t} a_i 2^i, \quad a_i \in \{-1, 0, 1\}$, e.g. the binary representation of k

  while $a_s a_{s+1} \neq 0$ for all $s, 0 \leq s \leq t-1$ **do**

    Let s be the least integer such that $a_s a_{s+1} \neq 0$.

    if $a_s \neq a_{s+1}$ **then**

      Set $a_s = -a_s$ and $a_{s+1} = 0$

    else

      Let r be the least integer such that $a_r \neq a_s$ and $a_{r-1} = \cdots = a_s$.

      if $a_r = 0$ **then**

        Set $a_t = a_s$.

      else

        Set $a_t = 0$ and $a_s = -a_s$ and $a_i = 0$ for all $s < i < t$.

      end if

    end if

  end while

Ensure: $NAF(k) = \sum_{i=0}^{t+1} a_i 2^i, \quad a_i \in \{-1, 0, 1\}$

After finding a NAF of k and applying the addition–subtraction method, we need $t+1$ doublings d and $w(NAF(k)) \approx \frac{t+1}{3}$ additions/subtractions a, where $t = \lfloor \log_2 k \rfloor$.

A further improvement is made using a base larger than 2 or 3 respectively. Then we get the **m-ary method**. Let

$$k = \sum_{i=0}^{t} a_i m^i, \qquad a_i \in \{0, 1, 2, \ldots, (m-1)\} \text{ for all } i, 0 \leq i \leq t,$$

and precompute $2P, 3P, \ldots, (m-1)P$ by the repeated doubling method.

Then we get the following algorithm:

m-ary method

Require: $P \in E(\mathbb{F}_q)$ and $k = \sum_{i=0}^{t} a_i m^i, \quad a_i \in \{0, 1, \ldots, (m-1)\}$

  Calculate $2P, 3P, \ldots, (m-1)P$ by a suitable method

  Set $R = P$

  for $i = t$ downto 0 **do**

    Evaluate R = mR by a suitable method above

    Compute $R = R + a_i P$

  end for

Ensure: $R = kP$

Especially if a very short addition–subtraction chain for m is known, e.g. $m = 2^l$ for l doublings or $m = 2^l \pm 1$, this method works fast for large k. For $m = 2^l$ we need 2^{l-1} additions a for precomputation, $\lfloor \log_2 k \rfloor$ doublings d and additionally $\lfloor \log_2 k \rfloor$ additions a for the worst case, i.e. all $a_i \neq 0$. Hence we can build an addition chain of length $2^{l-1} + (1 + 1/l)\lfloor \log_2 k \rfloor$. Minimizing this in l we get $l = \log_2 \log_2 k - 2 \log \log \log k$ for the minimum. Thus we have the upper bound for $l(k)$ in (4.3.1).

The m-ary method can also be extended using addition–subtraction chains and NAFs, but this seems to speed up the computation only slightly. See [43] for more details. For a survey over further improvements of this methods like the window method and precomputation, see [27].

The latest speedups were made by Solinas [87] for anomalous curves over $\mathbb{F}_{2^m}$, so-called anomalous binary curves (ABC). In this case the average number of elliptic additions is dropped to $\approx \lfloor \log_2 k \rfloor / 3$ additions a and no doublings. Since anomalous curves over odd prime finite fields may be insecure (subsection "Supersingular Curves" of Sect. 4.5.2) one has to choose these elliptic curves carefully. In Table 4.2 we summarized the running times of the methods above.

Hence the easiest way to speed up the scalar multiplication is done by an addition–subtraction chain with NAF representation of k.

Projective Coordinates Method

Because the addition formula in affine coordinates used in the last paragraph requires an inversion in $\mathbb{F}_q$, which is expensive in time, one can use projective coordinates to reduce the number of inversions in total. Since we can easily determine a point P in projective coordinates $(X_P : Y_P : Z_P)$ given the affine coordinates (x_P, y_P) by the rule $X_P = x_P,\ Y_P = y_P,\ Z = 1$ we can also do arithmetic in $\mathbb{P}^2(\mathbb{F}_q)$. The following addition formulas in projective space are obtained using the addition formulas from subsection "Curves over K, char$(K) \neq 2, 3$" of Sect. 4.2.2 for char$(\mathbb{F}_q) \neq 2, 3$.

Let $P = (X_P, Y_P, Z_P),\ Q = (X_Q, Y_Q, Z_Q)$. Assume that $P, Q \neq \mathcal{O}$ and $P \neq -Q$. We want to evaluate $R = P + Q = (X_R, Y_R, Z_R)$. If $P = Q$, i.e. $R = 2P$, we can apply Eq. (4.2.20) of Sect. 4.2 in order to evaluate $x_R = X_R/Z_R$. For simplicity define $N = 3X_P^2 + aZ_P^2$ and $D = 2Y_PZ_P$.

$$
\begin{aligned}
\frac{X_R}{Z_R} &= x_R = \left(\frac{3x_P^2 + a}{2y_P}\right) - 2x_P \\
&= \left(\frac{3\left(\frac{X_P}{Z_P}\right)^2 + a}{2\left(\frac{Y_P}{Z_P}\right)}\right)^2 - 2\left(\frac{X_P}{Z_P}\right)
\end{aligned}
$$

Table 4.2 Running times of several methods in d and a

Method	Worst case	Average case	Best case
Repeated Doubling	$t(a+d)$	$t(d+\frac{1}{2}a)$	td
Addition–Subtraction	$t(a+d)$	$\leq t(d+\frac{1}{2})a$	$\leq td$
Addition–Subtraction (NAF)	$(t+1)(a+d)$	$\leq (t+1)(d+\frac{1}{3}a)$	$\leq (t+1)d$
2^l – ary method	$(2^{l-1}+\frac{t}{l})a + td$	$(2^{l-1}+\frac{t}{2l})a + \frac{t}{2}d$	$2^{l-1}a + td$
Anomalous binary curve		$\approx \frac{t}{3}a$	

$t = \lfloor \log_2 k \rfloor$

$$
\begin{aligned}
&= \left(\frac{3X_P^2 + aZ_P^2}{2Y_PZ_P}\right)^2 - \frac{2X_P}{Z_P} \\
&= \frac{N^2}{D^2} - \frac{2X_P}{Z_P}. && (4.3.3)
\end{aligned}
$$

The same can be done for the y-coordinate substituting x_R by (4.3.3):

$$
\begin{aligned}
\frac{Y_R}{Z_R} &= y_R = \frac{3x_P^2 + a}{2y_P}(x_P - x_R) - y_P \\
&= \frac{3\left(\frac{X_P}{Z_P}\right)^2 + a}{2\left(\frac{Y_P}{Z_P}\right)}\left(\frac{X_P}{Z_P} - x_R\right) - \frac{Y_P}{Z_P} \\
&= \frac{N}{D}\left(\frac{X_P}{Z_P} - \frac{N^2}{D^2} + \frac{2X_P}{Z_P}\right) - \frac{Y_P}{Z_P} \\
&= \frac{3X_PN}{Z_PD} - \frac{N^3}{D^3} - \frac{Y_P}{Z_P}. && (4.3.4)
\end{aligned}
$$

When we now set $Z_R = D^3$, multiply (4.3.3) and (4.3.4) by Z_P, we obtain the following formulas for point doubling:

$$
\begin{aligned}
X_R &= (N^2 - 4X_PY_P)D^2 && (4.3.5) \\
Y_R &= 6X_PY_PND - N^3 - 2Y_P^2D^2 && (4.3.6) \\
Z_R &= D^3 && (4.3.7)
\end{aligned}
$$

We can do the doubling in 16 multiplications provided temporary variables are used.

If $P \neq Q$, we can apply Eq. (4.2.20) of Sect. 4.2. For simplicity define $N = (Y_QZ_P - Y_PZ_Q)$ and $D = (X_QZ_P - X_PZ_Q)$.

$$
\begin{aligned}
\frac{X_R}{Z_R} &= x_R = \left(\frac{y_Q - y_P}{x_Q - x_P}\right)^2 - x_P - x_Q \\
&= \left(\frac{\frac{Y_Q}{Z_Q} - \frac{Y_P}{Z_P}}{\frac{X_Q}{Z_Q} - \frac{X_P}{Z_P}}\right)^2 - \frac{X_P}{Z_P} - \frac{X_Q}{Z_Q} \\
&= \frac{(Y_QZ_P - Y_PZ_Q)^2}{(X_QZ_P - X_PZ_Q)^2} - \frac{X_P}{Z_P} - \frac{X_P}{Z_Q} \\
&= \frac{N^2}{D^2} - \frac{X_P}{Z_P} - \frac{X_Q}{Z_Q}. && (4.3.8)
\end{aligned}
$$

As above we can also evaluate y_R by substituting x_R by (4.3.8):

$$\begin{aligned}\frac{Y_R}{Z_R} &= y_R = \frac{y_Q - y_P}{x_Q - x_P}(x_P - x_R) - y_P \\ &= \frac{Y_Q Z_P - Y_P Z_Q}{X_Q Z_P - X_P Z_Q}\left(\frac{X_P}{Z_P} - x_R\right) - \frac{Y_P}{Z_P} \\ &= \frac{N}{D}\left(\frac{X_P}{Z_P} - \frac{N^2}{D^2} + \frac{X_P}{Z_P} + \frac{X_Q}{Z_Q}\right) - \frac{Y_P}{Z_P} \\ &= \frac{2X_P N}{Z_P D} - \frac{N^3}{D^3} + \frac{X_Q N}{Z_Q D} - \frac{Y_P}{Z_P} \end{aligned} \quad (4.3.9)$$

Hence defining $Z_R = Z_P Z_Q D^3$, multiplying (4.3.8) and (4.3.9) with Z_R, we get

$$X_R = Z_P Z_Q D N^2 - (X_P Z_Q + X_Q Z_P) D^3, \quad (4.3.10)$$
$$Y_R = (2X_P Z_Q + X_Q Z_P) N D^2 - Z_P Z_Q N^3 - Y_P Z_Q D^3, \quad (4.3.11)$$
$$Z_R = Z_P Z_Q D^3 \quad (4.3.12)$$

Observe that we can perform the addition within 16 multiplications provided, we have 9 variables to store calculation results.

Now we can use the repeated doubling method of the last paragraph to calculate kP in projective coordinates. After the computation in projective coordinates we can get the affine coordinates of R by $\left(\frac{X_R}{Z_R}, \frac{Y_R}{Z_R}\right)$.

Remark 39 In [32] projective methods for the cases $p = 2$ and $p > 3$ are given with even less multiplications needed as above.

p	Operation	Multiplications in $\mathbb{F}_{p^m}$	Squares in $\mathbb{F}_{2^m}$	Space
>3	Doubling d	≤10		5
	Addition a	≤16		≤7
2	Doubling d	5	5	4
	Addition a	≤15	5	≤9

Since $Z = 1$ in the conversion from affine to projective coordinates the methods can also be improved needing less time.

X-Coordinate Method

If only the x-coordinate of a product kP is needed, then we can apply the following method, if the curve is defined over $\mathbb{F}_p$, p prime.

Lemma 38 *Let* $P \in E_{a,b}(\mathbb{F}_p)$ *(or* $\overline{E_{a,b}(\mathbb{F}_p)}$*).*

(i) If $y_i \in \mathbb{F}_p^*$, *then*

$$x_{2i} = \frac{(x_i^2 - a)^2 - 8bx_i}{4(x_i^3 + ax_i + b)} \tag{4.3.13}$$

(ii) If $x_i \neq x_j$ *and* $y_i \in \mathbb{F}_p^*$ *then*

$$x_{i+j} = \frac{4b + 2(a + x_i x_j)(x_i + x_j)}{(x_i - x_j)^2} - x_{i-j}. \tag{4.3.14}$$

A proof for (i) is directly given by (4.2.13) and (4.2.8) of Sect. 4.2, setting $b_2 = b_8 = 0,\ b_4 = 2a,\ b_6 = 4b$. For (ii) see [11].

Setting $j = i + 1$ in (ii), we can quickly calculate x_{2i+1}. Hence if we want to calculate $x(kP)$, we apply the repeated doubling method only for x_i without using y_i:

Example 23 Let $k = 125 = 2^3(2(2(2(2+1)+1)+1)+1$ and $(x, y) \in E(\mathbb{Z}_p)$. Hence, computing $x_3 = x_{2+1},\ x_7 = x_{2\cdot 3+1},\ x_{15} = x_{2\cdot 7+1}, x_{31} = x_{2\cdot 15+1},\ x_{62} = x_{2\cdot 31}$ yields $x_{125} = x_{2\cdot 62+1}$ without calculating the y-coordinate in 6 steps. Note that this gets wrong, if there is an $i \in \{3, 7, 15, 31, 62, 125\}$ such that $y_i = 0 \bmod p$, since we can not use Lemma 38.

In oder to avoid $y_i = 0$ in the calculation we can use projective coordinates and therefore avoid division until the whole calculation as in subsection "Projective Coordinates Method" of Sect. 4.3.1: One can rewrite the Eqs. (4.3.13) and (4.3.14) to get a remainder part Z_{2i} and a divisor part X_{2i} of x_{2i}. The same is possible for x_{i+j}. For more details, see also [11].

4.3.2 *Finding Random Points in* $\mathrm{E}(\mathbb{F}_q)$

Points on an elliptic curve E given by a smooth Weierstrass equation can be found with the following probabilistic polynomial time algorithm. Assume q is odd (for $q = 2^m$ see [38]):

(i) Rewrite a given Weierstrass equation for E to $y^2 = f(x)$.
(ii) Choose a random $x \in \mathbb{F}_q$.

We have to determine, if $f(x)$ is a square in $\mathbb{F}_q^*$, because then there are (x, y) and $(x, -y)$ in $E(\mathbb{F}_q)$. By Theorem 69 we see that the probability that x is the x-coordinate of some point in $E(\mathbb{F}_q)$ is at least $\frac{1}{2} - \frac{1}{\sqrt{q}}$.

(iii) Calculate $f(x)$.
(iv) If $f(x) = 0$ or if $f(x)^{(q-1)/2} = -1$ go to step (ii).

No we have found x and $z = f(x)$, a square in $\mathbb{F}_q^*$. Finding a square root $y \in \mathbb{F}_q^*$ of z:

(v) If $q \equiv 3 \pmod 4$ then evaluate $y = z^{(q+1)/4}$. STOP. (since $y^2 = z^{(q+1)/2} = z \cdot z^{(q-1)/2}$)
(vi) if $q \equiv 1 \pmod 4$ then

(a) Find t odd, $s \geq 2$ such that $q - 1 = 2^s t$.
(b) Make a random search for $u \in \mathbb{F}_q^*$, such that $u^{(q-1)/2} = -1$ (i.e. u is a non-square in $\mathbb{F}_q^*$ by Eulers criterion).
(c) Evaluate $v = u^t$ (i.e. u is primitive (2^s)th root in $\mathbb{F}_q$)
(d) Search $l \in \mathbb{N}$ such that

$$v^{2l} = z^t. \tag{4.3.15}$$

(e) Compute $y = z^{(t+1)/2} v^{-l}$.

In order to find l in (4.3.15) rewrite $l = \sum_{i=0}^{s-2} l_i 2^i$. Then find $l_i,\ i = 0, \ldots, s-2$, by raising (4.3.15) inductively starting with $i = 0$ on both sides with 2^{s-2-i} and getting

$$l_i = \begin{cases} 0, & \text{if r.h.s. (4.3.15)} = 1 \\ 1, & \text{else} \end{cases}$$

4.3.3 Counting the Number of Points on $\mathbf{E}(\mathbb{F}_p)$

It is out of scope of this section to explain the methods of Schoof in order to count the number of rational points on an elliptic curve.

We will only give the main idea here, refering to [75].

Let E be an elliptic curve defined over $\mathbb{F}_p$, $\mathrm{char}(\mathbb{F}_p) > 3$, given by the Weierstrass equation

$$E : y^2 = x^3 + ax + b$$

and $\Phi_E \in \mathrm{End}(E)$ be the Frobenius endomorphism. As usual let $t = p + 1 - \#E(\mathbb{F}_p)$. Let $l < p$ be a prime. Then Φ_E induces an automorphism on the l-torsion group $E[l]$.

We know from Theorem 71 that $\mathbb{Z}[\Phi_E] \subseteq \mathrm{End}(E)$ and Φ_E satisfies the degree 2 equation.

$$\Phi_E^2 - t\Phi_E + p = 0. \tag{4.3.16}$$

We remark that (4.3.16) also is valid if we consider Φ_E as an element of the Galois group $G_{\overline{\mathbb{F}}_p/\mathbb{F}_p(E[l])}$.

Schoofs main idea is now to compute t modulo l by looking at the action of Φ_E on $E[l]$. This can be done using a special polynomial $f_l(x)$ of degree $(l^2 - 1)/2$. If we do this computation for primes l_i such that $\prod_i l_i > 4\sqrt{p}$, i.e. two times the Hasse bound, then we can compute the cardinality of E by the Chinese Remainder Theorem.

In practice the algorithm of Schoof gets infeasible for $l > 31$. But instead of calculating all $t \mod l_i$ Atkin and Elkies independently observed that not all f_{l_i} have to be computed. Both papers are unpublished by now. After further improvements of V. Müller [63] and Couveignes et al. [13, 14] using "isogeny cycles" $\#E(\mathbb{F}_q)$ can be computed in reasonable time in $O((\log q)^6)$.

4.4 Elliptic Curve Cryptosystems Based on Factorization

In this chapter we will present elliptic curve cryptoschemes where the security is based on factorization. The schemes are presented in the first section and attacks valid for these schemes in the second.

4.4.1 Cryptosystem Schemes

The first RSA-like elliptic curve cryptosystem was introduced 1991 [40], partially presented in subsection "KMOV Scheme". Then 1993 Demytko [15] proposed improvements of this scheme, see subsection "Demytko's Scheme". Here the main research was done in finding suitable TOFs in order to use the generalized cryptosystem in Sect. 4.1.3, subsection "RSA Cryptosystem". Recently Vanstone and Zuccherato [91], 1997, presented a new scheme with the security also based on factorization but the message held in the "exponent" and not in the group element.

Let $n = pq$, where p and q will denote large primes greater than 3.

KMOV Scheme

Three different TOFs based on elliptic curves over the ring $\mathbb{Z}_n$ were proposed in [40] by Koyama et al. The first class of functions, denoted a "type 0" TOF, can only be used in digital signature schemes. The second class, denoted "type 1" TOF, has the needed commutative property and will be used for the following KMOV scheme using restrictions on the primes p, q and the elliptic curves to be used. The third class, denoted "type 2", is the Rabin generalization of the "type 1" scheme.

Using Lemma 26 we get the following

KMOV Scheme

(i) **(Setup)** Each user i

(a) chooses two "large" primes p and q such that $p \equiv q \equiv 2 \pmod 3$ and computes $n = pq$,
(b) computes $N_n = \mathrm{lcm}(p+1, q+1)$ $(=\mathrm{lcm}(\#E_{0,b}(\mathbb{F}_p), \#E_{0,b}(\mathbb{F}_q)))$
(c) selects an integer e, $1 \leq e \leq N_n - 1$, such that $\gcd(e, N_n) = 1$
(d) evaluates an integer d, $1 \leq d \leq n - 1$, such that $ed = 1 \mod N_n$.
i's private-key is d and i's public-key is (n, e).

(ii) **(Communication)** User j wants to submit a plaintext $M = (x_M, y_M) \in \mathcal{M} = \mathbb{Z}_n \times \mathbb{Z}_n$ to person i, where $b = x_M^3 - y_M^2 \mod n$ is determined by M.

(enc) j sends $C = E(M) = eM \in E_{0,b}(\mathbb{Z}_n)$.
(dec) i recovers $M = D(C) = dC = deM \in E_{0,b}(\mathbb{Z}_n)$.

Note that b is never used in the encryption (enc) and decryption (dec) part, since the addition formulas do not need b, and furthermore $a = 0$. This will give a redundancy we will use in Sect. 4.4.2. Observe also that $E(M)$ in (enc) is a TOF using only the public key (e, n).

The main idea of this scheme is to use primes p, q such that $p \equiv q \equiv 2 \pmod 3$. Then we can apply the mentioned lemma and get $\#E_{0,b}(\mathbb{F}_p) = p + 1, \#E_{0,b}(\mathbb{F}_q) = q + 1$. This has the advantage that we do not have to count the number of points of two different elliptic curve groups. If we would take p and q of about 2^{500}, even the factorization methods described in Sect. 4.3.1 would take too long in practice.

Note that the addition in $E_{0,b}(\mathbb{Z}_n)$ may not be defined. In this case we would get an error. But the probability that this will occur is very small, see Sect. 4.2.4. Furthermore Koyama et. al. suggests to take $E_{a,0}$ instead of $E_{b,0}$ (cf. Lemma 27). For security considerations why the KMOV scheme is based on factorization, see [40], Chap. 7.

Demytko's Scheme

In 1993 Demytko [15] gave an improvement of the KMOV scheme, where the message is encrypted in the x-coordinate in order to overcome some of the attacks in Sect. 4.4.2.

Let $\left(\frac{w}{p}\right)$ denote the extended Legendre symbol.

Demytko Scheme

(i) **(Setup)** Each user i

(a) chooses two "large" primes p and q and computes $n = pq$,
(b) selects $a, b,\ 0 \le a, b \le n - 1$ such that $\gcd(4a^3 + 27b^2, n) = 1$.

To shorten notation we define $N_1 = \#E_{a,b}(\mathbb{F}_p)$, $N_2 = \#\overline{E_{a,b}(\mathbb{F}_p)}$, $N_3 = \#E_{a,b}(\mathbb{F}_q)$ and $N_4 = \#\overline{E_{a,b}(\mathbb{F}_q)}$, where $\bar{E}$ is the complementary group of E.

(c) Calculates N_1, N_2, N_3, N_4 by a method mentioned in Sect. 4.3.3.
(d) Chooses an integer $e,\ 0 \le e \le n - 1$ such that $\gcd(e, N_i) = 1$ for $i = 1, \dots, 4$.
(e) Computes d_1, d_2, d_3, d_4, determined by

$$ed_1 = 1 \pmod{\operatorname{lcm}(N_1, N_3)} \qquad ed_2 = 1 \pmod{\operatorname{lcm}(N_1, N_4)}$$
$$ed_3 = 1 \pmod{\operatorname{lcm}(N_2, N_3)} \qquad ed_4 = 1 \pmod{\operatorname{lcm}(N_2, N_4)}$$

i's public-key is (n, e, a, b) and i's private-key is (d_1, d_2, d_3, d_4).

(ii) **(Communication)** User j wants to submit a plaintext $x_M \in \mathbb{Z}_n \setminus \{0\}$ to i

(enc) j sends the x-coordinate $x_C \in \mathbb{Z}_n \setminus \{0\}$ of $C = e(x_M, y_M) \in E(\mathbb{Z}_n)$, where computation is done as in subsection "X-Coordinate Model" of Sect. 4.3.1, i.e. y_M is not needed.

(dec) i recovers the message x_M by computing $w = x_C^3 + ax_C + b \mod n$ and obtaining the plaintext $x_M = d_i(x_C, y_C)$, where

$$i = \frac{1}{2}\left(2\left(\frac{w}{p}\right) + \left(\frac{w}{q}\right) + 5\right).$$

Here again y_C may be unknown, see the aforementioned subsection for computation.

Note that only the first coordinate has to be calculated in the communication part, so many attacks on the KMOV scheme are useless in this case.

Here the main disadvantage is to calculate $N_i,\ i = 1, \ldots, 4$. We know by Lemma 32 that if $N_1 = p + 1 - t_p$ then $N_2 = p + 1 + t_p$. The same is valid for $N_3 = q + 1 - t_q$, $N_4 = q + 1 + t_q$. Hence using Schoofs method in Sect. 4.3.3 we obtain $t_p,\ t_q$.

A further improvement can be made choosing $a,\ b,\ p$ and q such that $t_p = t_q = 0$. Then $N_i = \operatorname{lcm}(p + 1, q + 1)$ for all $i = 1, \ldots, 4$. Hence, the Legendre symbols $\left(\frac{w}{p}\right)$ and $\left(\frac{w}{q}\right)$ are not needed for decryption in (dec), since $d = d_i$ for all $i = 1, \ldots, 4$.

Vanstone–Zuccherato Scheme

Vanstone and Zuccherato describe in their paper [91] two methods how to find primes p and q and curves

$$E_{-D,0} :\ y^2 = x^3 - Dx,$$

respectively

$$E_{0,D} :\ y^2 = x^3 + D$$

over $\mathbb{Z}_n$ such that $\#E_{a,b}(\mathbb{F}_p)$ and $\#E_{a,b}(\mathbb{F}_q)$ have small prime factors.

Then we can set up the following

Vanstone–Zuccherato Scheme

(i) **(Setup)** Each user i

(a) selects two "large" primes p and q and computes $n = pq$,

(b) chooses an elliptic curve $E_{a,b} :\ y^2 = x^3 + ax + b$ over $\mathbb{Z}_n$ with the property that $\#E_{a,b}(\mathbb{F}_p)$ and $\#E_{a,b}(\mathbb{F}_q)$ have small prime factors of about 15/16 digits,

(c) calculates $\#E_{a,b}(\mathbb{F}_p)$ using Schoofs method (cf. Sect. 4.3.3).

Then each user i sets (a, b, n, P, k), where $P \in E(\mathbb{Z}_n)$ is a point of order k, as the public-key and $(p, \#E_{a,b}(\mathbb{F}_p))$ as the private-key.

(ii) **(Communication)** User j wants to send a message $m \in \mathcal{M} = \mathbb{Z}_k$ to i

(enc) j sends $Q = mP \in E(\mathbb{Z}_n)$
(dec) i calculates $Q = [Q_p, Q_q] \in E(\mathbb{F}_p) \times E(\mathbb{F}_q)$ and solves the ECDLP $Q_p = mP_p$, cf. Definition 148, using the Silver–Pohlig–Hellman method in combination with the giant-step baby-step method. Therefore i needs $\#E_{a,b}(\mathbb{F}_p)$.

(iii) **(Cryptanalysis)** If an eavesdropper wants to use the Silver–Pohlig–Hellman method directly on $E_{a,b}(\mathbb{Z}_n)$, he has to know $\#E_{a,b}(\mathbb{Z}_n)$. But Schoofs method does not seem to generalize to $\mathbb{Z}_n$, so determining $\#E_{a,b}(\mathbb{Z}_n)$ is infeasible, unless p and q are known, which is a factorization problem.

There are further public-key schemes using elliptic curves known, where the security is based on factorization (cf. [39, 57]). We will not discuss them here, since they mainly vary only the elliptic curve and have almost the same properties than KMOV. At least the disadvantages mentioned in subsection "Requirements on the Modulus n" of Sect. 4.4.3 will be valid.

4.4.2 *Known Attacks on KMOV and Demytko*

In the next subsections we will follow Bleichenbacher [6], who analysed the KMOV scheme in detail.

Plaintext Attacks

For the KMOV cryptoscheme we have the following partially known plaintext attack.

Theorem 83 ([6]) *Let (n, e) be a public-key for KMOV and $C = (x_C, y_C)$ be the encryption of a message $M = (x_M, y_M)$. Then, given n, e, C and either x_M or y_M, the plaintext M can be computed efficiently.*

The proof of this theorem uses the redundancy given by

$$b \equiv y^2 - x^3 \pmod{n}. \tag{4.4.1}$$

The practical algorithm can easily be obtained from the following example.

Example 24 Let $(n, e) = (493, 16)$ be the public-key of KMOV. The ciphertext may be $C = (492, 77)$ and $y_M \equiv 109 \pmod{n}$. Then

$$b \equiv y_C^2 - x_C^3 \equiv 14 \pmod{493}$$

and

$$x_M^3 + b - y_M^2 \equiv x_M^3 + 458 \equiv 0 \pmod{493}.$$

Encrypting $P = (x, y_M) = (x, 109)$ over $\mathbb{Z}[x]/(x^3 + 458, 493)$ yields

$$\bar{C} \equiv (152x, 77) \pmod{493}.$$

Hence,

$$x_M \equiv x_C \cdot (x_{\bar{C}}/x)^{-1} \equiv 492 \cdot 152^{-1} \equiv 120 \pmod{493}.$$

Broadcast Attacks

In a broadcast application the same message M is encrypted with different public-keys. Then we can apply the following theorem:

Theorem 84 ([6]) *Let $t \geq 1$ and $(e_1, n_1), (e_2, n_2), (e_3, n_3)$ be different public keys. Let $M = (x_M, y_M) \in \{0, \ldots, \min\{n_i\} - 1\}^2$ be an unknown message. If there exist 3 ciphertexts encrypted with these 3 keys then M can be found in time $O(t^2 \log(n)^3)$ with probability $1 - 1/t$, where $n = \max_i\{n_i\}$.*

Proof Following the ideas of Bleichenbacher we get

$$-b_i \equiv x_M^3 - y_M^2 \pmod{n_i} \quad \textit{for all } i \in \{1, 2, 3\}.$$

Defining $\bar{n} = \min_i\{n_i\}$ we get

$$b \equiv x_M^3 - y_M^2 \pmod{n_1 n_2 n_3} \quad \textit{for } -\bar{n} \leq b < n_1 n_2 n_3 - \bar{n}^2.$$

Since $-\bar{n}^2 \leq x_M^3 - y_M^2 < n_1 n_2 n_3 - \bar{n}^2$, we get $b = x_M^3 - y_M^2$.

Assume $y_M^2 \ll x_M^3$, so $x_M \approx b^{\frac{1}{3}}$.

Let $x_0 = \lceil b^{\frac{1}{3}} \rceil$. If $(\hat{x}_M - b)$, $x_0 \leq \hat{x}_M \leq x_0 + (4/3)t^2$ is a square, then let $\hat{y}_M = \sqrt{\hat{x}_M - b}$ and test, if $e_i(\hat{x}_M, \hat{y}_M) = (x_C, y_C)$ for all $i \in \{1, 2, 3\}$. The test can be done for one $\hat{x}_M$ in $O(\log(n)^3)$. Hence testing every $\hat{x}_M$ in the given bounds needs $O(t^2 \log(n)^3)$.

Now assume that $x_M \geq \bar{n}/t$ and let $\gamma = (4/3)t^2$. Then $x_M \leq y_M/t$ and thus

$$y_M^2 \leq (3/4)\gamma x_M^2 \leq \gamma(3/4)x_m^2 + (\gamma - (3/2)x_M)^2 = x_M^3 - (x_M - \gamma)^3.$$

Hence

$$x_M^3 \geq x_M^3 - y_M^2 \geq (x_M - \gamma)^3$$

and therefore

$$x_0 \leq x_M \leq x_0 + (4/3)t^2.$$

So if $x_M \geq \bar{n}/t$ the attack succeeds. Therefore if $x_M < \bar{n}/t$, the attack fails with the probability $1/t$. □

Bleichenbacher also extended this theorem to linearly related messages. Another method, based on the Hastad Theorem 63, can be applied also to the Demytko scheme: Assume that a user i wants to send the same message M to k different users with the public keys (n_i, e, a_i, b_i), $i \in \{1, \ldots, k\}$ for the Demytko scheme and (n_i, e) for the KMOV scheme. Let $\{n_i\}_i$ be relatively prime. Then i sends $x_M^{(1)}, x_M^{(2)}, \ldots, x_M^{(k)}$, respectively $(x_M^{(1)}, y_M^{(1)}), (x_M^{(2)}, y_M^{(2)}), \ldots, (x_M^{(k)}, y_M^{(k)})$.

Theorem 85 ([41, 42]) *Let $n = \min_i\{n_i\}$. For the Demytko scheme the message x_M can be found from $x_C^{(1)}, \ldots, x_C^{(k)}$ in polynomial time, if*

(i) $e = 2$ *and* $k \geq 11, n \geq 2^{175}$,
(ii) $e = 3$ *and* $k \geq 49, n \geq 2^{482}$,
(iii) $e = 4$ *and* $k \geq 173, n \geq 2^{511}$,
(iv) $e = 5$ *and* $k \geq 664, n \geq 2^{723}$,
(v) $e = 5$ *and* $k \geq 428, n \geq 2^{1024}$.

The cases (iv) and (v) are also valid for the KMOV scheme.

Proof We will use the techniques of [41] proving (iv). The proofs for the other cases are similar. At first we will prove the validity of Theorem 63 for the Demytko scheme. Let $lM = (x_l, y_l)$. From Lemma 38,

$$x_2^{(i)} \equiv \frac{(x^2 - a_i)^2 - 8b_i x}{4(x^3 + a_i x + b_i)} \pmod{n_i} \tag{4.4.2}$$

$$x_3^{(i)} \equiv \frac{(a_i - x x_2^{(i)})^2 - 4b(x + x_2^{(i)})}{x(x - x_2^{(i)})^2} \pmod{n_i} \tag{4.4.3}$$

Hence,

$$x_C^{(i)} \equiv x_5^{(i)} \equiv \frac{4b + 2(a + x_2^{(i)} x_3^{(i)})(x_2^{(i)} + x_3^{(i)})}{(x_2^{(i)} - x_3^{(i)})^2} - x \pmod{n_i} \tag{4.4.4}$$

By substituting the Eqs. (4.4.2) and (4.4.3) into (4.4.4), we get

$$x_C^{(i)} \equiv \frac{h_i(x)}{g_i(x)} \pmod{n_i}$$

for some polynomials $h_i(x)$ and $g_i(x)$, $\deg h_i(x) = 25$ and $\deg g_i(x) = 24$.

Define $F_i(x) = x_C^{(i)} g_i(x) - h_i(x)$. Then $F_i(x) = 0$. In Theorem 63 we have now $h = 25$. Hence, $n^{h(h+1)/2} = n^{325}$. If $k = 664$ we get

$$\begin{aligned} n^{h(h+1)/2}(k+h+1)^{(k+h+1)/2} 2^{(k+h+1)^2/2} (h+1)^{(h+1)} &\leq n^{325} 2^{241630} \\ &< n^{325} (2^{723})^{339} \\ &\leq n^{664} \leq \prod_{i=1}^{664} n_i. \end{aligned}$$

Hence, Theorem 63 holds.

To give a proof for the KMOV scheme, remember that a plaintext M is given by (x_M, y_M) and the ciphertext is $C = (x_C, y_C)$. At first x_M can be found from $x_C^{(i)}$ as stated. Now we can apply the partially known plaintext attack from subsection "Plaintext Attacks". □

This theorem can also be extended to linearly related messages, e.g. if a timestamp is combined with the message. We will analyse this situation especially for the KMOV scheme in the rest of this section.

Assume in KMOV that we have the two unknown plaintexts $M = (x_M, y_M)$ and $\bar{M} = (x_{\bar{M}}, y_{\bar{M}})$ with the dependences

$$\begin{aligned} x_{\bar{M}} &\equiv \alpha x_M + \gamma \\ y_{\bar{M}} &\equiv \beta y_M + \delta \end{aligned}$$

and the known ciphertexts

$$\begin{aligned} C = (x_C, y_C) &\equiv e(x_M, y_M) \pmod{n} \\ \bar{C} = (x_{\bar{C}}, y_{\bar{C}}) &\equiv e(x_{\bar{M}}, y_{\bar{M}}) \pmod{n} \end{aligned}$$

Using (4.4.1) we get b and $\bar{b}$ such that $(x_M, y_M) \in E_{0,b}(\mathbb{Z}_n)$ and $(x_{\bar{M}}, y_{\bar{M}}) \in E_{0,\bar{b}}(\mathbb{Z}_n)$. Hence,

$$\begin{aligned} x_M^3 + b - y_M^2 &\equiv 0 \pmod{n} \\ (\alpha x_M + \gamma)^3 + \bar{b} - (\beta y_M + \delta)^2 &\equiv 0 \pmod{n} \end{aligned} \tag{4.4.5}$$

Defining the polynomials $f(x) = \frac{(\alpha x+\gamma)^3 - \beta^2 x^3 - \delta^2 + \bar{b} - \beta^2 b}{2\beta\delta}$ and $g(x) = x^3 - f(x)^2 + b$ we get

$$\begin{aligned} f(x_M) &\equiv \frac{(\alpha x_M + \gamma)^3 - \beta^2 x_M^3 - \delta^2 + \bar{b} - \beta^2 b}{2\beta\delta} \pmod{n} \\ &\equiv \frac{(\beta y_M + \delta)^2 - \beta^2 (y_M^2 - b) - \delta^2 - \beta^2 b}{2\beta\delta} \pmod{n} \\ &\equiv y_M \pmod{n} \end{aligned}$$

using (4.4.5) and

$$g(x_M) \equiv x_M^3 - y_M^2 + b \equiv 0 \pmod{n}$$

using also (4.4.5). Now compute the polynomials h and j by

$$(h(x), j(x)) \equiv e(x, f(x)) \text{ over } \mathbb{Z}[x]/(g(x), n).$$

Knowing (x_C, y_C) we have

$$\begin{aligned}(h(x_M), j(x_M)) &\equiv e(x_M, f(x_M)) \pmod{n}\\ &\equiv e(x_M, y_M) \pmod{n}\\ &\equiv (x_C, y_C) \pmod{n}.\end{aligned}$$

The described attack succeeds, if we find a linear polynomial $\lambda(x - x_M)$ in $\gcd(g(x), h(x) - x_C)$ in order to find x_M.

Note that Demytko's cryptosystem uses, contrary to KMOV, only one coordinate to represent messages. This difference seems to be crucial, as the attack above can not be applied to Demytko's scheme. However Bleichenbacher states that the attacks above work in almost the same way against other proposed elliptic curve cryptosystems based on factorization like [39], where the plaintext message is stored in both coordinates.

4.4.3 Integer Factorization

The security of all public-key schemes in this chapter are based on factorization. Beside the factorization methods presented in Sect. 4.1.6 we want to present a method which exploits the properties of elliptic curves next.

Elliptic Curve Method

The elliptic curve method (ECM) is a further integer factorization method proposed by H.W. Lenstra, Jr., in 1985 [46]. Brent [9] and Montgomery [60] have proposed practical improvements on the original method. With this improvements ECM provides the fastest means of finding factors of up to approximately 30 decimal digits. Recently the tenth Fermat number $F_{10} = 2^{2^{10}} + 1$ was factored completely by ECM [10].

Lenstra's original elliptic curve method can be briefly described as follows:

Elliptic Curve Method

Let $n \in \mathbb{N} \setminus \{0\}$ be an integer coprime to 6, $n \neq m^e$ with $m, e \in \mathbb{N} \setminus \{0\}$. Repeat the following steps until a non-trivial factor of N has been found:

(i) Select a random pair (E, P) , consisting of an elliptic curve $E = E_{a,b}$ defined over $\mathbb{Z}_n$ and a point $P \in E(\mathbb{Z}_n) \setminus \{\mathcal{O}_n\}$.
(ii) Select a suitable positive integer k and apply the pseudo-multiplication to compute $Q = kP$.

In oder to choose a random pair (E, P) in (i) there are several methods. The easiest way is to choose a triple $(a, x_P, y_P) \in \mathbb{Z}_n^3$ at random and let $b = y^2 - x^3 - ax \pmod{n}$. If $\gcd(4a^3 + 27b^2, n) > 1$ then a non-trivial factor of n is already found and we can stop. The next lemma determines the situation where we find a non-trivial factor of n is step (ii):

Lemma 39 *Let p and q be different prime divisors of n. Let $kP_p = \mathcal{O}_p$ in $E(\mathbb{F}_p)$ and $kP_q \neq \mathcal{O}_q$ in $E(\mathbb{F}_q)$. Then the pseudo-multiplication kP must yield a non-trivial divisor of n.*

Proof Assume we found $Q = kP \in E(\mathbb{Z}_n)$. Then $Q_p = kP_p = \mathcal{O}_p$ in $E(\mathbb{F}_p)$. So $Q = \mathcal{O}_n$. But now also $kP_q = Q_q = \mathcal{O}_q$ in $E(\mathbb{F}_q)$, which gives a contradiction. □

For a more detailed proof, see [46] Proposition 2.6.

Remark 40 Notice the similarity with Pollard's $p-1$ method (cf. subsection "The Pollard ρ-Method", Sect. 4.1.6). Instead of the group $\mathbb{Z}_p^*$, we are using the group $E(\mathbb{F}_p)$. However if the elliptic curve E seem to be a bad choice, i.e. for each prime $p|n$ $\#E(\mathbb{F}_p)$ is divisible by a large prime and so $kP_p \neq \mathcal{O}_p$ for given k, we can choose a new pair (E, P) at random. Since by Theorem 73 $\#E(\mathbb{F}_p)$ is nearly uniformal distributed between $p+1-\sqrt{p}$ and $p+1+\sqrt{p}$ we have a new chance of finding a factor. This is not possible in the Pollard method.

We will now discuss how k has to be chosen in order to achieve the situation of Lemma 39 with high probability:

Let p be the smallest divisor of n. Suppose that k is given as a product

$$k = \prod_{r \leq w, r prime} r^{e(r)}, \tag{4.4.6}$$

where $e(r)$ is the largest integer with $r^{e(r)} \leq p + 2\sqrt{p} + 1$. Lenstra showed using an unproved hypothesis on the smoothness of random integers in intervals and facts based on Theorem 73, that using ECM one may expect to find the smallest prime p dividing n in

$$B_1 = e^{(1+o(1))(\log^{1/2} p \log^{1/2} \log p)/\sqrt{2}} \quad (\text{for } p \to \infty) \tag{4.4.7}$$

trials with $w = B_1$. Each trial takes time $O((\log n)^2 B_1)$, which leads to the expected running time $O((\log n)^2 B_1^2)$. Since always $p \leq \sqrt{n}$ we get the running time $L_n(1/2, c)$.

Since p is unknown we define B_1 in practice by a suggested small prime number $\bar{p}$ and then increase k after each trial slightly. For instance we can choose a random B_0 and define $B_1 = B_0 \cdot 1,02^{t-1}$ at the tth trial.

Practical improvements Since the pseudo-multiplication kP is the most time consuming part of the ECM fast multiplication methods as described in Sect. 4.3.1 are very important in order to reduce the total running time.

As in the Pollard $p-1$ method the performance of ECM can be further improved by adding a second step to each trial:

(i) **Montgomery's improvement** [60]: Take primes $q_1, \ldots, q_l$ such that $q_i \nmid k$ for all $i = 1, \ldots, l$. If n has a prime divisor p such that it exists an $i \in \{1, \ldots l\}$ with $q_i = k \cdot \mathrm{ord}(P_p)$ then p will be detected with high probability.

(ii) **Brent's improvement** [9]: Simulate a random walk $Q_1, Q_2, \ldots$ in $< Q >$ as in the Pollard $\rho-$method. Then compute $\gcd(\prod(x_{Q_i} - x_{Q_j}), n)$. For a short review and some implementation details, see [8], Chap. 6.

Example 25 The author implemented the ECM without a second phase using a bound $B_1 = B_0 \cdot 1,02^{t-1}$ for the tth-trial. B_0 was set to 250, 500 and 1000. Factoring $n = 2^{203} - 1$, a 61-digit number, was a not so hard task. For every $B_0 = 250, 500$ and 1000 four attempts were made. The factors 127, 233, 1103 and 2089 were found for all B_0 at the first trial. 136417 was found independently of B_0 in maximal four trials. For the 9-digit factor 121793911 we needed 10/1/5/3 trials for $B_0 = 250$, 1/16/3/12 trials for $B_0 = 500$ and 13/1/3/7 trials for $B_0 = 1000$. The rest was observed to be a 38-digit prime. Hence using a large bound B_0 was not necessary. But before the computation started we did not know B_1 as given in (4.4.7). So the best strategy is to begin with a low bound B_1 and then increasing it.

Factoring $2^{213} - 1$ was more difficult. We made one attempt for $B_0 = 250$ and 1000 and three attempts for $B_0 = 500$. The prime factors 7, 66457 and 228479 were found in maximal 2 trials. The next two prime factors are more interesting: 48544121 was found in the 22th trial for $B_0 = 250$, in the $nf(100)/5$th/5th trial for $B_0 = 500$ and 19th trial for $B_0 = 1000$, where nf(k) means, that the factor was not found in maximal k trials. For 212885833 we made 11 trials for $B_0 = 250$, 31/31/31 trials for $B_0 = 500$ and nf(150) for $B_0 = 1000$. It is remarkable that we need three (respectively two) times the same number of trials for $B_0 = 500$ and we did not find the latter prime factor for $B_0 = 1000$. The last two prime factors 284988192114740679 and 4205268574191396793 were not found within 100 trials for $B_0 = 250$ and 500, and 150 trials for $B_0 = 1000$.

Thus we get these main advantages of ECM:

(i) Although the GNFS has heuristic running time $L_n(1/3, c)$, the expected running time of ECM depends on the prime factors p of n and thus ECM is able to find factors of up to 30 decimal digits faster.
(ii) ECM is useful to find integer factorizations of auxiliary numbers consisting of small primes in other factorization methods.
(iii) ECM has very small storage requirements and can be massive parallelized on multi-processor systems.

Requirements on the Modulus n

Since the security of all schemes in Sect. 4.4.1 is based on factorization, we get a necessary condition for the primes p and q to be used. Since the running times for various factorization methods take the worst case if $p \sim q$, the primes should be of nearly equal size.

Furthermore, A. Odlyzko estimated in 1995 the running times for the GNFS factoring $n = pq$ in practice (cf. Table 4.3). Recently A. Shamir announced the design for an electro-optical sieving device, called "TWINKLE" (The Weizmann INstitute Key Locating Engine). TWINKLE will execute sieve-based factoring algorithms

Table 4.3 Computing power required to factor n using the GNFS, 1995, [66]

Size of n in bits	MIPS years	Size of n in bits	MIPS years
512	3×10^4	768	2×10^8
1024	3×10^{11}	1280	1×10^{14}
1536	3×10^{16}	2048	1×10^{20}

approximately two to three orders of magnitude as fast as a conventional fast PC. Shamir estimates that the device can be fabricated for about 5000 dollar. Following an analysis of R.D. Silverman [84] the device will speed up the sieving process of the QS and the GNFS (cf. subsection "The Pollard $p-1$ Method", Sect. 4.1.6) but not the equation solving process, i.e. the problem of solving a large matrix is still a bottleneck for the whole integer factorization. Using 15–20 devices factoring a 512-bit modulus n would take 9–10 weeks in total. By now 200 fast PC's and a CRAY supercomputer would take 7–8 month. In Table 4.4 we have estimated the time and space, provided TWINKLE can be built and large matrix solving/storing can be realized.

Hence p and q should at least be about 2^{512}, i.e. about 155 decimal digits, in order to achieve minimal security against an eavesdropper. For digital signatures and authentication schemes a 2024-bit modulus is advisable, since digital signatures must be valid also in future.

4.4.4 Conclusion

Using elliptic curve public-key schemes over the ring $\mathbb{Z}_n$ is not recommended in practice, because of the following known deficiencies:

(i) The KMOV scheme and partly the Demytko scheme is not secure against various attacks mentioned in Sect. 4.4.2.

Table 4.4 Very rough estimate of time and space required to factor n using the GNFS in combination with TWINKLE, 1999

Size of n in bits	512	768	1024
#TWINKLE devices	20	1200	45,000
Factor base	3×10^6	2.4×10^8	6×10^{11}
Sieving time	5–6 weeks	6×10^2 years	5×10^5 years
Sieve space [Mbytes]	1.3×10^2	1.0×10^4	2.6×10^5
Matrix solving time	4 weeks	1.8×10^3 years	5×10^6 years
Matrix space [Mbytes]	2×10^3	6.4×10^4	10×10^6
Total time	9–10 weeks	2.4×10^3 years	5.5×10^6 years

(ii) As RSA a large modulus n is required in order to avoid fast factorization. Hence all schemes must have a large public/private key storage space and practical implementations must use large-number arithmetic (at least about 2^{1024}-bit numbers must be handled) (cf. Sect. 4.4.3).

(iii) Finally all elliptic curve schemes in Sect. 4.4.1 are not as efficient as RSA schemes in practice, since for an single elliptic curve addition over $\mathbb{Z}_n$ we need more multiplications and modular-inversions than for RSA (cf. Sect. 4.3.1).

Thus elliptic curve public-key schemes based on factorization have more theoretical use, since they do not overcome the advantages of common available public-key schemes (at least in encryption and authentication).

A big progress will be made by another kind of elliptic curve public-key schemes, explained in the next chapter.

4.5 Elliptic Curve Cryptosystems Based on the ECDLP

In 1987 Koblitz [36] and Miller [58] independently invented the use of elliptic curves in cryptosystems based on the following discrete logarithm in the group of points of an elliptic curve.

Let E be an elliptic curve defined over the finite field $\mathbb{F}_q$, where $q = p^m$ and p is a prime. This assumption should hold for the rest of this chapter.

Definition 148 Let $P, R \in E(\mathbb{F}_q)$ and the order n of P be given. The problem of finding the unique integer $l \in \mathbb{Z},\ 0 \le l \le n-1$, such that

$$R = lP, \tag{4.5.1}$$

provided l exists, is denoted **elliptic curve discrete logarithm problem (ECDLP)**. The elliptic curve discrete logarithm is the unique integer l, provided it exists.

Remark 41 Using Definition 120 the ECDLP is a DLP in the group $< P >$, which is a subgroup of $E(\mathbb{F}_q)$, to the base P.

Example 26 Let $E/\mathbb{Z}_{13}:\ y^2 = x^3 + x + 6$ as in Example 12. Take the generator $P = (2, 4)$, which generates $< P >= E(\mathbb{Z}_{13})$. Let $R = (11, 3)$ be given. To solve the ECDLP $(11, 3) = l(2, 4),\ 0 \le l \le 12$ seems easy. We know from Example 12 that $l = 10$.

In order to solve the ECDLP in the example, we used a precomputed list. Hence taking a generator P such that the order of P is about 10^{50}, this gets infeasible (cf. subsection "Arbitrary Curves", Sect. 4.5.2. By now no efficient, i.e. polynomial or even subexponential time algorithm is known to solve the ECDLP for arbitrary elliptic curves, although there are subexponential time algorithms for supersingular curves (cf. subsection "Frey/Rück Reduction", Sect. 4.5.2) and polynomial time algorithms for anomalous curves (cf. subsections "Supersingular Curves" and "Anomalous Curves—Algebraic Geometrical Method", Sect. 4.5.2). Hence, for finding "secure" elliptic curves, also denoted cryptographically good elliptic curves, which are elliptic curves, where the ECDLP seems to be computationally infeasible, see subsection "Cryptographically Good Elliptic Curves", Sect. 4.5.2. An survey how to construct curves, where the desired properties are given, is given in Sect. 4.5.3. Finally the author discusses the design of a new elliptic curve public-key cryptosystem.

In the first section we will introduce several public-key schemes, where the security is based on the ECDLP.

4.5.1 Public-Key Schemes

Note that calculating $R = lP$, given P and l takes only polynomial time in $\log q$ by applying the methods given in Sect. 4.3.1. Hence we will get a TOF function provided the ECDLP is computationally hard.

EC El Gamal Cryptoscheme

The El Gamal scheme in Sect. 4.3.3, Sect. 4.1 can be applied for $E(\mathbb{F}_q)$ as follows.

EC El Gamal cryptoscheme

(i) **(Setup)** Choose an elliptic curve E defined over $\mathbb{F}_q$, $q = p^m$, where p is a prime, and a point $P \in E(\mathbb{F}_q)$. Let n be the order of P in $E(\mathbb{F}_q)$. Each user i selects a private-key $l \in \mathbb{Z}$, $0 \leq l \leq n-1$ and a public-key $R = lP$.

(ii) **(Communication)** If i wants to send a message $M \in E(\mathbb{F}_q)$ to j, then:

(enc) i generates a random integer k and evaluates $C_1 = kP \in E(\mathbb{F}_q)$.
i computes $C_2 = kR + M \in E(\mathbb{F}_q)$ using j's public-key R.
i sends $(C_1, C_2) \in E(\mathbb{F}_q) \times E(\mathbb{F}_q)$

(dec) j uses his private-key l and recovers

$$M = C_2 - lC_1 = k\underbrace{R}_{lP} + M - lkP.$$

(iii) **(Cryptanalysis)** The security is based on the ECDLP as in the original El Gamal scheme.

The main disadvantage of this scheme is the fact that we have to take a message $M \in E(\mathbb{F}_q)$. In practice we often have only messages $m \in \mathbb{Z}_{\bar{m}}$. So we would further need an injective map $h : \mathbb{Z}_{\bar{m}} \to E(\mathbb{F}_q)$. Note that we have a message-expansion factor of 2.

EC MOV Cryptoscheme

In [52] Menezes and Vanstone proposed a cryptosystem based on El Gamal where the message (m_1, m_2) is in $\mathbb{F}_q \times \mathbb{F}_q$. Hence, an injective map $h : \mathbb{Z}_{\bar{m}} \to \mathbb{F}_q$ can easily be found.

If $P = (x_P, y_P) \in E(\mathbb{F}_q)$ define the projection $x(P) := x_P$ and $y(P) := y_P$.

EC MOV cryptoscheme

(i) **(Setup)** Same as in EC El Gamal (subsection "EC El Gamal Cryptoscheme") with the public-key l and the private-key $R = lP$.

(ii) **(Communication)** i wants to send a message $(m_1, m_2) \in \mathbb{F}_q \times \mathbb{F}_q$ to j.

(enc) i generates a random integer k such that $x(kP) \neq 0 \neq y(kR)$.
i computes $C_1 = kP$.
i evaluates $\bar{c}_1 = m_1 \cdot x(kR)$ and $\bar{c}_2 = m_2 \cdot y(kR)$.
i sends $(C_1, \bar{c}_1, \bar{c}_2) \in E(\mathbb{F}_q) \times \mathbb{F}_q \times \mathbb{F}_q$.

(dec) j recovers the plaintext using l by calculating lC_1. Then

$$m_1 = \bar{c}_1 \cdot x(lC_1)^{-1} = m_1 \cdot x(kR) \cdot x(lkP)^{-1} \in \mathbb{F}_q,$$
$$m_2 = \bar{c}_2 \cdot y(lC_1)^{-1} = m_2 \cdot y(kR) \cdot y(lkP)^{-1} \in \mathbb{F}_q.$$

Note that "$\cdot$" denotes the multiplication in the field $\mathbb{F}_q$.

(iii) **(Cryptanalysis)** The security is based on the ECDLP. If an eavesdropper knows m_1 (or m_2), he can easily evaluate m_2 (or m_1) using similar methods as in subsection "Broadcast Attacks", Sect. 4.4. To increase the security, it is possible to send only $(C_1, \bar{c}_1) \in E(\mathbb{F}_q) \times \mathbb{F}_q \subseteq \mathbb{F}_q^3$ as an encryption of $m_1 \in \mathbb{F}_q$ which would increase the message expansion factor from 2 to 3.

We can reduce the message expansion factor further by compressing the y-coordinate of a ciphertext point $C \in E(\mathbb{F}_q)$ to an one bit value for instance as follows: Let $C \in E_{a,b}(\mathbb{F}_p), p > 3$ prime. Then let $\bar{y} = y_C \mod 2$ be the compressed y-coordinate.

Decompression can then easily be done: Given x_C we can find a possible y-coordinate y from the point finding method (subsection "Number of Points", Sect. 4.2.3). So set $y_C = y$, if $\bar{y} = y_C \mod 2$, else $y_C = p - y$.

This reduces the message-expansion factor to $\approx 3/2$.

Remark 42 There are several further point-compression methods for different curves and underlying fields known, e.g. if we choose $E/\mathbb{F}_{2^m} :\ y^2 + xy = x^3 + a_2x^2 + a_6$, cf. [56], Sect. 6.4 for a suitable point-compression.

Assume now that the ECDLP gets computational infeasible if $q \sim 2^{160}$ (cf. the next section). Provided the same 'security' for RSA, El Gamal over $\mathbb{F}_q^*$ and EC MOV is given, we get Table 4.5 using point-compression for C_1 and only sending $(C_1, \bar{c}_1)$.

Table 4.5 Comparison of the encryption size of a 100-bits message

System	Encryption size in bits
RSA	1024
El Gamal	2048
EC MOV	321

Hence, the elliptic curve cryptoschemes are very interesting if short messages, e.g. money accounts, passwords and short signals, have to be encrypted and send. Note also that the field size is dramatically smaller compared to RSA and El Gamal. Even if the elliptic curve addition needs more modular operations than RSA and El Gamal, the underlying field is smaller and arithmetic can be done faster (about 8 times in contrast to RSA).

EC DSA Signature Scheme

Like for the El Gamal cryptosystem there is a variation of DSA using elliptic curves that might be even harder to break than the finite field DSA.

EC DSA Signature Scheme

(i) **(Setup)** Choose an elliptic curve E defined over $\mathbb{F}_q$, where q is a prime power, and a basepoint $P \in E(\mathbb{F}_q)$. Let n be the order of P in $E(\mathbb{F}_q)$.
Each user picks a random private-key $l,\ 0 \le l \le n-1$, and makes $R = lP$ public.

(ii) **(Signing)** i wants to sign a message $m \in \mathcal{M}$:

(a) i applies a hash function H to m to obtain $H(m),\ 0 < H(m) \le n-1$, see Definition 119.
(b) i picks a random integer $k,\ 0 < k \le n-1$, such that $\gcd(k,\ n) = 1$.
(c) i computes $T = kP$. If $x_T \equiv 0 \pmod{n}$ goto (a).
(d) i finds an integer s such that

$$sk \equiv H(m) + lx_T \pmod{n}. \tag{4.5.2}$$

(e) If $s \equiv 0 \pmod{n}$ goto (a).
(f) i's signature of m is $(x_T,\ s) \mod n$.

(iii) **(Verifying)** j wishes to verify the signature $(x_T,\ s)$ of a message m from i:

(a) j computes $u_1 \equiv s^{-1}H(m) \pmod{n},\quad u_2 \equiv s^{-1}x_T \pmod{n}$.
(b) j evaluates $V = u_1P + u_2R \in E(\mathbb{F}_q)$.
(c) j verifies if $x_V = x_T$.

In the signing step (ii)(c) we have to assure that $x_T \not\equiv 0 \pmod{n}$, because otherwise the signing equation in (ii)(d) does not involve the private-key l! Also if $s \equiv 0 \pmod{n}$ in step (ii)(e), we can not calculate s^{-1} in the verification part.

Observe that the values that are difficult to generate are the system parameters (q, E, P, n) which are public; thus their generation can be audited and independently checked for validity!

To prove the correctness of j's verification observe that by (4.5.2)

$$\begin{aligned} k &= s^{-1}H(m) + s^{-1}lx_T && \mod n \\ &= u_1 + u_2 l && \mod n \end{aligned}$$

Table 4.6 Comparison of the key and signature sizes of a 2000-bit message, which should be signed with the same security, in bits (approx)

System	System parameter	Public key	Private key	Signature size
RSA	–	1088	2048	1024
DSA	2208	1024	160	320
EC DSA	481	161	160	320

and multiplying by P gives

$$\begin{aligned} kP &= (u_1 + u_2 l)P \in < P > \subseteq E(\mathbb{F}_q) \\ \Longleftrightarrow \\ T &= u_1 P + u_2 R \\ \Longrightarrow \\ x_T &= x(u_1 P + u_2 R) \end{aligned}$$

Hence if $x_T \neq x(u_1 P + u_2 R)$ the signature must be false.

Remark 43 It is also possible to create for each user an own elliptic curve and an own base point P, which increases the public key to (E, P, n, R), but also increases the security, since if an ECDLP is solved for all $R \in < P >$, then the scheme is only corrupt for those users who have selected this curve and basepoint.

Hence, if we assume that the ECDLP is infeasible if $q \sim 2^{160}$, we get Table 4.6 which shows that EC DSA has a great advantage, since the key and signature sizes are really short in comparison to RSA and DSA.

So the EC DSA can be used for systems where the sizes of the signature and especially the private and public keys are crucial, e.g. in smart cards or the wireless communication.

With the advantages of Tables 4.5 and 4.6 the elliptic curve crypto- and signature schemes are very useful in commercial and non-commercial applications, e.g. internet banking and email. The IEEE P1363 group [32], which is responsible for the standardization of cryptoschemes and techniques, is just working on a standardization of these elliptic curve public-key schemes.

Note that Menezes, Vanstone and Zuccherato are members of the IEEE P1363 working group.

4.5.2 *Elliptic Curve Discrete Logarithm Problem*

In this section we want to give necessary conditions for the ECDLP in order to be computationally infeasible. As will be shown in this section, we have to choose the parameters for the elliptic curves used in Sect. 4.5.1 carefully.

For the rest of this section let $P \in E(\mathbb{F}_q)$ be a base point of the group $< P >$ generated by P, which is a subgroup of $E(\mathbb{F}_q)$. Let

$$n = \mathrm{ord}(P) = \min\{n : nP = \mathcal{O}\}$$

denote the order of P.

Remark 44 Let $\mathrm{Pic}^0(E)_n$ be the n-torsion subgroup of $\mathrm{Pic}^0(E)$, the group of divisor classes of degree 0 on E. Instead of solving the ECDLP in E we can apply the isomorphism given by Theorem 79 in order to solve the ECDLP in the divisor class subgroup $\mathrm{Pic}^0(E)_n$: Let $\bar{D}_1, \bar{D}_2 \in \mathrm{Pic}^0(E)_n$ be given. Determine $l \in \mathbb{Z},\ 0 \leq l \leq n-1$, such that $\bar{D}_2 = l\bar{D}_1$, provided such l exists.

We will assume further that $R \in< P >$, i.e. l exists. This can also be checked using the following lemma.

Lemma 40 *Let $E(\mathbb{F}_q)$ be an elliptic curve group with group type (n_1, n_2) and $n|n_1$. If $nR = \mathcal{O}$ and $e_n(P, R) = 1$ then $R \in< P >$.*

Proof Since $e_n(P, R) = 1 = e_n(P, P)$, we get from Lemma 35 that R and P are in the same coset of $< P >$. Hence, $R \in< P >$, since $\mathrm{ord}(R)|\mathrm{ord}(P)$. □

Arbitrary Curves

The baby-step giant-step method (cf. subsection "Square Root Methods", Sect. 4.1.5) requires time $O(\sqrt{n} \cdot \log(n))$ and space $O(\sqrt{n})$ in order to solve the ECDLP completely. This method becomes impractical if $n > 2^{90}$, since it needs more than 2^{45} bits, i.e. 2^{13} GByte.

The Pollard-ρ method has the same asymptotic running time, but requires only constant space. But if we assume that we can do $2^{20} \approx 10^6$ group additions in a second, which is by now impractical, and we perform computations using $2^{20} \approx 10^6$ processors in parallel, we need about 2^{45}s $\approx 10^{10}$ years to solve the ECDLP for $n \approx 2^{160}$.

P. van Oorschot and M. Wiener [90] provided a detailed study of the possibility to make a parallel search using the Pollard-ρ method. They estimated that if $n \approx 2^{120}$, then a machine with 325000 processors that could be built for 10 million dollar would solve a single ECDLP in about 35 days; Table 4.7 summarizes their work. Note that the difficulty of solving the ECDLP by the Pollard-ρ method raises exponentially in the field size q.

Table 4.7 Computing power needed to solve the ECDLP with the Pollard-ρ method

Field size q	Size of order n	MIPS years
$\sim 2^{155}$	$\sim 2^{150}$	3.8×10^{10}
$\sim 2^{210}$	$\sim 2^{205}$	7.1×10^{18}
$\sim 2^{239}$	$\sim 2^{234}$	1.6×10^{28}

The Silver–Pohlig–Hellman method (cf. Sect. 4.1.5) reduces the ECDLP in $< P >$ to ECDLPs in subgroups of $< P >$. Hence, the largest prime factor $\bar{n}$ of $n = \text{ord}(P)$ should satisfy the above restrictions, i.e. $\bar{n} > 2^{160}$.

Summing up, this gives the following necessary condition for secure elliptic curves:

Condition 1: $\#E(\mathbb{F}_q) = \bar{n} \cdot d,$ where $\bar{n} > 2^{160}$ is prime.

Index-Calculus Method

Due to Miller [58] there is no index-calculus method (cf. Sect. 4.1.5) which could be applied to the ECDLP, since index-calculus methods require a large number of free generators. For elliptic curves, or more generally, curves of non-zero genus, this seems to be not possible.

Recently J. Silverman [85] announced a new attack denoted Xedni Calculus Attack, on the ECDLP at the Second Elliptic Curve Cryptography Workshop, Sep. 1998:
Let $q = p$ be a prime, i.e. we want to solve the ECDLP (4.5.1) in $E(\mathbb{F}_p)$. Take r random linear combinations of the two points $P, R,\ 2 \leq r \leq 9$. Then consider points P_i with rational coordinates that reduce modulo p to these r points and elliptic curves $\hat{E}/\mathbb{Q}$ that pass through all of the P_i and reduce modulo p to the original curve $E/\mathbb{F}_p$. If those 'lifted' points P_i are linearly dependent, then the ECDLP is solved. But the probability of dependence is almost certain very low (cf. [38] for a nice illustration).

Silverman's idea is to fix a set of auxiliary conditions modulo l on the P_i and $\hat{E}$ for several small primes l, in order to increase the probability of success. These conditions guarantee that the elliptic curves will have fewer-than-expected points modulo l, and this presumably decreases the likelihood that the r $\mathbb{Q}$-points P_i will be independent. Mathematically most interesting is that Silverman's approach involves some ideas of arithmetic algebraic geometry that never before had any practical application, e.g. the Brich–Swinnerton–Dyer Conjecture.

J. Jacobson, et. al. [33] analysed the xedni calculus attack and proved using a conjecture of Lang (cf. [82], p. 233) that under certain plausible assumptions (cf. [33], Lemma 4.1) there exists an absolute constant C_0 such that the probability of success of the xedni algorithm in solving the ECDLP (4.5.1) is less than C_0/p. Hence for sufficiently large prime p, the xedni algorithm must be repeated $O(p)$ times (with different r) in order to find a discrete logarithm which yields an asymptotic running time of at least $O(p)$.

Using some heuristic arguments the constant $C_0(r)$ is supposed to increase with r ($C_0(2) \approx 2^{13}, \ldots, C_0(5) \approx 2^{125}, C_0(6) \approx 2^{180}, \ldots, C_0(9) \approx 2^{320}$). Hence r should be at least 6 in order to have a chance of finding the discrete logarithm in $E(\mathbb{F}_p),\ p \approx 2^{160}$. Nevertheless in practice also the discriminants of the elliptic curves over $\mathbb{Q}$ increase (for $r = 6$ at least 10000 digits) and an empirical analysis in the practical range for $r = 2, 3, 4$ shows that even the theoretical bounds $C_0(r)$ are chosen too optimistically.

Thus the main advantage of elliptic curve schemes over conventional public-key schemes using the finite field group $\mathbb{F}_q^*$ is still given, i.e. by now no practical index calculus method for elliptic curves with the Condition 1 is known.

Nevertheless, Adleman et. al. [2] give an index-calculus method of the Jacobians of hyperelliptic curves with large genus. Hence more detailed analysis on the ideas of Miller and Silverman is desired for further research.

MOV Reduction

This paragraph is mainly based on the paper of Menezes, Okamoto and Vanstone [55], presented in 1993.

Let $E/\mathbb{F}_q$ be an elliptic curve with

(i) group structure $\mathbb{Z}_{n_1} \oplus \mathbb{Z}_{n_2}$, where $n_2|n_1$,
(ii) $\gcd(\#E(\mathbb{F}_q), q) = 1$ and
(iii) $n|n_1$.

In order to determine n_1 and n_2 we can apply a probabilistic polynomial time algorithm proposed by Miller [59] (for a summarized work, see [51], Sect. 5.4). To apply this algorithm we need $\#E(\mathbb{F}_q)$, which we can compute in polynomial time by the Schoof method of Sect. 4.3 and the integer factorization of $\gcd(\#E(\mathbb{F}_q), q-1)$, which should be given.

The assumption (ii) determines $E[n_1] \cong \mathbb{Z}_{n_1} \oplus \mathbb{Z}_{n_1}$ by Theorem 76.

Let e_n be the Weil Pairing defined in Sect. 4.2.3. gives

MOV reduction
Require: $P \in E(\mathbb{F}_q)$ of order n and $R \in < P >$.
1: Find the smallest integer k such that $E[n] \subseteq E(\mathbb{F}_{q^k})$.
2: Find $Q \in E[n]$ such that $\alpha = e_n(P, Q)$ has order n
3: Compute $\beta = e_n(R, Q)$.
4: Compute l, the discrete logarithm of β to the base α ($l = \log_\alpha \beta$) in $\mathbb{F}_{q^k}$.
Ensure: $l, 0 \leq l \leq n-1$, such that $R = lP$.

Remark 45 By the MOV reduction we get a reduction of the ECDLP to the DLP in the finite extension field $\mathbb{F}_{q^k}$ of $\mathbb{F}_q$. In general the reduction takes exponential time in $\log q$, as k is exponentially large.

Theorem 86 *The MOV reduction works correctly.*

Proof In step 1 it is clear that k exists.

Let $\mu_n(\mathbb{F}_{q^k})$ denote the subgroup of the nth roots of unity in $\mathbb{F}_{q^k}$. In order to observe step 2, we show that there exists a $Q \in E[n]$ such that $e_n(P, Q)$ is a primitive nth root of unity: Let $Q \in E[n]$. Then

$$e_n(P, Q)^n = e_n(P, nQ) = e_n(P, \mathcal{O}) = 1,$$

by the bilinearity of the Weil-Pairing. Hence $e_n(P, Q) \in \mu_n(\mathbb{F}_{q^k})$. There are

$$|E[n]/ < P > | = \frac{|E[n]|}{| < P > |} = \frac{n^2}{n} = n$$

cosets of $< P >$ within $E[n]$.

Let $\bar{Q}_1, \ldots, \bar{Q}_n \in E[n]$ be the representatives of the n cosets of $< P >$ within $E[n]$. Since $\bar{Q}_i \neq \bar{Q}_j$ for all $i, j = 1, \ldots, n,\ i \neq j$, we know by Lemma 35(ii) that $e_n(P, \bar{Q}_i) \neq e_n(P, \bar{Q}_j)$. Hence we can identify

$$\{e_n(P, \bar{Q}_i) : i = 1, \ldots, n\} = \mu_n(\mathbb{F}_{q^k}).$$

Let ξ_n be a primitive element of $\mu_n(\mathbb{F}_{q^k}) \subseteq \mathbb{F}_{q^k}^*$. Thus it exists an $j,\ 1 \leq j \leq n$, such that $e_n(P, \bar{Q}_j) = \xi_n$, which determines $Q = \bar{Q}_j$.

To prove the rest of the algorithm, let $Q \in E[n]$ such that $e_n(P, Q) \in \mu_n(\mathbb{F}_{q^k})$ is primitive. We define the map

$$\begin{aligned} \rho_Q : < P > &\to \mu_n(\mathbb{F}_{q^k}) \\ S &\mapsto e_n(S, Q) \end{aligned}$$

and prove that ρ_Q is a group isomorphism:
By the bilinearity of e_n we can observe for all $S_1, S_2 \in < P >$

$$\rho_Q(S_1 + S_2) = e_n(S_1 + S_2, Q) = e_n(S_1, Q)e_n(S_2, Q) = \rho_Q(S_1)\rho_Q(S_2).$$

Now let $S \in < P >$ such that $S = l_S P,\ 0 \leq l_S \leq n - 1$. Thus

$$\rho_Q(S) = \rho(l_S P) = e_n(l_S P, Q) = e_n(P, Q)^{l_S} = \xi_n^{l_S} \in \mu_n(\mathbb{F}_{q^k}). \tag{4.5.3}$$

Since we can apply (4.5.3) for any $l_S,\ 0 \leq l_S \leq n - 1$ and know that $\mu_n(\mathbb{F}_{q^k}) = \{\xi_n^l : 0 \leq l \leq n - 1\}$ we get a group isomorphism.

Notice that we always used the Galois compatibility, i.e. for all $P_1, P_2 \in E[n]$ $e_n(P_1, P_2) \in \mathbb{F}_{q^k}$. □

Remark 46 By now the algorithm only works if $\gcd(n, p) = 1$. However if $\gcd(n, q) \neq 1$ we get $n = p^s m$ where $s > 0$ and $\gcd(m, p) = 1$. Consequently, the ECDLP in $< P >$ is reduced to a ECDLP in subgroups of order m and p by applying the Silver–Pohlig–Hellman method. For the subgroup of order m one can then apply the MOV reduction.

We will give a short running time consideration:
Let $q \geq 6$, i.e. $\ln q \geq \sqrt{3} \geq 1$.

Assume that k can be found in polynomial time, i.e. $E[n] \subseteq E(\mathbb{F}_{q^k})$, and Q can also be found in probabilistic polynomial time (since rational points on E can be found in probabilistic polynomial time using the method mentioned in Sect. 4.3.3).

Suppose further that the best algorithm to solve the DLP in $\mathbb{F}_{q^k}$ has running time $L_{q^k}(1/3, c)$ (cf. subsection "The Index-Calculus Method", Sect. 4.1.5). Notice that $\ln x,\ x > 0$, is straightly monotonically increasing and $1 \leq \ln x \leq \ln y$ if $e \leq x \leq y$. If $k \geq (\ln q)^2 \geq 3$, we get the following runtime estimate:

$$(\ln q)^2 \leq k \Longleftrightarrow (\ln q)^3 \leq k \ln q \Longleftrightarrow \ln q \leq (k \ln q)^{\frac{1}{3}} = (\ln q^k)^{\frac{1}{3}},$$

since $\ln q \geq 1$. Now since $k \geq 3$ and $\ln q \geq 1$

$$e \leq k \ln q \Longleftrightarrow 1 \leq \ln(k \ln q) \Longleftrightarrow 1 \leq (\ln(k \ln q))^{\frac{2}{3}} = (\ln \ln q^k)^{1-\frac{1}{3}},$$

Hence, using Definition 121

$$\begin{aligned} L_q(1, c) &= O(e^{c(\ln q)\cdot 1}) \\ &\leq O(e^{c(\ln q^k)^{\frac{1}{3}}\cdot}) \\ &\leq O(e^{c(\ln q^k)^{\frac{1}{3}}(\ln \ln q^k)^{1-\frac{1}{3}}}) \\ &= L_{q^k}(\frac{1}{3}, c), \end{aligned}$$

i.e. if $k \geq (\ln q)^2$ the DLP-solver in $\mathbb{F}_{q^k}$, which is subexponential in $\ln q^k$, gets fully exponential in $\ln q$ and thus the whole MOV algorithm gets exponential. The converse can also be shown, i.e. in order to get at least a probabilistic subexponential algorithm to solve the ECDLP with the MOV reduction we need $k < (\ln q)^2$.

In order to find a condition such that $E[n] \not\subseteq E(\mathbb{F}_{q^k})$ for all $k < (\ln q)^2$ in the MOV reduction we will use the following lemmas due to Schoof and Balasubramanian/Koblitz.

Lemma 41 ([76], Lemma 3.7) *Let* $\gcd(n, q) = 1$. *If* $E[n] \subseteq E(\mathbb{F}_q)$, *then* $n^2 | \#E(\mathbb{F}_q)$ *and* $n | q - 1$

Using further conditions Schoof also proved the converse.

Lemma 42 ([4]) *Let* $n = \bar{n}$, *the order of* P, *be a prime such that* $\bar{n} | \#E(\mathbb{F}_q)$ *and* $\bar{n} \not| (q - 1)$. *Then* $E(\mathbb{F}_{q^k})$ *contains* $\bar{n}^2$ *points of order* $\bar{n}$, *if and only if* $\bar{n} | (q^k - 1)$.

Observe that in the proof of the MOV reduction we mainly need n^2 points of order n. Since in practical applications we would avoid curves with the property $n | (q - 1)$ (see subsection "Arbitrary Curves" and the Hasse theorem which bounds $\#E(\mathbb{F}_q)$), $n | (q^k - 1)$ is both necessary and sufficient for the MOV reduction, if n is a prime dividing the order of $E(\mathbb{F}_q)$.

In order to assure that the MOV reduction in combination with a DLP method for $\mathbb{F}_{q^k}$ can not solve the ECDLP in subexponential time we get the following condition for the system parameter of an elliptic curve public-key scheme.

Let $\bar{n}$ be the largest prime dividing n. Then

Condition 2: $\bar{n} \not| (q^k - 1)$ for all $k,\ 1 \leq k \leq c,$ where $c \geq (\ln q)^2$

assures that the ECDLP gets infeasible with the MOV reduction. Notice that this condition is equivalent to

$$q^k \not\equiv 1 \bmod \bar{n}, \quad \text{for all } k,\ 1 \leq k \leq c,$$

which can be easily checked by a computer. Furthermore $c \approx 10$ is already sufficient in practice. For more specific values for c, see [32].

Example 27 Let $E/\mathbb{F}_{13^{19}} : y^2 = x^3 + x + 6$. From Example 13 we know the value of $\#E(\mathbb{F}_{13^{19}}) = 13 \cdot P21$, where $P21$ is a 21 digit prime. Note that $p = 13$ does not divide $t = 13^{19} + 1 - \#E(\mathbb{F}_{13^{19}}) = -37 \cdot 987678179$, i.e. E is non-supersingular (cf. subsection “Frei/Rück Reduction”). Assume $P \in E(\mathbb{F}_{13^{19}})$ of order $\bar{n} = P21$. Since the DLP in $\mathbb{F}_{(13^{19})^{10}}$ seems to be intractable, we could set $c_1 = 10$ or to be more secure $c_2 = (\ln 13^{19})^2 = (19 \ln 13)^2 \approx 2300$. Testing condition 2 especially for $c = c_2$ yields that the MOV reduction can not be used to solve the ECDLP in subexponential time even if the reduction itself is done in probabilistic polynomial time.

Remark 47 Semaev [77] also describes an algorithm for computing the Weil-Pairing and the reduction of the ECDLP to a finite field DLP. He notes that “... this result was obtained in 1990 and the method used can be applied to any Abelian manifold over a finite field.” In the literature always the paper of Menezes et. al. is mentioned, because they proposed additionally a special class of elliptic curves, supersingular curves, where k is small (cf. subsection “Supersingular Curves”).

Frey/Rück Reduction

In 1994 G. Frey and H.-G. Rück [23] proposed a method to reduce the discrete logarithm problem in a projective smooth curve of arbitrary genus to that in a finite field using the Tate pairing as an improvement of the MOV reduction. Since elliptic curves are projective smooth curves of genus 1 we can also apply this method.

Remark 48 In 1987 Koblitz [37] invented hyperelliptic curve cryptosystems which uses projective smooth curves of genus greater 1. Hence this method can be applied to those curves. For an introduction in hyperelliptic curves, see Appendix A of [38].

We will give a summary of the ideas of the Frey/Rück method specialized to the genus 1.

Let X be a projective smooth (irreducible) curve over the field K. Let $\mathrm{Div}_t^+(X) := \{A \in \mathrm{Div}X : A \geq 0,\ \deg A = t\}$ and P_0 be a rational point on X. Frey and Rück make two general assumptions for their method, which depend on the genus g of the curve X:

(i) The surjective map $c_g : \mathrm{Div}_g^+(X) \to \mathrm{Pic}(X),\ c_g(A) \mapsto \overline{A - gP_0}$ must be given.
(ii) Let $A_1, A_2 \in \mathrm{Div}_g^+(X)$. Then it must be possible to find $A_3 \in \mathrm{Div}_g^+(X)$ and $h \in K(X)$ such that $\div(h) = A_1 + A_2 - A_3 - gP_0$.

Since for elliptic curves E it is $g = 1$ we can choose $P_0 = \mathcal{O}$, since $\mathcal{O}$ is always a rational point on $E(\mathbb{F}_q)$, and
$\mathrm{Div}_1^+(E) = \{(P) \in \mathrm{Div}(E) :\ P \in E(\mathbb{F}_q)\}$. We can use Theorem 79(ii) to satisfy (i) and Corollary 10 to satisfy (ii).

Now let the ECDLP be given in the n-torsion point divisor class group $\mathrm{Pic}^0(E(\mathbb{F}_q))_n$, (cf. Remark 44). Note that if $n|q-1$, then n is prime to $\mathrm{char}(\mathbb{F}_q) = p$ and therefore $\mu_n(\mathbb{F}_q) \subseteq \mathbb{F}_q^*$.

Definition 149 Let D_P be a divisor with $\overline{D_P} \in \mathrm{Pic}^0(E)_n$,
$D_Q = \sum_{i=1}^r n_i(P_i) \in \mathrm{Div}^0(E)$ such that $\mathrm{supp}(D_P) \cap \mathrm{supp}(D_Q) = \emptyset$, i.e. D_P is relatively prime to D_Q. Let $f_P \in \overline{\mathbb{F}}_q(E)$ such that $\div(f_P) = nD_P$. Then we can define $f_P(D_Q) = \prod_{i=1}^r f_P(P_i)^{n_i}$.

Theorem 87 ([23]) *If* $n|q-1$ *then* $\{\overline{D_P}, \overline{D_Q}\}_n := f_P(D_Q)$ *defines a non-degenerate bilinear pairing*

$$\{\ ,\ \}_n :\ Pic^0(E)_n \times Pic^0(E)/nPic^0(E) \to \mathbb{F}_q^*/\mathbb{F}_q^{*n}.$$

The crucial part of this theorem is to prove that $\{\ ,\ \}_n$ is indeed a non-degenerate pairing. This can be done by deriving $\{\ ,\ \}_n$ from the Tate–Lichtenbaum pairing using algebraic geometry.

Now let n be prime to q. By Theorem 79 $E(\mathbb{F}_q)$ is isomorphic to $\mathrm{Pic}^0(E(\mathbb{F}_q))$ by mapping a point $Q \in E(\mathbb{F}_q)$ to the class of $(Q) - (\mathcal{O})$. Let $Q \in E(\mathbb{F}_q)$. Defining D_P and D_Q to be relatively prime divisors in $\overline{(P) - (\mathcal{O})}$ and $\overline{(Q) - (\mathcal{O})}$, respectively, we see that we can rewrite Theorem 87 to

Theorem 88 *If* $n|q-1$ *then* $\{P, Q\}_n := (f_P(D_Q))^{(q-1)/n}$ *defines a non-degenerate bilinear pairing*

$$\{\ ,\ \}_n :\ E[n](\mathbb{F}_q) \times E(\mathbb{F}_q)/nE(\mathbb{F}_q) \to \mu_n(\mathbb{F}_q).$$

Following [23] we give a method in order to evaluate the $\{\ ,\ \}_n$-pairing in almost $\log n$ elliptic curve additions, i.e. in $O((\log n)^3)$. Almost all of the ideas will be used and proved in a similar way in subsection "Anomalous Curves—Algebraic Geometrical Method". Hence we will give only a short survey:
Let $D_P = (P) - (\mathcal{O})$ and assume that D_Q is prime to all divisors
$(P_i), P_i \in < P >$. On $< P > \times \mathbb{F}_q^*$ we can define a group law

$$(A, a) \oplus (B, b) := (A + B, a \cdot b \cdot h_{A,B}(D_Q)),$$

where $h_{A,B}$ is a function such that

$$\div(h_{A,B}) = (A) + (B) - (A + B) - (\mathcal{O}).$$

We can easily compute $h_{A,B}$ (cf. Example 18 and (4.5.13)). Since D_Q is relatively prime to $\div(h_{A,B})$ we know that $h_{A,B}(D_Q) \neq 0$, and so $h_{A,B}(D_Q) \in \mathbb{F}_q^*$. Furthermore by induction we get

$$\underbrace{(P, 1) \oplus \cdots \oplus (P, 1)}_{n} = (\mathcal{O}, f_P(D_Q)).$$

See subsection "Anomalous Curves—Algebraic Geometrical Method" for a similar proof.

Observe that the evaluation of $h_{A,B}$ is almost done by the computation of $A + B$ (compare (4.5.13) and (4.2.20) (Sect. 4.2) in the case $\mathrm{char}(\mathbb{F}_q) = p > 3$). Hence using repeated doubling we can evaluate $f_P(D_Q)$ in $O(\log n)$ steps, where one step is mainly an elliptic curve addition in $E(\mathbb{F}_q)$. Note further that for evaluating $f_P(D_Q)$ we do not need the whole group $< P > \times \mathbb{F}_q^*$. Thus D_Q has not to be prime to all divisors $(P_i),\ P_i \in < P >$. Let

$$S(n) := \{i : 0 \leq i < n,\ iP \text{ needed to compute } nP = \mathcal{O} \text{ by repeated doubling}\}.$$

Hence we can precompute $S(n)$ and can choose D_Q such that D_Q is prime to all divisors $(P_i),\ i \in S(n)$.

By this construction we get the following

Lemma 43 *Let $P \in E[n](\mathbb{F}_q)$ and $Q \in E(\mathbb{F}_q)$, take divisors $D_P \in \overline{(P) - (\mathcal{O})}$ and $D_Q \in \overline{(Q) - (\mathcal{O})}$ with different support and let f_P be a function on E such that $\div(f_P) = nD_P$. Then $f_P(D_Q) \in \mathbb{F}_q^*$ can be evaluated in $\log n$ steps, where one step is nearly an elliptic curve addition.*

Now we are able to reduce the ECDLP:

Frey/Rück Reduction
Require: $P \in E[n](\mathbb{F}_q)$ and $R \in < P > \subseteq E[n](\mathbb{F}_q)$, where $n|(q-1)$.
1: Find a $Q \in E(\mathbb{F}_q)$ such that $\alpha = \{P, Q\}_n$ has order n in $\mathbb{F}_q^*$.
2: Compute $\beta = \{R, Q\}_n \in \mu_n(\mathbb{F}_q)$.
3: Solve the DLP $\beta = \alpha^l$ in $\mathbb{F}_q^*$.
Ensure: $l, 0 \leq l \leq n-1$, such that $R = lP$.

The algorithm fails if in step 1 no $Q \in E(\mathbb{F}_q)$ is found such that $\{P, Q\}_n$ is a primitive nth-root of unity, or the $\{\ ,\ \}_n$-pairing in step 2 can not be evaluated. If $n \not| (q-1)$ but $n|(q^k - 1)$ then we can easily extend the algorithm to the ECDLP in $E(\mathbb{F}_{q^k})$ since $E(\mathbb{F}_q)$ is a subgroup of $E(\mathbb{F}_{q^k})$ as done in the MOV reduction.

Corollary 12 ([23], Corollary 1) *If $n|q-1$ then the ECDLP in $E[n](\mathbb{F}_q)$ can be reduced to the corresponding DLP in $\mathbb{F}_q^*$ in probabilistic polynomial time in $\log q$.*

For simplicity we shall proof a weaker version:

Corollary 13 *Let* $P \in E[n](\mathbb{F}_q)$ *be a point of prime power order* $n = \bar{n}^r$. *Let* $r_{max} = \max\{r : \bar{n}^r | \#E(\mathbb{F}_q)\}$ *and* $E[\bar{n}^{r_{max}}](\mathbb{F}_q)$ *be cyclic. If* $n|(q-1)$ *then the ECDLP* $R = lP$ *can be reduced to the corresponding DLP in* $\mathbb{F}_q^*$ *in probabilistic polynomial time in* $\log q$.

Proof Observe that $\#E(\mathbb{F}_q) = (1-\alpha)(1-\overline{\alpha})$, where $\alpha, \overline{\alpha} \in \mathbb{C}$ and $|\alpha| = |\overline{\alpha}| = \sqrt{q}$. Hence $\log n = O(\log q)$.

Now we shall show that the probability to find a point $Q \in E(\mathbb{F}_q)$ with $\{\{P_i, Q\}_n | P_i \in < P >\} = \mu_n(\mathbb{F}_q)$ is positive. If $r = r_{max}$ then we can choose $Q = P$ and $\{P, P\}_n$ is primitive $\bar{n}^r$th root of unity. Otherwise if $r < r_{max}$ we have to find a point $Q \in E(\mathbb{F}_q)$ such that $\mathrm{ord}(Q) = \bar{n}^{r_{max}}$. Assume E is given by $E: \ y^2 = f(x)$. Then choose a random $x \in \mathbb{F}_q$:

(a) *f(x) is a square in* $\mathbb{F}_q$: Compute y by the method in Sect. 4.3.3 in probabilistic polynomial time in $\log q$. Then

$$Q = \frac{\#E(\mathbb{F}_q)}{\bar{n}^{r_{max}}}(x, y)$$

is a point of order $\bar{n}^j$, $j \le r_{max}$.

(a) *f(x) is not a square in* $\mathbb{F}_q$: Then $(x, \sqrt{f(x)}) \in E(\mathbb{F}_{q^2})$. Since $\#E(\mathbb{F}_{q^2}) = \#E(\mathbb{F}_q)(-\#E(\mathbb{F}_q) + 2q + 2)$ the p-primary parts of $E(\mathbb{F}_q)$ and $E(\mathbb{F}_{q^2})$ are the same. Then

$$Q = \frac{\#E(\mathbb{F}_{q^2})}{\bar{n}^{r_{max}}}(x, \sqrt{f(x)})$$

is again a point of order $\bar{n}^j$, $j \le r_{max}$ in $E(\mathbb{F}_q)$.

Thus for every random chosen $x \in \mathbb{F}_q$ we find $Q \in E(\mathbb{F}_q)$ of order $\bar{n}^j$, $j \le r_{max}$. Hence

$$\mathrm{Prob}(< Q >= E[\bar{n}^{r_{max}}](\mathbb{F}_q)) = \frac{\bar{n}^{r_{max}} - \bar{n}^{r_{max}-1}}{\bar{n}^{r_{max}}} = 1 - \frac{1}{\bar{n}^{r_{max}}} > 0.$$

Assume now that $Q \in E(\mathbb{F}_q)$ such that $\mathrm{ord}(Q) = \bar{n}^{r_{max}}$. Then

$$\alpha = \{P, Q\}_{\bar{n}^r} \in \mu_{\bar{n}^r}(\mathbb{F}_q)$$

is a primitive $\bar{n}^r$th root of unity. Calculating further

$$\beta = \{R, Q\}_{\bar{n}^r} \in \mu_{\bar{n}^r}(\mathbb{F}_q)$$

we get the DLP $\beta = \alpha^l$ since

$$\beta = \{R, Q\}_{\bar{n}^r} = \{lP, Q\}_{\bar{n}^r} = \{P, Q\}^l_{\bar{n}^r} = \alpha^l.$$

□

Recently Frey et al. proposed a further corollary which makes the difference to the MOV-Reduction/Weil-Pairing more clear:

Corollary 14 ([24]) *Let $E/\mathbb{F}_q$ be an elliptic curve containing a point of order n such that $\#E(\mathbb{F}_q) = q + 1 - t$. If*

$$t \equiv 2 \pmod{n} \tag{4.5.4}$$

then the ECDLP in $E[n](\mathbb{F}_q)$ can be reduced to the DLP in $\mathbb{F}_q^*$ probabilistically in polynomial time using the $\{\ ,\ \}_n$-pairing.

Proof $E(\mathbb{F}_q)$ contains a point of order n. Hence $n|\#E(\mathbb{F}_q) = q + 1 - t$. By (4.5.4) $t = a \cdot n + 2$, for any $a \in \mathbb{Z}$. Thus $n|(q - 1 - an)$ and we can use Corollary 12. □

When we compare the MOV reduction and the Frey/Rück reduction, we see that in order to use the Weil-Pairing (MOV) we need $E[n](\overline{\mathbb{F}_q}) \subseteq E(\mathbb{F}_{q^k})$ for a fixed k, which implies that $\mu_n(\mathbb{F}_{q^k}) \subseteq \mathbb{F}_{q^k}^*$ (the converse is not generally true). For the $\{\ ,\ \}_n$-pairing the proof of Theorem 87 shows that we only need $\mu_n(\mathbb{F}_q) \in \mathbb{F}_q^*$. Especially if the genus of a projective smooth curve is greater 1, it seems to be weaker to assume that $\mu_n(\mathbb{F}_q) \subseteq \mathbb{F}_q^*$ than $E[n] \subseteq E(\mathbb{F}_{q^k})$ (cf. [23]) and the computation of the $\{\ ,\ \}_n$-pairing is easier than the generalized Weil-Pairing. A further great difference is that $e_n(P, P) = 1$ for all $P \in E[n]$, but $\{P, P\}_n$ may be nontrivial as shown in the next example.

Example 28 Let $E/\mathbb{F}_q$ be an elliptic curve with $\#E(\mathbb{F}_q) = \bar{n} \cdot d$, where $\bar{n}$ is the prime order of the point $P \in E(\mathbb{F}_q)$. Let $\gcd(\bar{n}, d) = 1$ and $\bar{n}|(q - 1)$. Then we can compute $\alpha = \{P, Q\}_{\bar{n}} = (f_P(D_Q))^{(q-1)/\bar{n}}$ choosing $Q = P$. In order to evaluate the $\{\ ,\ \}_{\bar{n}}$-pairing we define $D_P = (P) - (\mathcal{O})$ and $D_Q = (kP) - ((k - 1)P)$, where $k, k - 1 \notin S(\bar{n})$. Then we can compute $\beta = \{R, Q\}_{\bar{n}}$ by choosing $D_R = (R) - (\mathcal{O})$ and $D_Q = (jP) - ((j - 1)P)$ such that $\mathrm{supp}(D_R) \cap \mathrm{supp}(D_Q) = \emptyset$. Thus for the cryptographically relevant case where $\mathrm{ord}(P)$ is prime we get a deterministic polynomial time reduction if $\mathrm{ord}(P)|(q - 1)$.

Nevertheless we get the same necessary

Condition 3: $\bar{n} \nmid (q^k - 1)$ for all k, $1 \leq k \leq c$, where c is a constant as in Condition 2, i.e. $c > (\ln q)^2$

for "secure" elliptic curves as in the MOV reduction. Again $\bar{n}$ is the largest prime dividing $n = \mathrm{ord}(P)$.

Supersingular Curves

In [55] Menezes et. al. state how to find a small k and Q for the MOV reduction under the assumption that E is a supersingular elliptic curve.

Let E be a supersingular curve of type (n_1, n_2) or (n_1), respectively, defined over $\mathbb{F}_q$, where $\#E(\mathbb{F}_q) = q + 1 - t$. By Corollary 8(iii) and Lemma 29 E lies in one of the curve classes of Table 4.8. Since we can count $\#E(\mathbb{F}_p)$ in polynomial time (see Sect. 4.3.3) we get t and can determine the class of the supersingular curve.

Note also that $n_1 = q + 1 - t$ if $E(\mathbb{F}_q)$ is cyclic. Since $n_1 | \#E(\mathbb{F}_q)$, i.e. $n_1 | q + 1 - t$, and E is supersingular, i.e. $p|t$, we get $\gcd(n_1, q) = \gcd(p^m + 1 - t, p^m) = \gcd(p^{p/t+m} + 1, p^m) = 1$, since p is a prime. Hence we satisfy the basic conditions for a MOV reduction of the ECDLP $R = lP$.

We shall discuss next how to determine the smallest $k \in \mathbb{N}$ such that $E[n] \subseteq E(\mathbb{F}_{q^k})$: Recall that n is the order of P. If $n = 2$ the ECDLP becomes trivial. Suppose that the order of P is greater 2. Then $n | n_1$ (cf. Table 4.8). Hence

$$E[n] = \{P \in E(\overline{\mathbb{F}}_q) : \; nP = \mathcal{O}\} \subseteq \{P \in E(\overline{\mathbb{F}}_q) : \; n_1 P = a \cdot nP = \mathcal{O}\} = E[n_1]$$

for some $a \in \mathbb{N} \setminus \{0\}$. Therefore if $E[n_1] \subseteq E(\mathbb{F}_{q^k})$ it follows that $E[n] \subseteq E(\mathbb{F}_{q^k})$ in step 1 of the MOV reduction.

Now we can use the Weil Theorem 72 in order to find the smallest k such that $E[n_1] \subseteq E(\mathbb{F}_{q^k})$, since we have all necessary parameters.

Example 29 Let E be a supersingular elliptic curve in the class (III), i.e. $t^2 = q$. From Table 4.8 we see that $E(\mathbb{F}_q)$ is cyclic of order $q + 1 \mp \sqrt{q}$. Let $\#E(\mathbb{F}_q) = n_1 = q + 1 + \sqrt{q}$ (the case $n_1 = q + 1 - \sqrt{q}$ is similar). Using the roots $\alpha = -\frac{\sqrt{q}}{2} + i\frac{\sqrt{3q}}{2}$ and $\overline{\alpha}$ of $1 + \sqrt{q}T + T^2$, we can apply the Weil Theorem in order to find

$$\#E(\mathbb{F}_{q^2}) = q^2 + 1 - \alpha^2 - \overline{\alpha}^2 = q^2 + 1 + q,$$
$$\#E(\mathbb{F}_{q^3}) = q^3 + 1 - \alpha^3 - \overline{\alpha}^3 = q^3 + 1 - 2\sqrt{q^3}.$$

Table 4.8 Supersingular elliptic curve classes, cf. also [55], Table 4.1

Class	t	Group structure	n_1
(I)	0	Cyclic	$q+1$
(II)	0	$\mathbb{Z}_{(q+1)/2} \oplus \mathbb{Z}_2$	$(q+1)/2$
(III)	$\pm\sqrt{q}$	Cyclic	$q+1 \mp \sqrt{q}$
(IV)	$\pm\sqrt{2q}$	Cyclic	$q+1 \mp \sqrt{2q}$
(V)	$\pm\sqrt{3q}$	Cyclic	$q+1 \mp \sqrt{3q}$
(VI)	$\pm\sqrt{4q}$	$\mathbb{Z}_{\sqrt{q}\mp 1} \oplus \mathbb{Z}_{\sqrt{q}\mp 1}$	$\sqrt{q} \mp 1$

Thus $E(\mathbb{F}_{q^2})$ is cyclic by Lemma 26(i), since $(t_{q^2})^2 = q^2$, and $E(\mathbb{F}_{q^3}) \cong \mathbb{Z}_{\sqrt{q^3}-1} \oplus \mathbb{Z}_{\sqrt{q^3}-1}$ by Lemma 26(ii), since $t_{q^3} = 2\sqrt{q^3}$. Therefore

$$E(\mathbb{F}_{q^2}) \cap E[n_1] = E(\mathbb{F}_q)$$

and

$$E[n_1] \subseteq E(\mathbb{F}_{q^3}) \cong \mathbb{Z}_{dn_1} \oplus \mathbb{Z}_{dn_1},$$

since $\sqrt{q^3} - 1 = (\sqrt{q} - 1)n_1 = dn_1$.

Calculating the smallest k and d such that $E[n_1] \subseteq E(\mathbb{F}_{q^k}) \cong \mathbb{Z}_{dn_1} \oplus \mathbb{Z}_{dn_1}$ can be done for all supersingular curve classes following the example. Menezes et. al. showed that $k \leq 6$ (cf. [55], Table 1). This yields the following probabilistic polynomial time reduction to a finite field DLP:

MOV reduction for supersingular elliptic curves
Require: $P \in E(\mathbb{F}_q)$ of order n, where $E/\mathbb{F}_q$ is supersingular, $R \in < P >$ and the supersingular elliptic curve class of E.
1: Determine k and d from [55], table 1 or compute it as in Example 29
2: Pick a random point $Q' \in E(\mathbb{F}_{q^k})$.
3: Calculate $Q = (dn_1/n)Q' \in E[n] \subseteq E(\mathbb{F}_{q^k})$.
4: Evaluate $\alpha = e_n(P, Q)$ and $\beta = e_n(R, Q)$.
5: Compute the discrete logarithm $l' = \log_\alpha \beta$ in $\mathbb{F}_{q^k}$.
6: **if** $l'P = R$ **then**
7: Set $l = l'$. STOP
8: **else**
9: Goto 2
10: **end if**
Ensure: $l, 0 \leq l \leq n - 1$, such that $R = lP$.

If $l'P \neq R$ in step 5 then $\mathrm{ord}(\alpha) < n$. Thus we have found a 'bad' point Q, i.e. a point Q such that $e_n(P, Q)$ is not primitive in $\mu_n(\mathbb{F}_{q^k})$, in step 3.

Theorem 89 ([55], Theorem 11) *Let $E/\mathbb{F}_q$ be a supersingular curve. The ECDLP in $E(\mathbb{F}_q)$ can be reduced to the DLP in $\mathbb{F}_{q^k}$ in probabilistic polynomial time in* $\log q$.

Proof Following [55] assume that a basis of $\mathbb{F}_q$ over its prime field and a irreducible polynomial $f(x)$ of degree k over $\mathbb{F}_q$ is given. Hence $\mathbb{F}_{q^k} \cong \mathbb{F}_q[x]/(f(x))$, $(f(x))$ the ideal generated by $f(x)$. Note that $\log n = O(\log q)$ as in the proof of Corollary 13 since $k \leq 6$. Now we can analyse the above algorithm:

2: Q' can be chosen at random in probabilistic polynomial time by the algorithm nqof Sect. 4.3 since $Q' \in E(\mathbb{F}_{q^k})$, where $k \leq 6$.
3: Q can be determined by repeated doubling in polynomial time in $\log q$. Note that Q is a random point in $E[n]$ since $E(\mathbb{F}_{q^k})$ is a abelian group of type (dn_1, dn_1).

4: α and β can be computed in probabilistic polynomial time (cf. Remark 37)
6: $l'P = R$ can be tested in polynomial time.

Let now Q be a random point in $E[n]$. Since there are n cosets of $< P >$ in $E[n]$ and by Lemma 35(ii) there are only n different values $e_n(P, Q) \in \mathbb{F}_{q^k}$ possible. Defining

$$\phi(n) := \#\{\alpha \in \mathbb{F}_{q^k} : \text{ord}(\alpha) = n\}$$

we get

$$\text{Prob}(\text{ord}(e_n(P, Q)) = n, Q \text{ a random point in } E[n]) = \phi(n)/n.$$

For $n > 5$ it can be shown that $n/\phi(n) \leq 6 \log \log n$ (cf. [55]). Hence the algorithm finds a $Q \in E[n]$ such that $\text{ord}(e_n(P, Q)) = n$ in expected $O(\log \log n)$ iterations. □

This immediately leads to

Corollary 15 ([55]) *Let $E/\mathbb{F}_q$ be a supersingular elliptic curve, q a prime power, and let $P, R \in E(\mathbb{F}_q)$ such that $R = lP$ and $n = ord(P)$. Then the MOV reduction for supersingular elliptic curves can determine l in probabilistic subexponential time in* $\log q$.

We only mention that the DLP in a small extension of the finite field $\mathbb{F}_q$ can be solved in probabilistic subexponential time using methods as in Sect. 4.1.5.

Note that k is also valid for the Frey/Rück reduction, so we could also apply this method for supersingular curves directly.

Example 30 Let $E/\mathbb{F}_{2^m} : y^2 + y = x^3$, m odd, be an elliptic curve. E was considered for the implementation of the elliptic curve cryptosystem in the pioneering paper of Koblitz [36]. It can be shown that t $= 0$ and E is a class (I) supersingular elliptic curve with $k = 2$ (cf. [51], Table 3.3). Hence the ECDLP in $E(\mathbb{F}_{2^m})$ is almost as difficult as the DLP in $\mathbb{F}_{2^{2m}}$.

Thus we get a further necessary condition for "secure" elliptic curves:

Condition 4: E should not be supersingular

Notice that condition 2 already implies this condition.

Anomalous Curves—Algebraic Geometrical Method

In 1995 Semaev [78] developed an algebraic geometrical method to compute the p-part of the ECDLP. In 1997 H.-G. Rück [71] extends this ideas to curves of arbitrary genus. We will follow Rück in order to reduce his algorithm to the genus one case.

Let

$$E : y^2 = x^3 + ax + b \tag{4.5.5}$$

be an elliptic curve defined over $\mathbb{F}_q$, where $q = p^m$ and $p > 3$ is a prime.

Assume that the basepoint P of an elliptic curve public-key scheme is of order p, i.e. $P \in E[p](\mathbb{F}_q)$. Let $t = t_{\mathcal{O}} = -x/y$ be the local parameter of the rational point $\mathcal{O}$.

In the following let D_Q be a divisor in the divisor class $\overline{(Q) - (\mathcal{O})}$, where $Q \in E[p]$, e.g. $D_Q = (Q) - (\mathcal{O})$. Define

$$\mathrm{lg} : Pic^0(E)_p \to \Omega(E) \tag{4.5.6}$$

$$\overline{D_Q} \mapsto \frac{df_Q}{f_Q}, \tag{4.5.7}$$

where f_Q is a function on E such that $\div(f_Q) = pD_Q$.

Note that by Corollary 9 f_Q exists, since pD_Q is principal. The key-point in constructing an isomorphic embedding of $< P >$ into $\mathbb{F}_q$ is the following lemma:

Lemma 44 *For all non-principal divisors D_Q $\underset{\cdot}{f}_Q/f_Q$ is holomorphic and non-vanishing.*

Proof We prove this lemma using ideas of Semaev [78], Lemma 1.

Let $Q \in E[p]$. Let $D_Q \in \overline{(Q) - (\mathcal{O})}$, i.e. $D_Q = \sum n_T(T)$ such that $Q = \sum n_T T$. Let t_T denote the local parameter at T (cf. Example 16). For convenience we write f instead of f_Q.

We shall show that $\div(\underset{\cdot}{\mathrm{f}}/\underset{\cdot}{\mathrm{x}}) = \div(f) - \div(y)$. Hence $\div(\underset{\cdot}{\mathrm{f}}/f) = \div(\underset{\cdot}{\mathrm{x}}/y) = 0$ and the lemma is proved (cf. Example 19).

Set $f = t_T^{l_T} f_1$, where f_1 is regular at T and $f_1(T) \neq 0$.

(a) Let $T \notin \mathrm{supp}\{\div(y)\}$, i.e. $T \notin E[2]$ and especially $T \neq \mathcal{O}$. Thus $t_T = x - x_T$.

$$\frac{df}{\underset{\cdot}{\mathrm{x}}} = \frac{df}{(\underset{\cdot}{x} - x_T)} = \frac{\underset{\cdot}{\mathrm{f}}}{\underset{\cdot}{\mathrm{t}}_T} = t_T^{l_T} \frac{\underset{\cdot}{\mathrm{f}}_1}{\underset{\cdot}{\mathrm{t}}_T}$$

Hence $\mathrm{ord}_T(\underset{\cdot}{\mathrm{f}}/\underset{\cdot}{\mathrm{x}}) = l_T + m_T$, where $m_T = \mathrm{ord}_T(\underset{\cdot}{\mathrm{f}}_1/dt_T) \geq 0$, since $\underset{\cdot}{\mathrm{f}}_1/\underset{\cdot}{\mathrm{t}}_T$ is regular at T by Theorem 80(ii).

(b) Let T a point of order 2, i.e. $T \in E[2] \setminus \{\mathcal{O}\}$: Thus $t_T = y$.

$$\frac{df}{\underset{\cdot}{\mathrm{x}}} = \left(\frac{\underset{\cdot}{\mathrm{f}}}{\underset{\cdot}{\mathrm{t}}_T}\right)\left(\frac{\underset{\cdot}{\mathrm{t}}_T}{\underset{\cdot}{\mathrm{x}}}\right) = \left(t_T^{l_T} \cdot \frac{\underset{\cdot}{\mathrm{f}}_1}{\underset{\cdot}{\mathrm{t}}_T}\right)\left(\frac{\underset{\cdot}{\mathrm{y}}}{\underset{\cdot}{\mathrm{x}}}\right) = y^{l_T} \frac{\underset{\cdot}{\mathrm{f}}_1}{\underset{\cdot}{\mathrm{t}}_T}\left(\frac{3x^2 + a}{2y}\right)$$

Since $T = (x_T, 0)$ $\mathrm{ord}_T((3x^2 + a)/(2y)) = -1$. Let $m_T = \mathrm{ord}_T(\underset{\cdot}{\mathrm{f}}_1/\underset{\cdot}{\mathrm{t}}_T)$. Then $m_T \geq 0$ and $\mathrm{ord}_T(\underset{\cdot}{\mathrm{f}}/\underset{\cdot}{\mathrm{x}}) = l_T + m_T - 1$.

(c) Let $T = \mathcal{O}$. Thus $t_T = x/y$.

$$\frac{df}{\underset{\cdot}{\mathrm{x}}} = \left(\frac{\underset{\cdot}{\mathrm{f}}}{\underset{\cdot}{\mathrm{t}}_{\mathcal{O}}}\right)\left(\frac{\underset{\cdot}{\mathrm{t}}_{\mathcal{O}}}{\underset{\cdot}{\mathrm{x}}}\right) = \left(t_{\mathcal{O}}^{l_{\mathcal{O}}} \cdot \frac{\underset{\cdot}{\mathrm{f}}_1}{\underset{\cdot}{\mathrm{t}}_{\mathcal{O}}}\right)\frac{\underset{\cdot}{(x/y)}}{\underset{\cdot}{\mathrm{x}}} = t_{\mathcal{O}}^{l_{\mathcal{O}}} \frac{\underset{\cdot}{\mathrm{f}}_1}{\underset{\cdot}{\mathrm{t}}_{\mathcal{O}}}\left(\frac{-x^3 + ax + b}{2y^3}\right).$$

Let $m_{\mathcal{O}} = \mathrm{ord}_{\mathcal{O}}(\underset{.}{\mathrm{f}}_1/\underset{.}{\mathrm{t}}_{\mathcal{O}})$. Then $m_{\mathcal{O}} \geq 0$ and $\mathrm{ord}_{\mathcal{O}}(\underset{.}{\mathrm{f}}/\underset{.}{\mathrm{x}}) = l_{\mathcal{O}} + m_{\mathcal{O}} + 3$, since

$$\mathrm{ord}_{\mathcal{O}}((-x^3 + ax + b)/(2y^3)) = \mathrm{ord}_{(0:1:0)}((-X^3 + aXZ^2 + bZ^3)/(2Y^3)) = 3.$$

Now let $\tilde{D} = \sum m_T(T)$. Thus $\tilde{D}$ is a positive divisor, i.e. $m_T \geq 0$ for all $T \in E$. Since

$$\begin{aligned}
\div(\underset{.}{\mathrm{f}}/\underset{.}{\mathrm{x}}) &= \sum \mathrm{ord}_T(\underset{.}{\mathrm{f}}/\underset{.}{\mathrm{x}})(T) \\
&= \sum_{T \notin E[2]} (l_T + m_T)(T) + \sum_{T \in E[2]\setminus\{\mathcal{O}\}} (l_T + m_T - 1)(T) \\
&\quad + (l_{\mathcal{O}} + m_{\mathcal{O}} + 3)(\mathcal{O}) \\
&= \sum_{T \in E(\mathbb{F}_q)} l_T(T) - \left[\sum_{T \in E[2]\setminus\{\mathcal{O}\}} 1(T) - 3(\mathcal{O}) \right] + \sum_{T \in E(\mathbb{F}_q)} m_T(T) \\
&= \div(f) - \div(y) + \tilde{D}.
\end{aligned}$$

Thus $\tilde{D}$ is principal and $\tilde{D} \in \mathrm{Div}^0(E)$. Hence $\tilde{D} = 0$. □

Thus the image of lg is in $\Omega^1(E)$. We define the following map

$$\begin{aligned}
\Phi : \; < P > &\to \mathbb{F}_q \\
Q &\mapsto c(\underset{.}{\mathrm{f}}_Q/f_Q), \\
\mathcal{O} &\mapsto 0,
\end{aligned}$$

where $c(f)$ is the constant term of the Laurent expansion of f around $\mathcal{O}$ with respect to the local parameter $t = t_{\mathcal{O}}$. More precisely, if we calculate the Laurent series

$$\frac{\partial f_Q/\partial t_{\mathcal{O}}}{f_Q} = \sum_{i=0}^{\infty} a_i t^i, \qquad a_i \in \mathbb{F}_q, \tag{4.5.8}$$

then $c(\underset{.}{\mathrm{f}}_Q/f_Q) = a_0$.

Lemma 45 *Φ is an isomorphic embedding of $< P >$ into $\mathbb{F}_q^+$.*

Proof By Theorem 79 we have an isomorphism of sets

$$\chi : E \to \mathrm{Pic}^0(E), P \mapsto \overline{(P) - (\mathcal{O})},$$

which can be reduced to an isomorphism $\chi_p : E[p] \to \mathrm{Pic}^0(E)_p$. Combining this with Lemma 44 gives the map

$$\Phi' := c \circ \lg \circ \chi_p : \; E[p] \to \mathrm{Pic}^0(E)_p \to \Omega^1(E) \to \mathbb{F}_q^+$$
$$Q \mapsto \overline{D_Q} \mapsto \underset{.}{f}_Q/f_Q \mapsto c(df_Q/f_Q),$$
$$\mathcal{O} \mapsto 0.$$

Φ' is well-defined:
Let $Q \in E[p]$. Let $D_Q,\ \tilde{D}_Q$ be linearly equivalent divisors, i.e. there is a $g \in \overline{\mathbb{F}_q}(E)$ such that $\div(g) = \tilde{D}_Q - D_Q$. Hence if $\div(f) = p\tilde{D}_Q$ then $g^p f = f_Q$. Therefore

$$\frac{\underset{.}{f}_Q}{f_Q} = \frac{\underset{.}{(g^p f)}}{g^p f} = \frac{\underset{.}{(g^p)}}{g^p} + \frac{\underset{.}{f}}{f} = p\frac{(g^{p-1})\underset{.}{g}}{g^p} + \frac{\underset{.}{f}}{f} = \frac{\underset{.}{f}}{f}, \tag{4.5.9}$$

since $\mathrm{char}(\mathbb{F}_q) = p$.

Let $Q_1, Q_2 \in E[p]$ and $\div(f_{Q_i}) = pD_{Q_i}, i = 1, 2$. Defining $D_{Q_1+Q_2} = D_{Q_1} + D_{Q_2}$ we get

$$\div(f_{Q_1} + f_{Q_2}) = pD_{Q_1+Q_2} = pD_{Q_1} + pD_{Q_2} = \div(f_{Q_1} f_{Q_2}), \tag{4.5.10}$$

i.e. $f_{Q_1} + f_{Q_2} = k \cdot f_{Q_1} f_{Q_2}$, where k is a multiplicative constant, and

$$c(f+g) = c((\sum f_i t^i)\underset{.}{t} + (\sum g_i t^i)\underset{.}{t})$$
$$= c(\sum (f_i + g_i)t^i\,\underset{.}{t}) = f_0 + g_0 = c(f) + c(g) \tag{4.5.11}$$

for functions f, g on E.

Hence using (4.5.10) and (4.5.11)

$$\begin{aligned}\Phi'(Q_1+Q_2) &= c \circ \lg \circ \chi_p(Q_1+Q_2) = c(\lg(\overline{D_{Q_1+Q_2}}))\\ &= c\left(\frac{\underset{.}{f}_{Q_1+Q_2}}{f_{Q_1+Q_2}}\right) = c\left(\frac{\underset{.}{(f_{Q_1} f_{Q_2})}}{f_{Q_1} f_{Q_2}}\right)\\ &= c\left(\frac{\underset{.}{f}_{Q_1}}{f_{Q_1}} + \frac{\underset{.}{f}_{Q_1}}{f_{Q_1}}\right)\\ &= c\left(\frac{\underset{.}{f}_{Q_1}}{f_{Q_1}}\right) + c\left(\frac{\underset{.}{f}_{Q_1}}{f_{Q_1}}\right)\\ &= \Phi'(Q_1) + \Phi'(Q_2).\end{aligned}$$

Therefore Φ' is a homomorphism. Reducing Φ' to Φ we take $Q \in < P >$, where $< P >$ is a subgroup of $E[p]$ in $E(\mathbb{F}_q)$. Hence $Q \neq \mathcal{O}$ is rational over $\mathbb{F}_q$. Therefore we can take D_Q also rational over $\mathbb{F}_q$ as well as $\underset{.}{f}_Q/f_Q$ and therefore $c(\underset{.}{f}_Q/f_Q)$.

Observe further that $\underset{.}{f}_Q(t)/f_Q(t) = (f_Q(t))^{-1}\partial f_Q(t)/\partial t\underset{.}{t}$. Hence $\underset{.}{f}_Q/f_Q$ determines $f_Q^{-1}\partial f_Q/\partial t$ uniquely by Theorem 80(i). Since $\underset{.}{f}_Q/f_Q$ is holomorph we can evaluate the power series expansion (4.5.8). By Corollary 11 (or more generally by the Riemann–Roch theorem) $\underset{.}{f}_Q/f_Q$ determines a_0 uniquely. Hence c is an isomorphism.

Finally Φ is an isomorphic embedding of $< P >$ into $\mathbb{F}_q^+$ because lg is non-vanishing on $< P >$ and an isomorphism (cf. [79], Proposition 10). □

For a more general construction of this isomorphism, see Serre [79] Proposition 10 for arbitrary genus curves.

Next we will show that $\Phi(Q)$, $Q \in < P > \subseteq E[p]$, can be evaluated in polynomial time. For convenience we define $\text{tlg}(f) := \dot{f}/f$.

For $G := < P > \times \mathbb{F}_q^+$ define the following operation

$$(A, a) \oplus (B, b) = (A + B, a + b + c(\text{tlg}(h_{A,B}))), \tag{4.5.12}$$

where $h_{A,B}$ is the line passing through the points A, B such that

$$\div(h_{A,B}) = (A) + (B) - (A + B) - (\mathcal{O}).$$

Following Example 18 we get $h_{A,B} : \frac{\delta(A,B)x+\alpha-y}{x-\gamma} = 0$, where

$$\delta(A, B) = \begin{cases} \frac{y_B - y_A}{x_B - x_A}, \ if A \neq \pm B, \\ \frac{3x_A^2 + a}{2y_A}, \ if A = B, \end{cases} \tag{4.5.13}$$

is the slope of $\overline{AB}$ and $\alpha, \gamma = x_{-(A+B)}$, are constants in $\mathbb{F}_q$ if $A \neq -B$. If $A = -B$ then we can take $h_{A,B} : \ x - c = 0$.

Lemma 46 *(i)* $(G, \oplus)$ *is an abelian group.*
(ii) Let $Q \in < P >$. *Then* $\underbrace{(Q, 0) \oplus \cdots \oplus (Q, 0)}_{p} = (\mathcal{O}, c(\text{lg}(\overline{D_Q})))$, *where calculation takes place in* $< P > \times \mathbb{F}_q^+$.

Proof (i) Let $(A, a), (B, b), (C, c) \in G$.

Identity element: If $A \neq \mathcal{O}$ observe that $c(\text{tlg}(h_{A,\mathcal{O}})) = 0$: Since $\div(h_{A,\mathcal{O}}) = 0$, i.e. $h_{A,\mathcal{O}} \in \mathbb{F}_q^*$ is constant, $\text{tlg}(h_{A,\mathcal{O}}) = 0$.
Hence $(A, a) \oplus (\mathcal{O}, 0) = (A, a)$.

Inverse element: $(A, a) \oplus (-A, -a) = (\mathcal{O}, 0)$.

As well the symmetry and thus the commutative law as the nonemptyness is given already by definition.

Note that

$$\begin{aligned}
\div (h_{A,B}h_{A+B,C}) &= [(A) + (B) - (A + B) - (\mathcal{O})] \\
&\quad +[(A + B) + (C) - (A + B + C) - (\mathcal{O})] \\
&= [(B) + (C) - (B + C) - (\mathcal{O})] \\
&\quad +[(A) + (B + C) - (A + B + C) - (\mathcal{O})] \\
&= \div(h_{B,C}h_{A,B+C}).
\end{aligned} \tag{4.5.14}$$

Hence using (4.5.14) we get

$$\begin{aligned}
c(\text{tlg}(h_{B,C})) + c(\text{tlg}(h_{A,B+C}) &= c(\text{tlg}(h_{B,C}) + \text{tlg}(h_{A,B+C})) \\
&= c(\text{tlg}(h_{B,C}h_{A,B+C})) \\
&= c(\text{tlg}(h_{A,B}h_{A+B,C})) \\
&= c(\text{tlg}(h_{A,B}) + \text{tlg}(h_{A+B,C})) \\
&= c(\text{tlg}(h_{A,B})) + c(\text{tlg}(h_{A+B,C}))
\end{aligned} \tag{4.5.15}$$

Using (4.5.14) and (4.5.15) we get the associative law:

$$\begin{aligned}
(A, a) \oplus [(B, b) \oplus (C, c)] &= (A, a) \oplus (B + C, b + c + c(\text{tlg}(h_{B,C}))) \\
&= (A + B + C, a + b + c + c(\text{tlg}(h_{B,C})) \\
&\quad +c(\text{tlg}(h_{A,B+C}))) \\
&= (A + B + C, a + b + c + c(\text{tlg}(h_{A,B})) \\
&\quad +c(\text{tlg}(h_{A+B,C}))) \\
&= (A + B, a + b + c(\text{tlg}(h_{A,B}))) \oplus (C, c) \\
&= [(A, a) \oplus (B, b)] \oplus (C, c).
\end{aligned}$$

(ii) Observe that $\text{tlg}(f_Q) = \text{lg}(\overline{D_Q})$. Defining $h_i(Q) \in \overline{\mathbb{F}_q}(E)$ such that

$$\div (h_i(Q)) = i(Q) - (iQ) - (i - 1)(\mathcal{O}), \quad i \geq 2, \tag{4.5.16}$$

it suffices to show by induction that

$$\underbrace{(Q, 0) \oplus \cdots \oplus (Q, 0)}_{i} = (iQ, c(\text{tlg}(h_i(Q)))) \text{ for all } i = 2, \ldots, p.$$

Define $D_Q = (Q) - (\mathcal{O})$. Then

$$\begin{aligned}
\div(h_p(Q)) &= p(Q) - (pQ) - (p - 1)(\mathcal{O}) = pD_Q \\
&= \div(f_Q)
\end{aligned}$$

and $h_p(Q), f_Q$ are equal up to a multiplicative constant. Hence

$$\mathrm{tlg}(h_p(Q)) = \frac{(kh_p(Q))^{\cdot}}{kh_p(Q)} = \frac{\dot{f}_Q}{f_Q} = \mathrm{tlg}(f_Q).$$

In the following we will use (4.5.10) and $\mathrm{tlg}(fg) = \mathrm{tlg}(f) + \mathrm{tlg}(g)$:

For $i = 2$ observe that $h_{Q,Q} = h_2(Q)$. Then

$$\begin{aligned} \div\,(h_i(Q)h_{iQ,Q}) &= \div(h_i(Q)) + \div(h_{iQ,Q}) \\ &= i(Q) - (iQ) - (i-1)(\mathcal{O}) \\ &\quad +(iQ) + (Q) - (iQ+Q) - (\mathcal{O}) \\ &= (i+1)(Q) - ((i+1)Q) - i(\mathcal{O}) \\ &= \div(h_{i+1}(Q)) \end{aligned} \tag{4.5.17}$$

Hence using (4.5.17)

$$\begin{aligned} \underbrace{(Q,0) \oplus \cdots \oplus (Q,0)}_{i+1} &= (iQ, c[\mathrm{tlg}(h_i(Q))]) \oplus (Q,0) \\ &= ((i+1)Q, c[\mathrm{tlg}(h_i(Q))] + c[\mathrm{tlg}(h_{iQ,Q})]) \\ &= ((i+1)Q, c[\mathrm{tlg}(h_i(Q)) + \mathrm{tlg}(h_{iQ,Q})]) \\ &= ((i+1)Q, c[\underbrace{\mathrm{tlg}(h_i(Q)h_{iQ,Q})}_{\mathrm{tlg}(h_{i+1}(Q))}]), \end{aligned}$$

since $h_i(Q)h_{iQ,Q}$ and $h_{i+1}(Q)$ are also equal up to an multiplicative constant. This is also valid for another representative D_Q of $\overline{(Q) - (\mathcal{O})}$.

Although $h_{A,B}$ has a pole at $\mathcal{O}$, $h_{A,B}$ is rational over $\mathbb{F}_q$ (cf. (4.5.13)). Hence $c(\mathrm{tlg}(h_{A,B}))$, is also rational over $\mathbb{F}_q$. See also the next lemma for details. □

Now we will give an algorithm for evaluating $\Phi(Q)$:

Semaev/Rück Method

Require: $Q = (x_Q, y_Q) \in < P > \subseteq E[p], p = \sum_{i=0}^{l} p_i 2^i, p_i \in \{0, 1\}$

1: **if** $Q = \mathcal{O}$ **then**

2: Set $s = 0$. STOP

3: **end if**

4: Extend the function

$$\delta(A,B) := \begin{cases} \frac{y_B - y_A}{x_B - x_A}, & \text{if } A \neq \pm B, \\ \frac{2x_A^2 + a}{2y_A^2}, & \text{if } A = B \neq \mathcal{O}, \\ 0, & \text{if } A = -B \end{cases}$$

Let (S,s)=(Q,0)
for $i = l$ downto 0 **do**
  Compute $(S, s) = (S, s) \oplus (S, s) = (S + S, s + s + \delta(S, S))$
  if $p_i = 1$ **then**
    Set $(S, s) = (S, s) \oplus (Q, 0) = (S + Q, s + \delta(S, Q))$
  end if
end for
Ensure: $\Phi(Q) = s$

Lemma 47 *Let $Q \in < P > \subseteq E[p]$. The Semaev/Rück method computes $\Phi(Q)$ in $O(\log p)$ elliptic curve additions.*

Proof If $Q = \mathcal{O}$ we are in the trivial case that $\Phi(\mathcal{O}) = 0$ by definition. So assume $Q \neq \mathcal{O}$. Using the local parameter $t = t_{\mathcal{O}}$ we can make a change of variables $t = -x/y,\ w = -1/y$. Thus (4.5.5) becomes

$$E:\ w = t^3 + atw + bw^3, \tag{4.5.18}$$

where $\mathcal{O} = (0, 0)$.

We can rewrite $w = w(t) = t^3 \sum_{i=0}^{\infty} a^i t^i$, since t is a local parameter by recursively rewriting w by (4.5.18):

$$\begin{aligned} w &= t^3 + atw + bw^3 \\ &= t^3 + at[t^3 + atw + bw^3] + b[t^3 + atw + bw^3]^3 \\ &\ \vdots \\ &= t^3 + at^4 + a^2t^5 + a^3t^6 + a^4t^7 + \cdots \\ &= t^3(1 + at + a^2t^2 + a^3t^3 + a^4t^4 + \cdots \\ &= t^3 \sum_{i=0}^{\infty} a^i t^i. \end{aligned}$$

Let $A \neq -B \neq \mathcal{O}$. Then

$$\begin{aligned} & h_{A,B}(x, y):\ \tfrac{\delta(A,B)x+\alpha-y}{x-\gamma} = 0 \\ \Longleftrightarrow\ & \\ & h_{A,B}(t, w):\ \tfrac{\delta(A,B)t+\alpha w+1}{t-\gamma w} = 0 \\ \Longleftrightarrow\ & \\ & h_{A,B}(t):\ \tfrac{\delta(A,B)t+1+\alpha t^3 \sum_{i=0}^{\infty} a^i t^i}{t-\gamma t^3 \sum_{i=0}^{\infty} a^i t^i}. \end{aligned}$$

Now evaluating the Laurent series (4.5.8) yields

$$\frac{\partial h_{A,B}(t)/\partial t}{h_{A,B}} = -\frac{1}{t} + \delta(A, B) + (2\alpha - \delta(A, B))t + \cdots$$

For $A = -B$ we can set $\delta(A, B) = 0$. Hence we can ease the group law (4.5.12) of G to

$$(A, a) \oplus (B, b) = (A + B, a + b + \delta(A, B)).$$

Since G is associative we can evaluate $(A, 0) \oplus \cdots \oplus (A, 0) = (\mathcal{O}, \Phi(Q))$ by $2\lfloor \log_2 p \rfloor$ computations of $\oplus$ using repeated doubling. Note that $\delta(A, B)$ is already computed by the elliptic curve addition $A + B$ (cf. (4.2.20)) and thus '$\oplus$' takes time of an addition in $E(\mathbb{F}_q)$. □

Theorem 90 *Let $E/\mathbb{F}_q$ be an elliptic curve, char$(\mathbb{F}_q) = p > 3$. If ord$(P) = p^e | q$, $e \in \mathbb{N}$, then the ECDLP (4.5.1) is solvable in polynomial time.*

Proof Assume $\text{ord}(P) = p$. Then we can set up the isomorphic embedding Φ. Since all points of $< P > \setminus\{\mathcal{O}\}$ are rational over $\mathbb{F}_q$ in the Weierstrass form we can evaluate $\Phi(P)$, $\Phi(R)$ by the Semaev/Rück method. Then $l = \Phi(R)/(\Phi(P))^{-1}$. Note that $\Phi(R) = 0$ if $R = \mathcal{O}$.

Now assume $\text{ord}(P) = p^e$, $e > 1$. Then we can use the Silver–Pohlig–Hellman method of Sect. 4.1.5:
There exists integers $l_0, \ldots, l_{e-1}$ satisfying $l \equiv \sum_{i=0}^{e-1} l_i p^i \pmod{p^e}$ and $0 \le l_i < p$. We put $R_0 := p^{e-1}R$ and $P_0 := p^{e-1}P$. Then $pP_0 = \mathcal{O}$ and $R_0 = l_0 P_0$. l_0 can be obtained by the Semaev/Rück method computing $l_0 = \Phi(R_0)/(\Phi(P_0))^{-1}$.

Assume now that we have obtained $l_0, \ldots, l_{k-1}$. Then
$R_k := p^{e-k-1}(R - (\sum_{i=0}^{k-1} l_i p^i)P)$ satisfies $R_k = l_k P_0$, which yields l_k by the same method. Finally we obtain $l \mod p^e$. This can be done in $O(e^2 \log p)$ elliptic curve additions. □

Corollary 16 *The ECDLP for a totally anomalous curve is solvable.*

We note only that $\#E(\mathbb{F}_q) = q$. Hence $\text{ord}(P)$ is a prime power of p.

Hence we get again a necessary condition for secure elliptic curves:

Condition 5: $n = \text{ord}(P)$ must not be divisible by $p = \text{char}(\mathbb{F}_q)$

At least $n = \text{ord}(P)$ must be divisible by a large prime $\bar{n}$ other than p to prevent a Silver–Pohlig–Hellman attack.

Example 31 Let $E : y^2 = x^3 + 444x + 7581$ be defined over $\mathbb{F}_p$, where $p = 30971$ is a prime of 15 binary bits. E is a non-supersingular anomalous elliptic curve, since $\#E(\mathbb{F}_p) = p$.

Let $P = (18784, 23524)$ be the basepoint of an elliptic curve public-key scheme of order $n = p$. Let $R = (18091, 4566)$ be the public known point. The private key l must exist, since $R \in< P >= E(\mathbb{F}_p)$. Using the computer, we can calculate

$$(P,0)\underbrace{\oplus\cdots\oplus}_{p\text{ times}}(P,0)=(\mathcal{O},973)$$

and

$$(R,0)\underbrace{\oplus\cdots\oplus}_{p\text{ times}}(R,0)=(\mathcal{O},7831).$$

Hence $l \equiv c(\mathrm{tlg}(f_R))(c(\mathrm{tlg}(f_P)))^{-1} \equiv 7831 \cdot 973^{-1} \equiv 11467 \pmod{p}$. The correctness can be easily checked by $R = lP$.

Anomalous Curves—Number Theoretical Method

In 1997 Satoh/Araki [72] and Smart [86] independently proposed a further method to solve the ECDLP in polynomial time for anomalous curves. The main difference of the so-called Fermat quotient method to the Semaev/Rück method is that we take a number theoretical instead of an algebraic geometrical approach. We shall only give an survey on the mathematical background of this attack.

We will denote by $\mathbb{Q}_p$ and $\hat{\mathbb{Z}}_p$ the ring of p-adic numbers. Note that in this paper $\hat{\mathbb{Z}}_p \neq \mathbb{Z}_p = \mathbb{Z}/p\mathbb{Z}$. For an introduction into p-adic numbers we refer to Mahler [49].

Let p be a prime and a an integer prime to p. Then we have the differential-like operator

$$L_p(a) := \frac{a^{p-1}-1}{p}$$

studied by Eisenstein in 1850. We call L_p the Fermat quotient of a to the base p. Then

$$\begin{aligned} L_p(ab) &= L_p(a) + L_p(b) \\ L_p(a+b) &= L_p(a) - ca^{-1}, \end{aligned}$$

where $a, b \in \mathbb{Z} \setminus p\mathbb{Z}$, $c \in \mathbb{Z}$ and a^{-1} is the inverse of a in $\mathbb{F}_p^*$. It can be shown that L_p induces an $\mathbb{F}_p$-valued logarithm defined over $(\mathbb{Z}/p^2\mathbb{Z})^*$. For details, see [72], § 2.

The idea of Smart/Satoh-Araki is to construct an elliptic curve version of the Fermat quotient.

Let $p \geq 3$be a prime and $\tilde{E}/\mathbb{F}_p : y^2 = x^3 + \tilde{a}x + \tilde{b}$ be an anomalous elliptic curve, i.e. $\#E(\mathbb{F}_p) = p$.

Choose any $a, b \in \mathbb{Z}$ satisfying $a \mod p = \tilde{a}$ and $b \mod p = \tilde{b}$ and define $E : y^2 = x^3 + ax + b$. Thus we get a lifting $\tilde{E}(\mathbb{F}_p) \to E(\mathbb{Q}_p)$. Note that there are many possible liftings. Now we fix a lifting $u : \tilde{E}(\mathbb{F}_p) \to E(\mathbb{Q}_p)$. If we denote ε to be the formal group associated to E we have the following isomorphism

$$\lg : \ker \pi \xrightarrow{\psi} \varepsilon(p\hat{\mathbb{Z}}_p) \xrightarrow{\log_\varepsilon} p\hat{\mathbb{Z}}_p,$$

where π is the reduction map $\pi : E(\mathbb{Q}_p) \to \tilde{E}(\mathbb{F}_p)$, i.e. $\pi \circ u = \mathrm{id}_{\tilde{E}(\mathbb{F}_p)}$ and $\psi(x, y) := x/y$, $\log_\varepsilon(t) := t - \frac{2a}{5}t^5 - \ldots$

For an introduction into the formal group of an elliptic curve and the defined logarithm in this group, we refer to Silverman [82], Chap. IV.

Remark 49 For anomalous elliptic curves the analogous of a^{p-1} in the Fermat quotient is pA for $A \in E$.

Define

$$\lambda_E : \tilde{E}(\mathbb{F}_p) \xrightarrow{u} E(\mathbb{Q}_p) \xrightarrow{\cdot p} \ker \pi \xrightarrow{\lg} p\hat{\mathbb{Z}}_p \xrightarrow{\bmod p^2} p\hat{\mathbb{Z}}_p/p^2\hat{\mathbb{Z}}_p \cong \mathbb{F}_p.$$

It can be shown ([72], Theorem 3.2) that λ_E is a group homomorphism independent of choice of u but depending on E. Furthermore, λ_E is either a zero-map or an isomorphism. In order to achieve the isomorphism we see that λ_E is surjective, since $\#\tilde{E}(\mathbb{F}_p) = p = \mathbb{F}_p^+$, provided $\tilde{E}$ is anomalous. Moreover the anomality of $\tilde{E}$ assures that $pE(\mathbb{Q}_p) \subseteq \ker \pi$ (cf. [82], proof of Proposition VII.2.1).

Let $p \geq 7$. First we give an algorithm to evaluate the isomorphism λ_E

Fermat Quotient Method Part I

Require Curve parameter of a lifted curve $E : y^2 = x^3 + ax + b$ of $\tilde{E}$, $A = (x_A, y_A) \in \tilde{E}(\mathbb{F}_p) \setminus \{\mathcal{O}\}$.

1: Find $\alpha, \beta \in \mathbb{Z}$ such that $\alpha \mod p = x_A$ and $\beta \mod p = y_A$.

2: Compute $x_1 = \alpha \mod p^2$, $\gamma = \frac{x_1^3+a+x_1+b-\beta^2}{2\beta p} \mod p$, $y_1 = (\beta + p\gamma) \mod p^2$
 Note that $S := (x_1, y_1) \in E(\mathbb{Z}/p^2\mathbb{Z}) = E(\mathbb{F}_{p^2})$.

3: Compute $(x_{p-1}, y_{p-1}) = (p-1)(x_1, y_1) \in E(\mathbb{Z}/p^2\mathbb{Z})$ by repeated doubling

4: **if** $x_{p-1} \neq x_1 \mod p^2$ **then**

5: Compute $\lambda_{a,b}(A) = \frac{x_{p-1}-x_1}{p(y_{p-1}-y_1)} \mod p$. STOP

6: **else**

7: Set $\lambda_{a,b} = 0$.

8: **end if**

Ensure: $\lambda_E(A) = \lambda_{a,b}(A)$

If we have a lifted elliptic curve E we find the lifted point $S = u(A)$ by step 1 and 2. The proof of [72], corollary 3.6 shows that $S \in E(\mathbb{Z}_{p^2}) = E(\mathbb{Z}/p^2\mathbb{Z})$!

Thus all computations in step 3 take place in $E(\mathbb{Z}/p^2\mathbb{Z})$ and can be performed in $2\lfloor \log_2 p^2 \rfloor$ additions (cf. Sect. 4.3.1). Hence $O((\log p)^3)$ basic operations are needed to evaluate $\lambda_{a,b}(A)$. Since $\lambda_{a,b}(A)$ can also be a zero-map we use [72], theorem 3.5(ii) to get a condition for a non-zero map λ_E. Note also that by [72], theorem 3.5(iii) the formula for λ_E is well-defined.

Fermat Quotient Method Part II

Require: $P, R \in \tilde{E}(\mathbb{F}_p) \setminus \{\mathcal{O}\}$, where $\tilde{E}_{\tilde{a},\tilde{b}}$ is anomalous, $p \geq 7$ prime and $R \in < P >$.

1: Choose integers $a, b \in \mathbb{Z}$ such that $a \mod p = \tilde{a},\ b \mod p = \tilde{b}$.

2: Find $(\delta, i) \in \mathbb{Z} \times \{1, 2, 3\}$ with $\delta \mod p = x(iP)$ and $3\delta^2 \not\equiv a \pmod p$.

3: Compute $\lambda_{a,b}(iP)$ and $\lambda_{a,b}(iR)$.

4: **if** $\lambda(iP) \neq 0$ **then**

5: Compute $l = \lambda_{a,b}(iR)(\lambda_{a,b}(iP))^{-1}$. STOP

6: **else**

7: Set $a = a + p \in \mathbb{Z}$ and $b = b - p\delta \in \mathbb{Z}$. Goto 3

8: **end if**

Ensure: $l,\ 0 \leq l \leq n - 1$ such that $R = lP$

In step 1 we choose a lifting of $\tilde{E}$ to E. Reference [72], theorem 3.7 assures us, that we can get a non-zero map λ_E or $\lambda_{E'}$, if we satisfy the condition $3x(u(P))^2 \not\equiv a \mod p$. Since by [72], theorem 3.5(ii) we can satisfy this for at least one of the three points $1P,\ 2P$ or $3P$, we can easily satisfy this condition.

Then either λ_E or $\lambda_{E'}$, where E' is a further lifting of $\tilde{E}$ calculated in step 7, is a non-zero map, i.e. must be an isomorphism.

Assume now that $\lambda_{a,b}$ is the isomorphism, then step 5 calculates l correctly, since

$$\begin{aligned} R &= lP \\ \Longleftrightarrow \quad iR &= liP \\ \Longleftrightarrow \quad \lambda_{a,b}(iR) &= l\lambda_{a,b}(iP) \\ \Longleftrightarrow \quad l &= \lambda_{a,b}(iR)(\lambda_{a,b}(iP))^{-1}. \end{aligned}$$

Hence we get a deterministic polynomial time algorithm in $\log p$ ($O((\log p)^3)$).

Since for fixed i the probability that $3\delta^2 \equiv a \mod p$ is $\approx\frac{1}{p}$, i.e. $\text{Prob}(\lambda_E(P) = 0) \approx \frac{1}{p}$, we could also use the following probabilistic method: Choose a lifting as in step 1, compute $\lambda_{a,b}(P)$. If $\lambda_{a,b}(P) = 0$ choose a new random lifting until $\lambda_{a,b}(P) \neq 0$. Then compute $\lambda_{a,b}(R)$ and l as in step 5.

Example 32 Let $\tilde{E} :\ y^2 = x^3 + 444x + 7581$ as in Example 31 be an anomalous elliptic curve defined over $\mathbb{F}_p,\ p = 30971$ prime.

Let $P = (18784, 23524)$ and $R = (18091, 4566)$. Choosing

$$a = 444 \text{ and } b = 7581$$

in step 1 we can take $(\delta, i) = (18784, 1)$, since $\delta^2 \equiv 20101 \not\equiv a \pmod p$. Choosing $\alpha = x_P,\ \beta = y_P$, we obtain

$$S = (x_1, y_1) = (18784, 97396348)$$

Table 4.9 Running times of Semaev/Rück and Fermat quotient Method for 10 different anomalous elliptic curves with $\#E(\mathbb{F}_p) = p = \text{ord}(P)$ and the curve/key construction time for the 10 curves in seconds

Bit size of p	Semaev/Rück	Fermat quotient	Curve/key constr.
100	3.0	11.4	13.1
160	8.3	26.2	34.3
200	13.7	44.1	63.5
300	30.5	113.5	143.9
400	59.4	231.6	287.0
512	105.1	494.5	583.6

following the algorithm in order to compute $\lambda_{a,b}(P)$. Then we can compute

$$(x_{p-1}, y_{p-1}) = (p-1)S = (332461498, 734453741)$$

by repeated doubling in the group $E(\mathbb{F}_{p^2})$ (not $E(\mathbb{Z}/p\mathbb{Z})$)! Hence $\lambda_{a,b}(P) = 13962$ and we can evaluate $\lambda_{a,b}(R) = 13155$ in the same way. This yields as in the Semaev/Rück algorithm the correct value $l \equiv \lambda_{a,b}(R)/\lambda_{a,b}(P) \equiv 11467 \pmod p$.

The author implemented the Semaev/Rück and Fermat quotient attack on a MAPLE V system using a common home computer (Celeron 400, 128 MB RAM) and achieved the results given in Table 4.9. The (totally) anomalous elliptic curves defined over $\mathbb{F}_p$ were found using an implementation of the complex multiplication method by the author (cf. subsection "Complex-Multiplication Method" of Sect. 4.5.3). The running times for the curve and private/public key constructions are given in the last column.

We get the same necessary condition for elliptic curves as in the last paragraph since this attack can also be extended to elliptic curves over $\mathbb{F}_q$ by the Silver–Pohlig–Hellman method if $p|\#E(\mathbb{F}_q)$.

Quantum Computing

D. Boneh and R.J. Lipton [7] showed 1995 that beside factoring of composite numbers (RSA) and the DLP in $\mathbb{F}_q^*$ (El Gamal) also the ECDLP can be computed in 'random quantum polynomial time'. Referring to Boneh and Lipton we give the following

Definition 150 Let $h : \mathbb{Z} \to G$ be a function.

(i) h has **period** p_h if for all $x \in \mathbb{Z} : h(x + p_h) = h(x)$.
(ii) h has **order** m_h if for all $g \in G : |h^{-1}(g) \bmod p_h| \le m_h$.

Let $f : \mathbb{Z}^k \to \bar{G}$ be a function with $\bar{G} \subseteq G$. f has a **hidden linear structure over** p_f if there exist $a_2, \ldots, a_k \in \mathbb{Z}$ and a function h with period $p_f = p_h$ such that

$$f(x_1, \ldots, x_k) = h(x_1 + a_2x_2 + \cdots + a_kx_k)$$

for all $x_1, \ldots, x_k \in \mathbb{Z}$. The order of f is given by the order of h.

Using quantum computing we get the following theorem which enables us to recover the hidden linear structure or the smallest period of a function. For details on quantum complexity theory we refer to [5].

Theorem 91 ([7])

(i) Suppose $f : \mathbb{Z}^k \to G$ to be a function which has hidden linear structure over p_f and is of order m_f. Let p_s be the smallest prime dividing p_f. If $m_f, k \le (\log p_f)^{O(1)}$ and $m_f < p_s$ then the values of $a_2, \ldots, a_k$ can be recovered modulo p_f in random quantum polynomial time in $\log p_f$ from an oracle for f.

(ii) Suppose $h : \mathbb{Z} \to G$ be a periodic function where p_h is the smallest period of h and h has order m_h. Let p_s be the smallest prime dividing p_h. If $m_h \le (\log p_h)^{O(1)}$ and $m < p_s$ then the period p_h of h can be recovered in random quantum polynomial time.

This leads directly to the

Corollary 17 *The ECDLP (4.5.1) can be solved in random quantum polynomial time even if n, the order of P, is not known.*

Proof Define the homomorphism

$$\begin{aligned} h : \mathbb{Z} &\to E(\mathbb{F}_q) \\ i &\mapsto iP. \end{aligned}$$

Since n is the order of P the function h does not map more than 1 element of $\mathbb{Z}_n$ to one. Thus the order of h is $m_h = 1$ and we can apply Theorem 91(ii) to find n, the smallest period of h.

Assume now that $R = h(l')$, where $l' \in \mathbb{Z}$ is unknown. Defining

$$\begin{aligned} f : \mathbb{Z}^2 &\to E(\mathbb{F}_q) \\ (x, y) &\mapsto h(x + l'y) = xP + yR. \end{aligned}$$

we obtain an function which has a hidden linear structure over n. Since the order of f is also 1 we can apply Theorem 91(i) to find an integer $l < n$ such that $l \equiv l' \bmod n$. This proves the lemma. □

Remark 50 Observe that we can easily extend the proof to the DLP in $\mathbb{F}_q^*$.

By now quantum computing is not (sufficiently) possible in practice. Nevertheless much research is done in physics and information theory concerning this area, so there might be a practical device in future.

Thus we get again a necessary condition for secure elliptic curves:

Condition 6: Quantum computing must not be practical

Further Research Areas

If an elliptic curve E is defined over $\mathbb{F}_q = \mathbb{F}_{p^m}$, then V. Müller and S. Paulus [64] suggest to take $m = 1$ or m prime, since every finite field extension $\mathbb{F}_{p^m}$ over $\mathbb{F}_p$ admits the action of the Galois group $G_{\mathbb{F}_{p^m}/\mathbb{F}_p}$. Hence if $G_{\mathbb{F}_{p^m}/\mathbb{F}_p}$ has small factors, i.e. $m > 1$ not prime, then there could be a method which exploits this fact.

Another attack Müller and Paulus thought of is to exploit "the number field over which some curve has a reduction modulo a prime ideal isomorphic to the given curve." [64]. This could be used for an index-calculus attack. They suggest to set 10 as the minimal number field degree.

By now no such methods are known, so we will not consider these ideas, but research in this area is needed. If some party wants to set up a "very secure" public-key scheme even for the future, they might consider these ideas and use elliptic curves which have the desired properties.

Cryptographically Good Elliptic Curves

Summarizing the necessary conditions deduced in this section we get the following definition for cryptographically good elliptic curve parameter, i.e. curve parameter where the ECDLP *seems* to be infeasible:

Definition 151 Let $E = E_{a,b}$ be an elliptic curve defined over $\mathbb{F}_q$, where $q = p^m$, $p > 3$ prime. E is denoted to be a **cryptographically good elliptic curve** if

(i) $\#E(\mathbb{F}_q) = \bar{n} \cdot d$, where $\bar{n} > 2^{160}$ is prime and $\gcd(\bar{n}, p) = 1$.
(ii) $q^k \not\equiv 1 \mod \bar{n}$ for $1 \leq k \leq c$, c as in Condition 2.
(iii) If $m > 1$ then $a, b \notin \mathbb{F}_p$.
(iv) Quantum computing is not practical.

Let $P \in E(\mathbb{F}_q)$ such that $\mathrm{ord}(P) = \bar{n}$ and l, $0 \leq l \leq \bar{n} - 1$, an statistically unique and unpredictable integer. Then $(E_{a,b}, \bar{n}, P, l, R = lP)$ is called a **cryptographically good elliptic curve parameter (G-ECP)** for the finite field $\mathbb{F}_q$.

The number d in (i) should be not too big, since the infeasibility of the mentioned ECDLP $R = lP$ is not effected by d, but the finite field $\mathbb{F}_q$ may be chosen needless too 'big', since $\#E(\mathbb{F}_q)$ and q are connected by the Hasse inequality. Note that the case (iii) is clear, since we should use the whole field $\mathbb{F}_q$ for defining the elliptic curve if possible.

Remark 51 The above definition can easily be extended to elliptic curves defined over $\mathbb{F}_{2^m}$.

As far as the author knows the sufficient conditions for elliptic curve cryptosystems to be secure is not known yet. So note that even if we use cryptographically good elliptic curves the ECDLP could be 'easy' since we only prevent (a few or all?) necessary conditions to solve the ECDLP.

4.5.3 Elliptic Curve Construction

Before implementing an elliptic curve public-key system, we have to choose if the field $\mathbb{F}_{2^m}$ or $\mathbb{F}_{p^m},\ p > 3$ should be used. The binary case 2^m is easier to implement in hardware systems, since that is the 'natural' computer arithmetic. For $\text{char}(\mathbb{F}_q) > 3$ the software implementation is often easier. We will restrict ourselves to the odd prime case, but the following approaches can also be extended to the binary case.

In order to construct elliptic curves over an finite field $\mathbb{F}_q$ there are these main methods known (we will also state unsolved mathematical questions concerning these methods refering to [38]):

(i) **Weil Theorem Approach**: If we have an elliptic curve E defined over $\mathbb{F}_q$, then find a suitable r such that $E(\mathbb{F}_{q^r})$ has the necessary properties. This approach has the advantage that we can quickly derive $\#E(\mathbb{F}_{q^r})$ from $\#E(\mathbb{F}_q)$ by the Weil Theorem 72, but we can not satisfy (the not so relevant) Definition 151(iii) since $a, b \in \mathbb{F}_q$ but $a, b \notin \mathbb{F}_{q^r}$. Since $E(\mathbb{F}_{q^{r'}})$ is a subgroup of $E(\mathbb{F}_{q^r})$ whenever $r'|r$, large prime factors of $\#E(\mathbb{F}_{q^r})$ are more likely to occur when r is prime then when r is composite. In the case of prime r, the best one can hope for is that

$$\frac{\#E(\mathbb{F}_{q^r})}{\#E(\mathbb{F}_q)} = \left|\frac{\alpha^r - 1}{\alpha - 1}\right|^2 \tag{4.5.19}$$

is prime. So we get the following natural questions: For fixed $E/\mathbb{F}_q$, what is the probability as r varies that (4.5.19) is prime? Can one ever prove that there are infinitely many r such that (4.5.19) is prime? Nothing is known on these questions by now. A short computer calculation shows the following: Let $S := \{q \text{ prime} : 1000 < q < 3000\}$, $R := \{11, 13, 17, 19, 23, 29, 31\}$. For each $q \in S$ we selected 20 different elliptic curves defined over $\mathbb{F}_q$ at random and tested for all $r \in R$ if (4.5.19) is prime. We got the following

$$\text{Prob}\left(\frac{\#E(\mathbb{F}_{q^r})}{\#E(\mathbb{F}_q)} \text{ is prime}\right) \approx 0.0474.$$

Thus a rough estimate shows that this method may be feasible.

(ii) **Randomization Approach**: Choose for a given finite field $\mathbb{F}_q$ the parameters (a, b) of an elliptic curve $E_{a,b}$ at random and check the desired properties. The main problem is, that we have to calculate $\#E(\mathbb{F}_q)$ lying in the small interval

$$[q+1-2\sqrt{q}, q+1+2\sqrt{q}], \tag{4.5.20}$$

which is the most time consuming part in this approach. Since the running time for the Schoof method has been improved dramatically in the last time, this becomes practical. Mathematically there is another question: We know already from Theorem 73 that as E varies over all elliptic curves defined over $\mathbb{F}_q$, q prime, $\#E(\mathbb{F}_q)$ is fairly uniformly distributed in (4.5.20). This is still true for prime powers $q = p^m$ except that the density drops off near the endpoints of the interval (4.5.20). The probability that an elliptic curve $E/\mathbb{F}_q$ has a prime factor greater than some lower bound B_1, is essentially the same as the probability that a random integer in the interval

$$[q, q+c\sqrt{q}], \quad c \text{ a constant}, \tag{4.5.21}$$

has this property. But nothing is proved about the number or distribution of primes in the interval (4.5.21). Not even whether there exists a c such that (4.5.21) contains at least one prime for $p \to \infty$ is known.

(iii) **Global Elliptic Curve Reduction**: Reduce a given elliptic curve $E_{a,b}$ over $\mathbb{Q}$ or $\mathbb{C}$ to an elliptic curve over $\mathbb{F}_p$, and vary the prime p until $E(\mathbb{F}_p)$ has the desired properties. For example choose $E: y^2 = x^3 + ax^2 + b$ defined over $\mathbb{Q}$. For many primes p we can reduce E to $E \bmod p$ defined over $\mathbb{F}_p$. $E \bmod p$ will always contain as a subgroup the image of the torsion subgroup E_{tors} of the curve over $\mathbb{Q}$. But one expects that in many cases

$$\frac{\#E \bmod p}{\#E_{tors}} \tag{4.5.22}$$

is a prime. Although by now the probability that (4.5.22) is prime as p varies for fixed $E/\mathbb{Q}$ is unknown.

(iv) **Complex-Multiplication Approach**: Choose a suitable curve order $N = \bar{n} \cdot d$ and construct $E/\mathbb{F}_q$ satisfying $\#E(\mathbb{F}_q) = N$ by 'complex-multiplication'. It is also possible to fix an underlying field $\mathbb{F}_q$ and then find a suitable curve order $N = \bar{n} \cdot d$ under some restrictions such that $\#E(\mathbb{F}_q) = N$. This turns out to be a special case of (iii).

The cryptographically best method is the randomization approach (ii), since we do not use possible isomorphism classes or other restrictions than those of Definition 151. In the past several isomorphism classes like supersingular curves and anomalous curves were considered, since these curves have fast addition properties. But as shown above especially those curves are vulnerable against attacks.

Nevertheless the randomization method is not so easy to implement, since we have to implement an algorithm which counts the number of rational points. The same is true for the general idea of (iii): We have to check if $\#E(\mathbb{F}_q)$ has the desired properties.

The author used the CM method described in the next section in order to find cryptographically good elliptic curve parameter.

Complex-Multiplication Method

In 1991 Morain [62] proposed a method to build elliptic curves modulo large primes. This was used in the Goldwasser–Killian–Atkin primality proving algorithm, implemented by Morain [3]. Frey et. al. [88] and Lay et al. [45] independently adapted this algorithm for determing elliptic curves of prescribed order in cryptology. We will present the idea of the algorithm. For a more algebraic number theoretical view, see [45].

Let $p > 3$ be a given prime, i.e. we fix the underlying finite field $\mathbb{F}_p$. We want to construct an cryptographically good elliptic curve E over $\mathbb{F}_p$ for a given integer t such that $\#E(\mathbb{F}_p) = p + 1 - t = \bar{n} \cdot d$. By the Hasse inequality t is restricted to $|t| \leq 2\sqrt{p}$. Assuming further that E is non-supersingular, i.e. $t \neq 0$, we get

$$4p - (p + 1 - \#E(\mathbb{F}_p))^2 \; = \; 4p - t^2 > 0.$$

Thus there exists a unique factorization

$$4p - t^2 = \bar{D}v^2,$$

where $\bar{D}$ is a squarefree positive integer.

By Theorem 75 we know that $\mathrm{End}(E)$ is an order of an imaginary field. Hence there exist a squarefree positive integer D and a positive integer f such that

$$\mathrm{End}(E) = \mathbb{Z} + f\omega_D\mathbb{Z}, \quad \text{where} \quad \omega_D = \begin{cases} \frac{1+\sqrt{-D}}{2}, & \text{if } (-D) \equiv 1 \bmod 4, \\ \sqrt{-D}, & \text{if } (-D) \equiv 2, 3 \bmod 4. \end{cases}$$

D is denoted *CM-discriminant*, while $-D$ with some further restrictions is also called fundamental discriminant in literature.

In $\mathrm{End}(E)$ the pth Frobenius endomorphism Φ_E satisfies $\Phi_E^2 - t\Phi_E + p = 0$. Since $\Phi_E \in \mathrm{End}(E)$ we can observe that $\bar{D} = D$ and v' is a multiple of f, if we define

$$v' = \begin{cases} v, & \text{if } (-D) \equiv 1 \bmod 4, \\ v/2, & \text{if } (-D) \equiv 2, 3 \bmod 4. \end{cases}$$

Now assume $E/\mathbb{F}_p$ is a non-supersingular elliptic curve such that

$$\mathrm{End}(E) = \mathbb{Z} + f\omega_D\mathbb{Z}$$

and

$$4p - t^2 = Dv^2 \quad \text{with } f | v'. \tag{4.5.23}$$

For $D = 1$ or $D = 3$ the unit group of $\mathbb{Z} + \omega_D\mathbb{Z}$ is not $\{\pm 1\}$. So for simplicity we assume further $D \neq 1, 3$. Since $\deg \Phi_E = p$ we can define the dual isogeny $\hat{\Phi}_E$ of Φ_E, i.e. $\hat{\Phi}_E\Phi_E = p$ (cf. [82]). From (4.5.23) we get

$$
\begin{aligned}
\hat{\Phi}_E\Phi_E = p &= \frac{1}{4}(t^2 + v^2D) \\
&= \begin{cases} \frac{1}{4}(t^2 - v^2 + 2v^2(1+\sqrt{-D}) - v^2(1 + 2\sqrt{-D} + (\sqrt{-D})^2) \\ \frac{1}{4}(t^2 - v^2(-D)) \end{cases} \\
&= \begin{cases} \frac{t-v}{2}\frac{t+v}{2} + v\frac{1+\sqrt{-D}}{2}\frac{t+v}{2} - v\frac{1+\sqrt{-D}}{2}\frac{t-v}{2} - v^2\frac{(1+\sqrt{-D})^2}{4} \\ \frac{t^2}{4} - \frac{v^2}{4}\sqrt{-D}^2 \end{cases} \\
&= \begin{cases} \left(\frac{t-v}{2} + v\frac{1+\sqrt{-D}}{2}\right)\left(\frac{t+v}{2} - v\frac{1+\sqrt{-D}}{2}\right) \\ \left(\frac{t}{2} + \frac{v}{2}\sqrt{-D}\right)\left(\frac{t}{2} - \frac{v}{2}\sqrt{-D}\right) \end{cases} \\
&= \begin{cases} \left(\frac{t-v}{2} + v\omega_D\right)\left(\frac{t+v}{2} - v\omega_D\right), & \text{if } (-D) \equiv 1 \bmod 4, \\ \left(\frac{t}{2} + \frac{v}{2}\omega_D\right)\left(\frac{t}{2} - \frac{v}{2}\omega_D\right), & \text{if } (-D) \equiv 2, 3 \bmod 4, \end{cases}
\end{aligned}
$$

i.e. p splits into a product of two principal prime ideals of $\mathrm{End}(E)$. Since $\Phi_E \notin \mathbb{Z}$ and the uniqueness of prime ideal decomposition,

$$\mathrm{tr}(\Phi_E) = \pm t.$$

If $\mathrm{tr}(\Phi_E) = t$ then $\#E(\mathbb{F}_p) = p + 1 - t$. Otherwise twist $\hat{E}$ of E satisfies $\hat{E}(\mathbb{F}_p) = p + 1 - (-t)$.

Using complex analytic theory of elliptic curves, we can construct an elliptic curve $\mathcal{E}/\mathbb{C}$ such that $\mathrm{End}(\mathcal{E}) = \mathbb{Z} + f\omega_D\mathbb{Z}$ (cf. [16] for details, also [31] for elliptic curves over $\mathbb{C}$ with complex multiplication). Let $\mathcal{E}_{\bmod p}$ be the modulo p reduction of $\mathcal{E}$. Then also $\mathrm{End}(\mathcal{E}_{\bmod p}) = \mathbb{Z} + f\omega_D\mathbb{Z}$.

The main idea now is to construct an elliptic curve E isomorphic to $\mathcal{E}_{\bmod p}$ without constructing $\mathcal{E}$ such that $\mathrm{End}(\mathcal{E}_{\bmod p}) = \mathrm{End}(E)$: The j-invariant $j(\mathcal{E})$ of $\mathcal{E}$ is an algebraic integer. We can compute the minimal Hilbert class polynomial $H_D(x)$ of $j(\mathcal{E})$ by an algorithm of Atkin, et al. [3]. The algorithm uses the connection between the CM-discriminant D, and reduced quadratic forms in order to work in the imaginary quadratic field $\mathbb{Q}(\sqrt{-D})$. Using Weber's and Dedekind's functions it is possible to express the j-invariant of $\mathcal{E}$ and to compute the minimal polynomial efficiently in $\mathbb{R}[x]$ with coefficients which are much smaller than for the Hilbert class polynomial. Provided the computation takes place with the necessary precision, we can round it to $\mathbb{Z}[x]$.

Now let j_0 be a root of

$$H_D(x) \equiv 0 \bmod p. \tag{4.5.24}$$

It can be shown that $j_0 \in \mathbb{F}_p$, i.e. that $H_D(x)$ splits completely over $\mathbb{F}_p$. Furthermore it is easy to see that

$$E/\mathbb{F}_p = \begin{cases} y^2 = x^3 - 1, & \text{if } j_0 = 0 \\ y^2 = x^3 - x, & \text{if } j_0 = 1728 \\ y^2 = x^3 + 3cx + 2c, & \text{otherwise with } c = \frac{j_0}{1728 - j_0} \end{cases} \tag{4.5.25}$$

has j-invariant j_0.

So we can easily compute an elliptic curve $E/\mathbb{F}_p$ with $j(E) = j_0$ and $\mathrm{End}(E) = \mathbb{Z} + f\omega_D\mathbb{Z}$. The case $j_0 = 0$ or $j_0 = 1728$ will actually occur if $D = 1$ or 3, respectively. Hence if $D = 1$ for example, we can immediately set $E = E_{-1,0}$.

If $E(\mathbb{F}_p) = p + 1 - t = \bar{n} \cdot d$ we are done, otherwise we construct its twist $\hat{E}$.

We get the following

CM-Method for CG-ECP construction

Require: lower bound B_1 ($\approx 2^{160}$) for $\bar{n}$ and upper bound B_2($\approx$256) for d

1: Choose a random prime p of about B_1.
2: Find a CM-discriminant D in $\mathbb{F}_p$, i.e. a positive squarefree integer
3: Find a pair $(t, v) \in \mathbb{Z}^2$ such that (4.5.23) is satisfied
4: **if** NOT ($p + 1 + t = d \cdot \bar{n}$ or $p + 1 - t = d \cdot \bar{n}$), where $\bar{n} > B_1$ is a prime, $\gcd(\bar{n}, p) = 1$ and $1 \leq d \leq B_2$. **then**
5: Goto step 1 or 2
6: **end if**
7: **if** there exists k, $1 \leq k \leq (\ln k)^2$ such that $p^k \equiv 1 \pmod{\bar{n}}$ **then**
8: Goto step 1 or 2
9: **end if**
10: Construct an elliptic curve $E_{a,b}$ with the CM discriminant D using the algorithm of [3]
11: Choose a random basepoint P such that $\mathrm{ord}(P) = \bar{n}$
If no basepoint of prescribed order exists,
use the twisted curve $\hat{E}$ instead of E.
12: Choose a random private-key l, $0 \leq l \leq \bar{n} - 1$.
13: Compute the public-key $R = lP$.

Ensure: $(E_{a,b}, \bar{n}, P, l, R)$ is a CG-ECP for the finite field $\mathbb{F}_p$.

The bound B_1 should be chosen to meet the necessary security, i.e. at least $B_1 > 2^{160}$. We fixed the field p first, since there are certain finite fields where modular arithmetic can be done faster (e.g. normal bases representations, etc.) Note that already in the steps 4 and 7 we can determine if the elliptic curve which will be constructed in step 10 has the desired properties!

Remark 52 The algorithm can easily be extended to finite fields over $q = p^m$, p prime (even 2). See [45] for the necessary algorithm in step 10.

Remark 53 Finding an anomalous elliptic curve over $\mathbb{F}_p$, $p > 2^{40}$ prime, by setting $B_2 = 1$ and searching until $\mathrm{ord}(P) = \bar{n} = p = \#E(\mathbb{F})$ gets infeasible, since the density of anomalous elliptic curves over $\mathbb{F}_p$ is at most $O(\frac{1}{\sqrt{p}} \log p \log\log p)$ and there are too many possible prime orders $\bar{n} \neq p$ between the Hasse bounds $p + 1 - 2\sqrt{p}$

and $p + 1 + 2\sqrt{p}$ found by the above algorithm. So we used another strategy to find anomalous elliptic curves for Table 4.9: Set $t = -1$ and $v \approx 4B_1$. By increasing v and D test if $p = (t^2 - Dv^2)/4$ is a positive integer and prime. If further D is squarefree in $\mathbb{F}_p$, we have found a CM-discriminant. Then either $E(\mathbb{F}_p)$ or twist $\hat{E}(\mathbb{F}_p)$ has order p, thus also $\text{ord}(P) = p$.

The last remark shows that it is also possible to prescribe the order of the elliptic curve E, but then we have restrictions on p and thus the underlying finite field $\mathbb{F}_p$.

Practical Implementation

The author implemented the CM-method using the strategy above on a Maple V system. Although some special speedups were made as described in [32] there will be further improvements in native code possible. In Table 4.10 the running times for the construction of CG-ECP are given. For every fieldsize 10 runs on a usual home computer (Celeron 300, 128 MB) were made in order to find 10 different cryptographically good elliptic curve parameter. Since it is possible to construct even CG-ECP with field size $p \approx 2^{160}$ in about 10 s, every user can choose an own elliptic curve as his public key. So if the curve E_i of user i is 'broken', i.e. the ECDLP in E_i has become feasible (in at most probabilistic subexponential time), the ECDLP in the curve E_j of user j may still be hard. Although this 'increases' the security, this would also increase the total length of the public-key.

Example 33 Let $p = 1070134007$ be a randomly selected prime in order to fix the underlying finite field $\mathbb{F}_p$. Then we can choose $D = 19$ as a CM-discriminant for the field p, i.e. D has no squares in $\mathbb{F}_p$. Selecting $t = 65423$ and $v = 139$ we can satisfy (4.5.23) (set $f = 1$). Furthermore

$$p + 1 + t = 7 \cdot 152885733 = d \cdot \bar{n},$$

where $\bar{n}$ is prime and $\bar{n} \nmid p$. Testing condition 2 yields

$$p^k \not\equiv 1 \pmod{\bar{n}} \quad \text{for all } 1 \leq k \leq 433.$$

Now we can construct an elliptic curve $E_{a,b}$ with the given CM-discriminant 19 and prescribed order $d \cdot \bar{n}$. Solving the minimal polynomial (4.5.24) we get the

Table 4.10 Constructions times for 10 different CG-ECP over $\mathbb{F}_p$ with $\#E(\mathbb{F}_p) = \bar{n} \cdot d$ such that $\text{ord}(P) = \bar{n}$ and $1 \leq d < 256$

Size of p in bit ($\approx B_1$)	Worst case (s)	Average case (s)	Best case (s)
100	31.9	22.3	16.3
150	137.6	88.6	61.6
160	184.3	99.7	58.2
200	383.3	223.0	125.2
250	514.2	360.2	236.6

j-invariant $j_0 = 1069249271$. Thus we can construct the elliptic curve $E_{526088222,32775985}$ by (4.5.25). Moreover the 'pseudo'-randomly chosen point $P = (938800742, 1020685579)$ has order $\bar{n}$. Choosing a 'pseudo'-random private-key $l, 0 \leq l \leq \bar{n} - 1$ is now trivial, yielding $R = lP$. Hence we get the following CG-ECP:

$$(E_{a,b}, \bar{n}, P, l, R) = (E_{526088222,32775985}, 152885633, (938800742, 1020685579),$$
$$93059745, (598183944, 412604570))$$

for the finite field $\mathbb{F}_p$. Note that there are many isomorphic elliptic curves with the same j-invariant j_0. For example we could also choose

$$(E_{a,b}, \bar{n}, P, l, R) = (E_{441448886,691574612}, 152885633, (354112736, 657172669),$$
$$120419930, (746398119, 847702797)).$$

Observe that we did not satisfy the important condition $\bar{n} > 2^{160}$ and only pseudo-random private-keys l were used in this example.

In the next section the author discusses the idea for a new public-key scheme he thought of using many techniques described in this paper.

4.5.4 Designing New Public-Key Cryptosystems

It is very difficult to design new public-key schemes with a verifiable security, i.e. the underlying trapdoor one-way function is computational infeasible and the scheme itself does not yield any further possibilities to gain knowledge about the plaintext.

The Scheme

In order to design a **new** scheme for a public-key cryptosystem the author thought of the following scheme using ideas of the Vanstone–Zuccherato scheme presented in Sect. 4.4.

A (broken) Public-Key Cryptoscheme Based on Factorization

(i) **(Setup)** Each user i

(a) selects two "large" primes p and q and an elliptic curve $E_{a,b}(\mathbb{Z}_n)$, $n = pq$, such that

$$\#E_{a_p,b_p}(\mathbb{F}_p) = p \text{ and } \#E_{a_q,b_q}(\mathbb{F}_q) = p.$$

Note that $E_{a_p,b_p}/\mathbb{F}_p$ is an anomalous elliptic curve.

(b) Chooses a point $P \in E(\mathbb{Z}_n)$ of order p.
(c) Chooses a random integer $k \in \mathbb{Z}_p$.
Then each user i sets (a, b, n, P, k) as the public-key and (p) as the private-key.

(ii) **(Communication)** User j wants to send a message $l \in \mathcal{M} = \{1, \ldots, k\}$ to i

(enc) j computes $C = lP \in E(\mathbb{Z}_n)$ (pseudo-multiplication).
j calculates $c = \Phi(C)/(\Phi(P))^{-1} \in \mathbb{Z}_n$ using the Semaev/Rück method
j sends $c \in \mathbb{Z}_n$.
(dec) i calculates

$$l = c \bmod p. \tag{4.5.26}$$

Using the Semaev/Rück method in $E_{a,b}(\mathbb{Z}_n)$ j obtains in the communication part (ii) the p-part of the ECDLP in

$$\tilde{E}_{a,b}(\mathbb{Z}_n) := E_{a_p,b_p}(\mathbb{F}_p) \times E_{a_q,b_q}(\mathbb{F}_q),$$

until the Semaev/Rück method does not yield a non-trivial divisor of n, which immediately 'breaks' the scheme. Since $l \in \mathbb{Z}_p$ and the ECDLP $R_p = lP_p$ is already solved modulo p by j using the Semaev/Rück method i gets the whole plaintext message l by computing Eq. (4.5.26).

Note that we get a message expansion factor of at least ≈ 2, depending on the relation of k and n. So one should not choose k too small at 'random'. Observe further that decryption can be done very fast.

Example 34 Let n be the product of $p = 1373$ and $q = 1423$ and

$$E_{a,b}/\mathbb{Z}_n : y^2 = x^3 + 825x + 952$$

be the elliptic curve over $\mathbb{Z}_n$. Further let

$$P = (490669, 449857) \quad \text{and} \quad k = 1212.$$

If j wants to send the message $194 \in \{1, \ldots, k\}$ to i, j computes

$$C = 194P = (556275, 1192351)$$

by repeated doubling in $\mathbb{Z}_n$ and

$$c = \Phi(C)/(\Phi(P))^{-1} = 1812554$$

by the Semaev/Rück method (also in $\mathbb{Z}_n$). Receiving c, i can decipher the message $l = c \bmod p = 194$. Observe that $nP = \mathcal{O}_n = (\mathcal{O}_p, \mathcal{O}_q) = (pP_p, pP_q) = pP$, but $qP = (1554011, 1429400)$.

Curve Construction

The main problem is to construct the desired elliptic curve over $\mathbb{Z}_n$. We used the following two strategies:

At first construct for a given prime p an anomalous elliptic curve E_{a_p,b_p} over $\mathbb{F}_p$ by the CM method. Let

$$S(p) := \{q \text{ a prime } : p - 2\sqrt{p} + 1 \leq q \leq p - 2\sqrt{p} + 1, q \neq p\}$$

be a set of possible primes for q given by the Hasse inequality.

(i) For any $q \in S(p)$ let $E_{a,b}/\mathbb{Z}_{pq}$ be the 'lifted' curve of $E_{a_p,b_p}/\mathbb{F}_p$. Find a point $P \in E(\mathbb{Z}_n)$ and test if $pP_q = \mathcal{O}_q$ in $E_{a_q,b_q}(\mathbb{F}_q)$. If the test succeeds count $\#E(\mathbb{F}_q)$ by Schoof's algorithm. If $\#E(\mathbb{F}_q) \neq p$ then choose a new prime $q \in S(p)$ and try again. Otherwise we have found the necessary curve $E_{a,b}/\mathbb{Z}_n$ and the point P. If for all $q \in S(p)$ no curve were found select a new prime p.

(ii) If a $q \in S(p)$ and a squarefree CM discriminant D in $\mathbb{F}_q$ exists such that

$$4p - (q + 1 - p)^2 = Dv^2$$

for some $v \in \mathbb{Z}$ then we can construct the elliptic curve $E_{a_q,b_q}/\mathbb{F}_q$ using the CM method. Then calculate $n = pq$ and $a \bmod n$, $b \bmod n$ by the Chinese Remainder Theorem. If no such $q \in S(p)$ exists select a new prime p.

The first approach already becomes computationally infeasible for $p \approx 2^{10}$. For greater primes $p \approx 2^{15}$ the second attempt succeeds, but it was not possible for the author to construct an elliptic curve with $n > 2^{50}$ by now.

An Analysis

In order to analyse the system we want to give a clearer encryption/decryption part:

(ii)' **(Communication)** User j wants to send a message $l \in \mathcal{M} = \{1, \ldots, k\}$ to i

(enc) j computes $C = lP \in E(\mathbb{Z}_n)$ (pseudo-multiplication).
j sends $C \in E(\mathbb{Z}_n)$.

(dec) i calculates $(C_p, C_q) \in E_{a_p,b_p}(\mathbb{F}_p) \times E_{a_q,b_q}(\mathbb{F}_q)$.
i solves the ECDLP $C_p = lP_p$ in $E_{a_p,b_p}(\mathbb{F}_p)$ using the Semaev/Rück method (cf. Sect. 4.5.2).

Note that now the message expansion factor is at least 4.

In (ii)' the decryption takes place in the anomalous elliptic curve group $E_{a_p,b_p}(\mathbb{F}_p) = E_{a_p,b_p}[p] \simeq \mathbb{Z}_p$ (cf. Theorem 76), since p is prime. Furthermore $P_q \in E_{a_q,b_q}[p] \simeq Z_p \oplus Z_p$. Hence if the pseudo-multiplication is well-defined in $E_{a,b}(\mathbb{Z}_n)$ then

$$C = lP = [(lP)_p, (lP)_q] = [lP_p, lP_q] \in E_{a_p,b_p}[p] \times E_{a_q,b_q}[p] \simeq Z_p \times Z_p \oplus Z_p.$$

If the communication is done by the scheme (ii)' then an eavesdropper could also calculate $c = \Phi(C)/(\Phi(P))^{-1} \in \mathbb{Z}_n$ and this will yield the private-key p as shown below in Lemma 50. In the first discussion we will assume that the two schemes (ii)' and (ii) are equivalent and will discuss scheme (ii)' in order to explain that no further free parameters are possible.

Lemma 48 *Let (a, b, n, P, k) be the public-key of the above cryptoscheme. If the order of $P_q \in E(\mathbb{F}_q)$ does not divide n, then we can factor n in $O(\log n)$ elliptic curve additions.*

Proof Let $h = \text{ord}(P_q),\ P_q \in E(\mathbb{F}_q), h \nmid n$. Since $p \neq q$ and $q \neq h \neq p$

$$nP = n(P_p, P_q) = (nP_p, nP_q) = (q(pP_p), nP_q) = (\mathcal{O}_p, nP_q) \neq (\mathcal{O}_p, \mathcal{O}_q) = \mathcal{O}_n.$$

Hence by Lemma 39 we must get a non-trivial divisor of n, i.e. p or q, in the pseudo-multiplication nP. The evaluation takes $O(\log n)$ elliptic curve additions by repeated doubling. □

Lemma 49 *Let (a, b, n, P, k) be the public-key of the above cryptoscheme. If the order of $E(\mathbb{F}_q)$ is q (and not p as required) then an eavesdropper can solve the ECDLP $C = lP$ in $\mathbb{Z}_n$ completely.*

Proof Since $qP_q = \mathcal{O}_q,\ P_q \in E_{a_q,b_q}[q](\mathbb{F}_q) \setminus \{\mathcal{O}_q\}$. Hence $E_{a_q,b_q}(\mathbb{F}_q) = E_{a_q,b_q}[q]$ $(\mathbb{F}_q)$ is also an anomalous elliptic curve.

Assume p and q are known. Hence we can use the Semaev/Rück method to solve the following ECDLPs

$$C_p = l_p P_p \text{ in } E_{a_p,b_p}[p](\mathbb{F}_p) = E_{a_p,b_p}(\mathbb{F}_p) \simeq \mathbb{Z}_p, \tag{4.5.27}$$
$$C_q = l_q P_q \text{ in } E_{a_q,b_q}[q](\mathbb{F}_q) = E_{a_q,b_q}(\mathbb{F}_q) \simeq \mathbb{Z}_q. \tag{4.5.28}$$

(4.5.27) yields $l \equiv l_p \pmod p$ and (4.5.28) $l \equiv l_q \pmod q$. Thus we can determine l by the Chinese Remainder Theorem.

Assume p and q are unknown. If we use the Semaev/Rück method in $E_{a,b}(\mathbb{Z}_n)$ we can obtain a non-trivial divisor of n, i.e. p or q, if the denominator of (4.5.13) has no modular inverse in $\mathbb{Z}_n$. Otherwise the algorithm works in the group

$$\tilde{E}_{a,b}(\mathbb{Z}_n) := E_{a_p,b_p}(\mathbb{F}_p) \times E_{a_q,b_q}(\mathbb{F}_q)$$

yielding directly $l \in \mathbb{Z}_n$, since the method solves the two ECDLPs (4.5.27) and (4.5.28) simultaneous. □

Hence it is very important how to choose the elliptic curve E used in the system above. The only free parameter we get from Lemmas 48 and 49 is to choose the prime q such that $\#E_{a_q,b_q}(\mathbb{F}_q) = p$ and thus $pP_q = \mathcal{O}_q$. Now $nP = \mathcal{O}_n$ and the elliptic curve pseudo-multiplication does not yield a non-trivial factor of n as in Lemma 48.

Nevertheless by a remark due to H.-G. Rück it is possible to break the scheme:

Lemma 50 *Let (a, b, n, P, k) be the public-key of the above scheme. Then p can be computed in probabilistic polynomial time in* $\log n$.

Proof Let D be an divisor in E_{a_p,b_p} such that $p \cdot D = \div(f)$. In the isomorphic embedding $\Phi :< P_p >\rightarrow \mathbb{F}_p$ we use the map $\lg : \overline{D} \rightarrow \mathfrak{f}/f$, where lg is independent of the representant of the divisor class $\overline{D}$ (cf. Eq. (4.5.9)). For example in the worked out Semaev/Rück algorithm we chose $D_Q = (Q) - (\mathcal{O})$. But this is not valid if the characteristic of the field is not p. Hence extending Φ to a map $\Phi :< P >\rightarrow \mathbb{Z}_n$, where Φ operates on $E[p](\mathbb{F}_p) \times E[p](\mathbb{F}_q)$ with $p \neq q$, we can choose two representants D_1, D_2 of the same divisor class in $\mathrm{Pic}^0(E_{a_p,b_p}(\mathbb{F}_p))_p \times \mathrm{Pic}^0(E_{a_q,b_q}(\mathbb{F}_q))_p$ different in the second component. Now if we encrypt any message $m \in \{1, \ldots, k\}$ with the communication part (ii) we obtain $c_1 \bmod n$ and $c_2 \bmod n$ according to D_1 and D_2, respectively. Thus c_1 and c_2 will be the same modulo p, but with high (at least positive) probability different in the second component. Hence

$$p = \gcd(c_1 - c_2, n).$$

□

Thus the above scheme does not lead to a new public-key cryptosystem.

4.5.5 Conclusion

In this section we described several public-key cryptosystems which exploit the propositions of elliptic curves. Even if the implementation and encryption/decryption of all of these schemes can be done without much knowledge about the mathematical theory of elliptic curves we presented various attacks due to the recent research using several mathematical areas and the theory of elliptic curves.

Especially elliptic curve public-key schemes based on the ECDLP discussed in the last chapter have many advantages over other known public-key schemes like:

(i) Shorter public and private key length.
(ii) Shorter digital signature and encrypted message length.
(iii) Faster arithmetic, since the underlying field $\mathbb{F}_q$ can be choosen smaller.

Although various mathematical attacks are possible to solve the ECDLP in polynomial or at least probabilistic subexponential time for special classes of elliptic curves this class of public-key schemes has the property to achieve the most 'security per key bit' by now compared with commercially available public-key schemes if we use the cryptographically good elliptic curves developed in this section. Nevertheless further research is necessary concerning the ECDLP in order to find a sufficient definition for cryptographically secure elliptic curves, i.e. for curves where the ECDLP **is** in fact computational infeasible.

Furthermore we have shown as well the efficient construction of cryptographically good elliptic curves using the structure of curves over different fields as the efficient

m-fold addition. So elliptic curve public-key schemes based on the ECDLP can be efficiently implemented in commercial software systems and because of this will become a standard by the IEEE and ANSI Standards groups in the near future. Finally an idea due to Vansone–Zuccherato is discussed for a new elliptic curve cryptoscheme based on factorization using the properties of anomalous curves, curve construction and curves over the ring $\mathbb{Z}_n$. Nevertheless it turned out that this scheme can be broken in probabilistic polynomial time. This shows that designing a public-key cryptosystem as well the underlying trapdoor one-way function as the protocol scheme must have a 'mathematical' and 'computational' verifiable security.

References

1. L.M. Adleman, A subexponential algorithm for the discrete logarithm problem with applications to cryptology, in *20th Annual Symposium on the Foundations of Computer Science* (1979), pp. 55–60
2. L.M. Adleman, J. DeMarrais, M.D. Huang, A subexponential algorithm for discrete logarithms over the rational subgroup of the Jacobians of large genus hyperelliptic curves over finite fields, *Algorithmic Number Theory*. LNCS, vol. 877 (Springer, Berlin, 1994)
3. A.O.L. Atkin, F. Morain, Elliptic curves and primality proving. Math. Comput. **61**(205), 29–68 (1993)
4. R. Balasubramanian, N. Koblitz, The improbability that an elliptic curve has subexponential discrete log problem under the Menezes-Okamoto-Vanstone algorithm. J. Cryptol. **11**, 141–145 (1998)
5. E. Bernstein, U. Vazirani, Quantum complexity theory, in *Proceedings of 26th ACM Symposium on Theory of Computation* (1993)
6. D. Bleichenbacher, On the security of the KMOV public key cryptosystem, in *Advances in Cryptology - CRYPTO '97*. LNCS, vol. 1294 (Springer, Berlin, 1997), pp. 235–247
7. D. Boneh, R.J. Lipton, Quantum cryptanalysis of hidden linear functions, in *Advances in Cryptology - CRYPTO '95*. LNCS, vol. 963 (Springer, Berlin, 1995), pp. 424–437
8. W. Bosma, A.K. Lenstra, An implementation of the elliptic curve integer factorization method, in *Mathematics and its Applications*, vol. 325 (Kluwer Academic Publishers, Dordrecht, 1995)
9. R.P. Brent, Some integer factorization algorithms using elliptic curves, Research Report CMA-R32-85 (The Australian National University, Canberra, 1985)
10. R.P. Brent, Factorization of the tenth fermat number. Math. Comput. **68**(225), 429–451 (1999)
11. D.M. Bressoud, *Factorization and Primality Testing* (Springer, New York, 1989)
12. C.C. Cocks, A note on non-secret encryption, CESG Report (1973), www.cesg.gov.uk/about/nsecret.htm
13. J.M. Couveignes, F. Morain, Schoofs algorithm and isogeny cycles, in *Algorithmic Number Theory*. LNCS, vol. 877 (Springer, Berlin, 1994), pp. 43–58
14. J.M. Couveignes, L. Dewaghe, F. Morain, Isogeny cycles and the Schoof-Elkis-Atkin algorithm, Research Report LIX/RR/96/03, LIX (1999)
15. N. Demytko, A new elliptic curve cryptosystem based analogue of RSA, in *Advances in Cryptology - EUROCRYPT '93*. LNCS, vol. 765 (Springe, Berlin, 1994), pp. 41–49
16. M. Deuring, Die Typen der Multiplikatorenringe elliptischer Funktionskörper. Abh. Math. Sem. Hamburg **14**, 197–272 (1941)
17. W. Diffie, M.E. Hellman, New directions in cryptography. IEEE Trans. Inf. Theory **22**, 644–654 (1976)
18. P. Downey, B. Leong, R. Sethi, Computing sequences with addition chains. SIAM J. Comput. **10**, 638–646 (1981)

19. T. El Gamal, A public key cryptosystem and a signature scheme based on discrete logarithms. IEEE Trans. Inform. Theory **31**, 469–472 (1985)
20. J.H. Ellis, The possibility of secure non-secret digital encryption, CESG Report (1970), www.cesg.gov.uk/about/nsecret.htm
21. P. Erdös, Remarks on number theory, III. On addition chains. Acta Arith. **6**, 77–81 (1960)
22. Final report on Project C43, Bell Telephone Laboratory (1944), p. 23
23. G. Frey, H.G. Rück, A remark concerning m-divisibility and the discrete logarithm in the divisor class group of curves. Math. Comput. **62**(206), 865–874 (1994)
24. G. Frey, M. Müller, H.G. Rück, The tate pairing and the discrete logarithm applied to elliptic curve cryptosystems. IEEE Trans. Inf. Theory **45**(5), 1717–1719 (1999)
25. D.M. Gordon, Discrete logarithms in $GF(p)$ using the number field sieve. J. Discrete Math. **6**(1), 124–138 (1993)
26. D.M. Gordon, Discrete logarithms in $GF(p^n)$ using the number field sieve, preprint (1995)
27. D.M. Gordon, A survey of fast exponentiation methods. J. Algorithms **27**, 127–146 (1998)
28. J. Guajardo, C. Paar, Efficient algorithms for elliptic curve cryptosystems, in *Advances in Cryptology - CRYPTO '97*. LNCS, vol. 1294 (Springer, Berlin, 1997), pp. 342–355
29. J. Hastad, On using RSA with low exponent in a public key network, in *Proceedings of CRYPTO '85* (1985), pp. 403–408
30. M.E. Hellman, S. Pohlig, An improved algorithm for computing logarithms over GF(p) and its cryptographic significance. IEEE Trans. Inf. Theory **24**, 106–110 (1978)
31. D. Husemöller, *Elliptic Curves* (Springer, Berlin, 1986)
32. IEEE P1363 Standards Draft, www.ieee.com
33. M.J. Jacobson, N. Koblitz, J.H. Silverman, A. Stein, E. Teske, Analysis of the xedni calculus attack. Des. Codes Cryptogr. **20**(1), 41–64 (2000)
34. D. Kahn, *The Codebreakers - The Story of Secret Writing* (MacMillan Publishing Co., New York, 1979). (ninth printing)
35. B.S. Kalinski, A chosen message attack on Demytko's elliptic curve cryptosystem. J. Cryptol. **10**, 71–72 (1997)
36. N. Koblitz, Elliptic curve cryptosystems. Math. Comput. **48**(177), 203–209 (1987)
37. N. Koblitz, Hyperelliptic cryptosystems. J. Cryptol. **1**, 139–150 (1989)
38. N. Koblitz, *Algebraic Aspects of Cryptography* (Springer, Berlin, 1998)
39. K. Koyama, Fast RSA-type schemes based on singular cubic curves $y^2 + axy = x^3 \ (mod\ n)$, in *Advances in Cryptology - EUROCRYPT '95*. LNCS, vol. 921 (Springer, Berlin, 1995), pp. 329–340
40. K. Koyama, U. Maurer, T. Okamoto, S. Vanstone, New public-key schemes based on elliptic curves over the ring $\mathbb{Z}_n$, in *Advances in Cryptology - CRYPTO '91*. LNCS, vol. 576 (Springer, Berlin, 1992), pp. 252–266
41. K. Kurosawa, K. Okada, S. Tsujii, Low exponent attack against elliptic curve RSA. Inf. Process. Lett. **53**, 77–83 (1995)
42. H. Kuwakado, K. Koyama, Security of RSA-type cryptosystems over elliptic curves against Hastad attack. Electron. Lett. **30**(22), 1843–1844 (1994)
43. C.S. Laih, W.C. Kuo, Speeding up the computations of elliptic curves cryptoschemes. Comput. Math. Appl. **33**(5), 29–36 (1997)
44. S. Lang, *Fundamentals of Diophantine Geometry* (Springer, Berlin, 1983)
45. G.J. Lay, H.G. Zimmer, *Constructing elliptic curves with given group order over large finite fields*. LNCS, vol. 877 (Springer, Berlin, 1994), pp. 250–263
46. H.W. Lenstra, Factoring integers with elliptic curves. Ann. Math. **126**, 649–673 (1987)
47. A.K. Lenstra, H.W. Lenstra, *The Development of the Number Field Sieve*, Lecture Notes in Mathematics, vol. 1554 (Springer, Berlin, 1991)
48. R. Lercier, Finding good random elliptic curves for cryptosystems defined over F_{2^n}, in *Advances in Cryptology - EUROCRYPT '97*. LNCS, vol. 1233 (Springer, Berlin, 1997), pp. 379–391
49. K. Mahler, *p-adic Numbers and their Functions* (Cambridge University Press, Cambridge, 1981)

50. J. McKee, Subtleties in the distribution of the numbers of points on elliptic curves over a finite prime field. J. Lond. Math. Soc. **59**(2), 448–460 (1999)
51. A.J. Menezes, *Elliptic Curve Public Key Cryptosystems* (Kluwer Academic Publishers, Boston, 1993)
52. A.J. Menezes, S.A. Vanstone, The implementation of elliptic curve cryptosystems, in *Proeedings of AUSCRYPT '90*. LNCS, vol. 453 (Springer, Berlin, 1990), pp. 2–13
53. A.J. Menezes, S.A. Vanstone, Elliptic curve cryptosystems and their implementation. J. Cryptol. **6**, 209–224 (1993)
54. A.J. Menezes, I.F. Blake, X.H. Gao, R.C. Mullin, S.A. Vanstone, T. Yaghoobian, *Applications of Finite Fields* (Kluwer Academic Press, Boston, 1993)
55. A.J. Menezes, T. Okamoto, S.A. Vanstone, Reducing elliptic curve logarithms to logarithms in a finite field. IEEE Trans. Inf. Theory **39**(5), 1639–1647 (1993)
56. A.J. Menezes, P. van Oorschot, S.A. Vanstone, *Handbook of Applied Cryptography* (CRC Press, Boca Raton, 1996)
57. B. Meyer, V. Müller, A public key cryptosystem based on elliptic curves over $\mathbb{Z}$, $n\mathbb{Z}$ equivalent to factoring, in *Advances in Cryptology - EUROCRYPT '96*. LNCS (Springer, Berlin, 1997), pp. 49–59
58. V. Miller, Use of elliptic curves in cryptography, in *Advances in Cryptology - CRYPTO '85*. LNCS, vol. 218 (Springer, Berlin, 1986), pp. 417–426
59. V. Miller, Short programs for functions on curves, unpublished paper (1986)
60. P.L. Montgomery, Speeding the Pollard and elliptic curve methods of factorization. Math. Comput. **48**(177), 243–264 (1987)
61. F. Morain, J. Olivos, Speeding up the computations on elliptic curves using addition-subtraction chains. Inf. Theory Appl. **24**, 531–543 (1990)
62. F. Morain, Building cyclic elliptic curves modulo large primes, in *Advances in Cryptology - EUROCRYPT '91*. LNCS, vol. 547 (Springer, Berlin, 1991), pp. 328–336
63. V. Müller, Ein Algorithmus zur Bestimmung der Punktanzahl elliptischer Kurven über endlichen Körpern der Charakteristik grösser drei, PhD thesis, Technische Fakultät der Universität des Saarlandes (1995)
64. V. Müller, S. Paulus, On the generation of cryptographically strong elliptic curves (1997, to appear)
65. National Securtity Action Memorandum 160, http://www.research.att.com/~smb/
66. A.M. Odlyzko, The future of integer factorization, CryptoBytes: The Technical Newsletter. RSA Laboratories, Summer (1995)
67. J.M. Pollard, A Monte Carlo method for factorization. BIT **15**, 331–334 (1975)
68. J.M. Pollard, Monte Carlo methods for index computation mod p. Math. Comput. **32**, 918–924 (1978)
69. C. Pomerance, *The Quadratic Sieve Factoring Algorithm*. LNCS, vol. 209 (Springer, Berlin, 1985), pp. 169–182
70. R. Rivest, A. Shamir, L.M. Adleman, A method for obtaining digital signatures and public-key cryptosystems. Commun. ACM **21**, 120–126 (1978)
71. H.G. Rück, On the discrete logarithm in the divisor class group of curves. Math. Comput. **68**(226), 805–806 (1999)
72. T. Satoh, K. Araki, Fermat quotients and the polynomial time discrete log algorithm for anomalous elliptic curves. Commentarii Mathematici Univ. St. Pauli **47**, 81–92 (1998)
73. B. Schneier, *Applied Cryptography: Protocols, Algorithms, and Source Code in C* (Wiley, New York, 1995)
74. C.P. Schnorr, Efficient signature generation by smart cards. J. Cryptol. **4**, 161–174 (1991)
75. R. Schoof, Elliptic curves over finite fields and computation of square roots mod p. Math. Comput. **44**(170), 483–494 (1985)
76. R. Schoof, Nonsingular plane cubic curves over finite fields. J. Comb. Theory **A 46**, 183–211 (1987)
77. I.A. Semaev, On computing logarithms on elliptic curves. Discrete Math. Appl. **6**, 69–76 (1996)

78. I.A. Semaev, Evaluation of discrete logarithms in a group of p-torsion points of an elliptic curve in characteristic p. Math. Comput. **67**(221), 353–356 (1998)
79. J.P. Serre, Sur la topologie des varietes algebriques en caracteristique p, in *Symposium Internacional de Topología Algebraica* (Mexico City, 1956), pp. 24–53
80. D. Shanks, Class number, a theory of factorization, and genera, (1969) Number Theory Institute. Proc. Symp. Pure. Math. **20**, 415–440 (1971)
81. C.E. Shannon, Communication theory of secrecy systems. Bell Syst. Tech. J. **28**, 656–715 (1949)
82. J.H. Silverman, *The Arithmetic of Elliptic Curves* (Springer, Berlin, 1986)
83. R.D. Silverman, The multiple polynomial quadratic sieve. Math. Comput. **48**, 329–340 (1987)
84. R.D. Silverman, An analysis of Shamir's factoring device, RSA Laboratories (1999), www.rsa.com/rsalabs/html/twinkle.html
85. J.H. Silverman, The xedni calculus and the elliptic curve discrete logarithm problem. Des. Codes Cryptogr. **20**(1), 5–40 (2000)
86. N.P. Smart, The discrete logarithm problem on elliptic curves of trace one. J. Cryptol. **12**(3), 193–196 (1999)
87. J.A. Solinas, An improved algorithm for arithmetic on a family of elliptic curves, in *Advances in Cryptology - CRYPTO '97*. LNCS, vol. 1294 (Springer, Berlin, 1997), pp. 357–371
88. A. Spallek, Konstruktion einer elliptischen Kurve über einem endlichen Körper zu gegebener Punktgruppe, Master Thesis, Institut für experimentelle Mathematik, Essen (1992)
89. J.H. van Lint, G. van der Geer, Introduction to Coding Theory and Algebraic Geometry, in *DMV Seminar*, vol. 12 (Birkhäuser, Berlin, 1988)
90. P. van Oorschot, M. Wiener, Parallel collision search with cryptanalytic applications. J. Cryptol. **12**(1), 1–28 (1999)
91. S. Vanstone, R.J. Zuccherato, Elliptic curve cryptosystems using curves of smooth order over the ring Z_n. IEEE Trans. Inf. Theory **43**(4), 1231–1237 (1997)
92. A.E. Western, J.P. Miller, Tables of indices and primitive roots, *Royal Mathmatical Tables*, vol. 9 (Cambridge University Press, Cambridge, 1968)
93. M.J. Williamson, Non-secret encryption using a finite field, CESG Report (1974), www.cesg.gov.uk/about/nsecret.htm
94. M.J. Williamson, Tougths on cheaper non-secret encryption, CESG Report (1976), www.cesg.gov.uk/about/nsecret.htm

Chapter 5
Founding Cryptography on Oblivious Transfer

5.1 Introduction[1]

In cryptography, an oblivious transfer protocol (abbreviated OT) is a fundamental protocol (see [5]) in which a sender transfers one of potentially many pieces of information to a receiver, but remains oblivious as to what piece has been transferred.

The first form of oblivious transfer was introduced in 1981 by Michael O. Rabin [10]. The sender sends a message to the receiver with probability $\frac{1}{2}$, while the sender remains oblivious as to whether or not the receiver received the message. A more useful form of oblivious transfer called 1-2 oblivious transfer or "1 out of 2 oblivious transfer," was developed later by Shimon Even, Oded Goldreich, and Abraham Lempel, in order to build protocols for secure multiparty computation. It is generalized to "1 out of n oblivious transfer" where the user gets exactly one database element without the server getting to know which element was queried, and without the user knowing anything about the other elements that were not retrieved. The latter notion of oblivious transfer is a strengthening of private information retrieval, in which the database is not kept private. In this chapter, unless stated otherwise, OT means "1-2 oblivious string transfer": Alice has two length-k binary strings K_0 and K_1 and Bob has a single bit Z as inputs; an OT protocol should let Bob learn K_Z while Alice remains ignorant of Z and Bob of $K_{\bar{Z}}(\bar{Z} = 1 - Z)$. The Shannon-theoretic approach is used, thus ignorance means negligible amount of information; formal definitions are given in Sect. 5.2.

Both source and channel models of OT are considered. In a source (or noisy correlations) model, a discrete memoryless multiple source (DMMS) with two component sources is given, whose outputs $X^n = (X_1, \ldots, X_n)$ and $Y^n = (Y_1, \ldots, Y_n)$

[1]This text was written by Rudolf Ahlswede and Imre Csiszar in 2007. In 2013 Imre Csiszar wrote a new version of this text, which appeared in the Book "Information Theory, Combinatorics, and Search Theory, In Memory of Rudolf Ahlswede, Lecture Notes in Computer Science, Vol. 7777, Springer, 2013".

A. Ahlswede et al. (eds.), *Hiding Data – Selected Topics*,
Foundations in Signal Processing, Communications
and Networking 12, DOI 10.1007/978-3-319-31515-7_5

are available to Alice respectively Bob. In a channel model, a discrete memoryless channel (DMC) is given, Alice selects the inputs X^n and Bob observes the outputs Y^n. In both models, Alice and Bob may use a public noiseless channel for unrestricted communication. The cost of OT is measured by the number n of observed DMMS outputs or of DMC transmissions; the public channel use is considered free.

The OT capacity C_{OT} of a DMMS of DMC is the limit as $n \to \infty$ of $\frac{1}{n}$ times the largest k for which OT is possible with "cost" n. This concept has been introduced by Nascimento and Winter [8] who also proved $C_{OT} > 0$ under general conditions. For previous results, showing that a DMMS or DMC makes OT possible for any k if n is sufficiently large (but not that $\frac{k}{n}$ may be bounded away from 0 while the conditions (5.2.1)–(5.2.3) below are satisfied) see the references in [8]. A related concept of commitment capacity has been introduced and characterized in [9].

In the literature, much of the effort is devoted to designing OT protocols that prevent a dishonest Alice from learning about Bob's bit Z, or a dishonest Bob from obtaining information also about $K_{\bar{Z}}$, if they violate the agreed upon protocol. This issue is not addressed here, both Alice and Bob are assumed to honestly follow the protocol. This simplification facilitates gaining basic insights, expected to be relevant also in dealing with more practical but more difficult situations where protection against cheating is also required. Of course, upper bounds derived for the simpler case are even more valid in those situations.

We report here on the paper [2]. We give a general upper bound to C_{OT} and show the tightness of this bound for a class of channels. For other cases, we give lower bounds to C_{OT} which do not coincide with the upper bound. A necessary and sufficient condition for $C_{OT} > 0$ is also given which is similar to but not the same as the condition in [8]; the difference is due to our not dealing with "distrustful cryptography".

5.2 Upper and Lower Bounds on the Oblivious Transfer Capacity

5.2.1 *Statement of Results*

An (n, k) *protocol* for OT via a DMC is described as follows. Let K_0, K_1, Z, and M, N be independent random variables (RVs), K_0 and K_1 uniformly distributed on $\{0, 1\}^k$, and Z on $\{0, 1\}$, while M and N (serving for randomization for Alice resp. Bob) are arbitrary. At times $t = 1, \ldots, n$ Alice transmits a RV X_t over the DMC, Bob receiving Y_t. Here X_t is chosen as a function of K_0, K_1, M, and of the previous public communication $F^{t-1} = F_1 \ldots F_{t-1}$ where F_i denotes the public communication in the time interval $(i, i+1)$ that may be interactive: F_i is a sequence of messages sent alternatingly by Alice and Bob, those by Alice are functions of K_0, K_1, M, and of the messages previously received by her, those by Bob are functions of Z, N, and of the messages previously received by him, including $Y^i = Y_1 \ldots Y_i$. Finally, Bob produces an estimate $\hat{K}_Z$ of K_Z where $\hat{K}_0$ and $\hat{K}_1$ are functions of Y^n, N, and of the total public communication $\mathbf{F} = F_1 \dots F_n$.

An (n, k) protocol for OT via a DMMS is similar but simpler: then X^n, Y^n are the length-n outputs of the two component sources, independent of K_0, K_1, Z, M, N, and the public communication takes place after Alice and Bob have observed X^n resp. Y^n, thus $\mathbf{F} = F_n$.

A positive number R is an *achievable OT rate* for a DMMS of DMC if, for n sufficiently large, there exist (n, k) protocols with $\frac{k}{n} \to R$ letting Bob learn K_Z, that is

$$Pr\{\hat{K}_Z \neq K_Z\} \to 0 \tag{5.2.1}$$

while Alice remains ignorant of Z:

$$I(K_0 K_1 M X^n \mathbf{F} \wedge Z) \to 0 \tag{5.2.2}$$

and Bob learns nothing about $K_{\bar{Z}}$:

$$I(N Y^n \mathbf{F} \wedge K_{\bar{Z}} \mid Z) \to 0. \tag{5.2.3}$$

The dependence on n of the RVs in (5.2.1)–(5.2.3) has been suppressed, to keep the notation transparent.

The *OT capacity* C_{OT} of a DMMS or DMC is the largest achievable OT rate, or 0 if no $R > 0$ is achievable.

Remark 54 An alternative definition requires convergence with exponential speed in (5.2.1)–(5.2.3). The results in this paper hold also with that definition.

Theorem 92 *The OT capacity of a DMMS with generic RVs X, Y is bounded above by*

$$\min\left[I(X \wedge Y), H(X \mid Y)\right]. \tag{5.2.4}$$

The OT capacity of a DMC is bounded above by the maximum of (5.2.4) for RVs X, Y connected by this DMC.

Remark 55 This bound holds also for a weaker concept of OT, requiring Bob to learn or remain ignorant about a single length-k string of Alice according as Z equals 0 or 1, Alice remaining ignorant of Z. Also, the "strong secrecy" postulated in (5.2.3), see [7], could be relaxed to "weak secrecy", dividing the mutual information by k.

Theorem 93 *For a binary erasure channel with erasure probability p*

$$C_{OT} = \min(1 - p, p),$$

thus the bound in Theorem 92 is tight.

A DMC $\{W : \mathcal{X} \to \mathcal{Y}\}$ will be called a *generalized erasure channel* (GEC) if the output alphabet $\mathcal{Y}$ can be decomposed as $\mathcal{Y}_0 \cup \mathcal{Y}^*$ such that $W(y \mid x)$ does not depend on $x \in \mathcal{X}$ if $y \in \mathcal{Y}^*$. For a GEC, we denote $W_0(y \mid x) = \frac{1}{1-p^*} W(y \mid x), \quad x \in \mathcal{X}$, $y \in \mathcal{Y}_0$, where p^* is the sum of $W(y \mid x)$ for $y \in \mathcal{Y}^*$ (not depending on x). The Shannon capacity of the DMC $\{W_0 : \mathcal{X} \to \mathcal{Y}_0\}$ is denoted by $C(W_0)$.

Theorem 94 *For a GEC, the bound in Theorem 92 is tight if $p^* \geq 1/2$, then $C_{OT} = (1 - p^*)C(W_0)$. If $0 < p^* < 1/2$, a lower bound is $C_{OT} \geq p^* C(W_0)$.*

Remark 56 The latter bound is not tight in general, see Example 36.

Theorem 95 *The OT capacity of a DMMS or DMC is positive if and only if there exist x' and x'' in $\mathcal{X}$ such that the joint probabilities $P_{XY}(x', y)$ and $P_{XY}(x'', y)$, respectively the conditional probabilities $W(y \mid x')$ and $W(y \mid x'')$, are not equal for all $y \in \mathcal{Y}$, and are simultaneously positive for some $y \in \mathcal{Y}$.*

Remark 57 Theorem 95 says that the positivity of the upper bound in Theorem 92, after merging identical rows (if any) of the matrix of joint respectively conditional probabilities, is necessary and sufficient for positive OT capacity, see Sect. 5.2.2 .

5.2.2 The Proofs

Lemma 51 *For arbitrary RVs U, V, Z with values in finite sets $\mathcal{U}, \mathcal{V}, \mathcal{Z}$, and any z_0, z_1 in $\mathcal{Z}$ with $Pr\{Z = z_0\} = p > 0, \quad Pr\{Z = z_1\} = q > 0$,*

$$
\begin{aligned}
&|H(U \mid V, Z = z_0) - H(U \mid V, Z = z_1)| \\
&\leq c\sqrt{I(UV \wedge Z)} \log_2 |\mathcal{U}| + h\left(\min\left[c\sqrt{I(UV \wedge Z)}, \frac{1}{2}\right]\right)
\end{aligned}
$$

where $h(t) = -t \log_2 t - (1 - t) \log_2(1 - t)$, and c is a constant depending on p and q.

The proof, whose details are omitted, uses the Pinsker inequality to bound the variation distance of the two conditional distributions of UV, given $Z = z_0$ respectively $Z = z_1$. Then the conditional entropy difference is bounded as in [3]. Though the value of c is not relevant here, by careful calculation (including an improvement of the bound in [3]) we have shown that $c = 3\sqrt{\frac{(p+q)\ln 2}{2pq}}$ suffices. Thus, for the case $p = q = 1/2$ used below, a suitable constant factor is $c = 3\sqrt{2 \ln 2}$.

Proof of Theorem 92 Concentrating on channel models, we sketch the proof of the following stronger result: if there exist (n, k) protocols with $\frac{k}{n} \to R$ and

$$Pr\{\hat{K}_0 \neq K_0 \mid Z = 0\} \to 0 \tag{5.2.5}$$

$$I(K_0 X^n \mathbf{F} \wedge Z) \to 0 \tag{5.2.6}$$

$$\frac{1}{k} I(NY^n\mathbf{F} \wedge K_0 \mid Z = 1) \to 0 \tag{5.2.7}$$

then R does not exceed the maximum of (5.2.4).

Now, (5.2.6) implies by Lemma 51 that

$$H(K_0 \mid X^n\mathbf{F}, Z = 0) - H(K_0 \mid X^n\mathbf{F}, Z = 1) = o(k) \tag{5.2.8}$$

$$H(K_0 \mid \mathbf{F}, Z = 0) - H(K_0 \mid \mathbf{F}, Z = 1) = o(k). \tag{5.2.9}$$

From (5.2.9) and the consequence $I(\mathbf{F} \wedge K_0 \mid Z = 1) = o(k)$ of (5.2.7), it follows due to $H(K_1 \mid Z = 0) = H(K_1 \mid Z = 1) = k$ that

$$I(K_0 \wedge \mathbf{F} \mid Z = 0) = o(k). \tag{5.2.10}$$

If (5.2.5) and (5.2.10) held without conditioning on $Z = 0$ then K_0 would be a secret key for Alice and Bob, with (weak sense) security from an eavesdropper observing the public communication $\mathbf{F}$. The rate $\frac{k}{n}$ of such secret key is asymptotically bounded [1, 6] as

$$\frac{k}{n} \leq \frac{1}{n} \sum_{t=1}^{n} I(X_t \wedge Y_t) + \varepsilon_n, \quad \varepsilon_n \to 0. \tag{5.2.11}$$

The actual (5.2.5) and (5.2.10) imply the analogue of (5.2.11) with $I(X_t \wedge Y_t)$ replaced by $I(X_t \wedge Y_t \mid Z = 0)$. This replacement, however, has an asymptotically negligible effect since, due to the consequence $\max_t I(X_t \wedge Z) \to 0$ of (5.2.6), the conditional distribution of X_t on the condition $Z = 0$ differs negligibly from the unconditional distribution. Thus, (5.2.5)–(5.2.7) imply (5.2.11).

It is not hard to show that $K_0 \to X^n\mathbf{F} \to NY^n\mathbf{F}Z$ is a Markov chain. This, (5.2.5), and Fano's inequality give

$$H(K_0 \mid X^n\mathbf{F},\ Z = 0) \leq H(K_0 \mid NY^n\mathbf{F},\ Z = 0) = o(k). \tag{5.2.12}$$

Then

$$\begin{aligned} k = H(K_0 \mid Z = 1) &\overset{(i)}{=} H(K_0 \mid NY^n\mathbf{F},\ Z = 1) + o(k) \\ &\leq H(K_0 \mid X^nY^n\mathbf{F}, Z = 1) + H(X^n \mid NY^n\mathbf{F},\ Z = 1) + o(k) \\ &\overset{(ii)}{\leq} H(X^n \mid Y^n,\ Z = 1) + o(k) \leq \sum_{t=1}^{n} H(X_t \mid Y_t,\ Z = 1) + o(k) \end{aligned}$$

where (i) follows from (5.2.7) and (ii) from (5.2.8) and (5.2.12). In the last sum, the conditioning on $Z = 1$ has an asymptotically negligible effect as before, thus we have

$$\frac{k}{n} \leq \frac{1}{n} \sum_{t=1}^{n} H(X_t \mid Y_t) + \varepsilon_n, \quad \varepsilon \to 0. \tag{5.2.13}$$

Finally the main term in (5.2.11) is $\leq I(X_T \wedge Y_T)$ and the main term in (5.2.13) is $\leq H(X_T \mid Y_T)$ where T is a RV uniformly distributed on $\{1, \ldots, n\}$, independent of the RVs X_t, Y_t. Hence, the claim follows from (5.2.11), (5.2.13).

Proof of Theorem 93 Theorem 92 gives the upper bound $C_{OT} \leq \min(1-p, p)$. The following protocol shows that each $R < \min(1-p, p)$ is an achievable OT rate.

(i) Alice transmits over the DMC n independent equiprobable bits X^n.
(ii) Bob determines the set $G \subset \{1, \ldots, n\}$ of "good" positions where no erasure occurred, and selects from G a random subset of size $k = nR$, and similarly from the "bad" set G^c. Denoting by S_0 the set of positions selected from G or G^c according as $Z = 0$ or $Z = 1$, and by S_1 the other set, Bob tells Alice S_0 and S_1, not leaking any information on Z.
(iii) Alice adds her strings K_i to $\{X_t : t \in S_i\}$, $i = 0, 1$, bitwise mod 2, and she reports the sums to Bob.

As Bob knows X_t for $t \in G$, he can recover K_Z, but remains ignorant of $K_{\bar{Z}}$, not knowing X_t for $t \in G^c$.

Proof of Theorem 94 Due to Theorem 92, it suffices to show that $C_{OT} \geq \min(1-p^*, p^*)C(W_0)$, that is, that $R = R'R''$ is an achievable OT rate if $R' < \min(1-p^*, p^*)$, $R'' < C(W_0)$. To this, a DMMS secrecy result [1, 6] will be used: Suppose Alice and Bob observe l outputs of the component sources of a DMMs whose generic RVs have mutual information larger than R. Then, for l sufficiently large, Alice can securely transmit $k = lR$ bits to Bob via sending a public message, with negligible probability of error and negligible leak of information to an eavesdropper who sees the public message alone.

Now, Alice transmits over the DMC n i.i.d. RVs X_t that achieve Shannon capacity (of both channels W and W_0). Then Bob selects $l = nR$ positions at random from the good set $G = \{t : Y_t \in \mathcal{Y}_0\}$, as well as from the bad set $G^c = \{t : Y_t \in \mathcal{Y}^*\}$. Calling the resulting sets S_0 and S_1 as in the previous proof, Bob tells Alice S_0 and S_1, leaking no information on Z.

Under the condition $Z = 0$, the RVs $\{(X_t, Y_t) : t \in S_0\}$ represent l output pairs of a DMMS whose generic RVs have mutual information $C(W_0)$, while under the condition $Z = 1$ these X_t and Y_t are independent. The joint distributions of $\{(X_t, Y_t) : t \in S_1\}$ under the same conditions coincide with those of $\{(X_t, Y_t) : t \in S_0\}$ as above, reversing $Z = 0$ and $Z = 1$. Hence, by the cited result and the assumption $R'' < C(W_0)$, there exists a function f on $\{0, 1\}^k \times \mathcal{X}^l$, where $k = lR'' = nR$, with the following properties: If Alice sends the public messages $f(K_0, \{X_t : t \in S_0\})$, $f(K_1, \{X_t : t \in S_1\})$ then, in case $Z = 0$ when Bob knows $\{Y_t : t \in S_0\}$, Bob can recover K_0 but remains ignorant of K_1 regarding which he observes, in effect, the public message only. Similarly, in case $Z = 1$ Bob can recover K_1 remaining ignorant of K_0.

Proof of Theorem 95 If some rows of the matrix of joint or conditional probabilities are equal then merging the corresponding elements of $\mathcal{X}$ does not change OT capacity. The necessity part of the assertion follows applying Theorem 92 after this merging.

For sufficiency, concentrate on channel models. Consider the two-block extension of the given channel $\{W\}$ and restrict its input alphabet $\mathcal{X}^2$ to $\tilde{\mathcal{X}} = \{(x', x''), (x'', x')\}$, for x', x'' in the hypothesis. It follows by that hypothesis and Theorem 94 that the so obtained channel $\{\tilde{W} : \tilde{\mathcal{X}} \to \mathcal{Y}^2\}$ which is a GEC, has positive OT capacity. Hence so does also the channel $\{W\}$.

5.2.3 *Discussion and Examples*

The OT capacity of discrete memoryless source and channel models has been studied. A general upper bound, and a lower bound for generalized erasure channels were given, determining the OT capacity of binary erasure channels, and of any GEC with erasure probability $\geq 1/2$. For the general case, lower bounds were shown to follow from those for GECs. While in proving the upper bound very complex protocols were admitted, the achievability (lower bound) results use simple protocols. It remains open whether OT capacity can be achieved in general with protocols of comparable simplicity, similarly, for example, to multiterminal secrecy capacities [4].

Protection against cheating has not been addressed. Still, it is worth noting that while the protocols in Theorems 93 and 94 are vulnerable to cheating by Bob if the erasure probability is less than $1/2$ (when Bob, in addition to learning K_Z, can gain information about $K_{\bar{Z}}$ via dishonest choice of $S_{\bar{Z}}$), such cheating could be prevented by a modified protocol achieving the same OT rate. To this, sets S_0 and S_1 both of size $\frac{n}{2}$ are taken, with $S_Z \subset G$; then, as $S_{\bar{Z}}$ intersects G, a stronger DMMS secrecy result has to be invoked in which the eavesdropper knows more than the public message alone.

The approach in this paper easily extends to other versions of OT, one mentioned in connection with Theorem 92. For example, Alice may have m stings $K_1, \ldots, K_m$ and Bob may chose either of them ("1 of m OT") or any subset of them, while Bob has to remain ignorant of the other stings, and Alice of Bob's choice. Our reason for stating Lemma 51 for a not necessarily binary RV Z has been to make it suitable to prove analogues of Theorem 92 for such OT problems, too.

We conclude by three examples.

Example 35 Consider a binary symmetric channel (BSC) with crossover probability p, and define a channel $\{\tilde{W} : \tilde{\mathcal{X}} \to \mathcal{Y}^2\}$ by restricting the input alphabet of the two-block extension of this BSC to $\tilde{\mathcal{X}} = \{(0, 1), (1, 0)\}$.Then $\tilde{W}$ is a GEC with $\mathcal{Y}^* = \{(0, 0), (1, 1)\}$, and the corresponding $\tilde{W}_0$ is a BSC with crossover probability $p^2/(1 - p^*)$ where $p^* = 2p(1 - p) < 1/2$.

Hence, Theorem 94 implies for the OT capacity of a BSC

$$C_{OT} \geq \frac{1}{2} p^* C(\tilde{W}_0) = p(1 - p) \left(1 - h \left(\frac{p^2}{1 - 2p(1 - p)}\right)\right).$$

Example 36 Consider the GEC with $\mathcal{X} = \{0, 1\}, \mathcal{Y} = \{0, 1, *\}$,

$$W = \begin{pmatrix} (1-p)(1-\varepsilon) & p(1-\varepsilon) \; \varepsilon \\ p(1-\varepsilon) & (1-p)(1-\varepsilon) \; \varepsilon \end{pmatrix}.$$

For its OT capacity, if $0 < \varepsilon < 1/2$, Theorem 94 gives $C_{OT} \geq \varepsilon C(W_0)$ where W_0 is the BSC with crossover probability p. Another lower bound in $C_{OT} \geq \frac{1}{2}p^* C(\tilde{W}_0)$ where $\{\tilde{W} : \tilde{\mathcal{X}} \to \mathcal{Y}^2\}$ is the GEC defined similarly as in Example 35, with $\mathcal{Y}^* = \{(0,0), (1,1), (**)\}$ and $p^* = 2p(1-p)(1-\varepsilon)^2 + \varepsilon^2$. If $\varepsilon \to 0$, the latter bound approaches that in Example 35, while the previous bound goes to 0. This shows that the lower bound in Theorem 94 is not tight, in general.

Example 37 Consider the additive DMC with $\mathcal{X} = \mathcal{Y} = \{0, 1, 2, 3\}$, $Y = X + N$ (mod 4), N binary $1/2 - 1/2$. This is not a GEC but the bound in Theorem 92 is tight for it, $C_{OT} = 1$. Indeed, the following simple (1, 1) protocol achieves perfect OT. (i) Alice transmits over the channel a uniformly distributed rv X. (ii) Bob receives $Y = X + N$ (mod4), and tells Alice $\phi = 0$ or 1 according as $Y + Z$ is even or odd. (iii) Alice reports the mod2 sums $K_0 + i_\phi(X)$ and $K_1 + i_{1-\phi}(X)$ were i_0 and i_1 are the indicator functions of the sets $\{1, 2\}$ and $\{2, 3\}$. This unambiguously tells Bob the bit K_Z, keeping him fully ignorant of $K_{\bar{Z}}$, because an even or odd value of Y uniquely determines $i_0(X)$ respectively $i_1(X)$ but provides 0 information about $i_1(X)$ respectively $i_0(X)$.

References

1. R. Alhswede, I. Csiszár, Common randomness in Information Theory and Cryptography, Part I. IEEE Trans. Inf. Theory **39**, 1121–1132 (1993)
2. R. Alhswede, I. Csiszár, On the oblivious transfer capacity, in *Proceedings of the IEEE International Symposium on Information Theory, ISIT* (2007), pp. 2061–2064
3. R. Alicki, M. Fannes, Continuity of quantum conditional information. J. Phys. A: Math. Gen. **37**, L55–L57 (2004)
4. I. Csiszár, P. Narayan, Secrecy capacities for multiterminal channel models. IEEE Trans. Inf. Theory **54**(6), 2437–2452 (2008)
5. J. Kilian, Founding cryptography on oblivious transfer, in *Proceedings of the STOC 1998* (1988), pp. 20–31
6. U. Maurer, Secret key agreement by public discussion. IEEE Trans. Inf. Theory **39**, 733–742 (1993)
7. U. Maurer, The strong secret key rate of discrete random triples, in *Communications and Cryptography: Two sides of One Tapestry*, ed. by R.E. Blahut, et al. (Springer, Boston, 1994), pp. 271–285
8. A. Nascimento, A. Winter, On the oblivious transfer capacity of noisy correlations, in *Proceedings of the ISIT 2006* (Seattle, 2006), pp. 1871–1875
9. A. Winter, A. Nascimento, H. Imai, Commitment capacity of discrete memoryless channels, *Cryptography and Coding*, vol. 2898, LNCS (Springer, Berlin, 2003), pp. 35–51
10. M.O. Rabin, How to exchange secrets by oblivious transfer, Technical Report TR-81, Aiken Computation Laboratory, Harvard University, (1981)

Obituary for Rudi

My name is Beatrix Ahlswede Loghin. I was married to Rudi Ahlswede from 1970 until 1984. Rudi and I are the parents of a son, Alexander Ahlswede.

Rudi's death was sudden. There was no warning, no time to consider, to right wrongs, to express love and thanks. He left us quickly and undramatically. Through the power of our remembrance, we evoke Rudi back into our world for this brief moment. Or, to quote T.S. Eliot: "History is now and England, with the drawing of this love, and the voice of this calling".

Preparing this obituary I found myself pondering the question, again and again: how to go about this? A human being is so complex. Of all the myriad possibilities, moments, experiences, selves, of which we consist, which ones do we choose to share? What does one write? Isn't anything that we write a reduction, a limiting of this particular human being's complexity? Is not our life a great work of algebra, in which we ponder the great X, the mystery of our lives? And so I realized that I cannot write about Rudi, because I don't know "Rudi". Even after all these years of experience with him, living with him, being in a family with him, I don't really know Rudi. All I know is my Rudi, my experience of him.

The Canadian writer, Margaret Atwood, gave this advice to young writers: "Say what is yours to tell". That is all we can do, but also all we need to do: Say what is ours to tell.

"I come to bury Caesar, not to praise him". No sooner are these words spoken, than Marc Antony of course begins to do just that' praise Caesar, in Shakespeare's historical drama. Nevertheless, I pondered the distinction. How does one speak of the dead? If we praise, we end up speaking only of the "nice", "pleasant" attributes. A kind of "Rudi Ahlswede lite" version. Those of us who spent time with Rudi know that this was not his way. Rudi's interaction with life was passionate. He loved "not wisely, but too well". He was not given to strategic behavior, even though it would

---

This obituary was hold during the conference at the ZiF in Bielefeld by Beatrix Ahlswede Loghin.

A. Ahlswede et al. (eds.), *Hiding Data – Selected Topics*,
Foundations in Signal Processing, Communications
and Networking 12, DOI 10.1007/978-3-319-31515-7

perhaps have been wiser at times. On the other hand, the dead are defenceless, they relinquish to us the power of definition, for we are still alive to tell the tale. Looking into my heart, I asked myself, "What is it really that you want to tell?" The answer that I found was this: I want to honor Rudi's life here, I want to honor the complexity of his being. I want to acknowledge the difference Rudi made in my life.

But what does it mean to acknowledge someone? The Oxford dictionary states that to acknowledge means to take something which has been previously known to us and which we now feel bound to lay open or make public. It means to recognize a particular quality or relationship which we forgot or did not consciously see. And it means to own with gratitude.

What did I know then, and wish to lay open now? Which qualities did I forget or not consciously see? What can I own with gratitude? Of the rich tapestry of Rudi's life, where do I begin to acknowledge? We cannot remember the entire sequence of life. We remember moments, special moments which—for some reason—stayed in our memory. So this is what I really want—to share with you some of these moments.

Thinking of Rudi, an image of a great mountain range comes to my mind, with invincible summits, terrifying plunges and depths, and a smattering of meadows in between. This image has been the defining core of my relationship with Rudi, beginning with our first meeting in the summer of 1967 in Columbus, Ohio. I was 18 years old and had just begun my freshman year at Ohio State University. Rudi was 29 years old and starting his first job in the US as an assistant professor in the Department of Mathematics.

At this time explosions were rocking the social and political fabric of American society. Afro-Americans, Latinos, Asian Americans and other groups were claiming their rightful place in American society, and protest against the Vietnam War was flaming up everywhere, even in politically conservative Ohio. I frequented a bar known as Larry's in Columbus, on High Street, refuge to those who considered themselves left-wing, or at least to the political left of the mainstream. In this bar, classical music, jazz and soul music was played, people of different races and nationalities congregated in cheerful bawdiness, and of course chess was played.

A mutual friend at Larry's Bar introduced us, and between long silences, in which he scrutinized his chess partners' moves, Rudi told me a little about himself, his fascination with his research, information theory, and the discoveries he was making about life in the United States. The more I became embroiled in the political demonstrations against the Vietnam War, the more Rudi became interesting for me. My fellow demonstrators and I quoted Ho Chi Minh, Mao Tse Tung and Marx, but Rudi had actually read some of Karl Marx's writings, and he was able to put these writings into a philosophical context, showing the evolution of Hegel's and Feuerbach's ideas. The great breadth of his knowledge left me stunned. I began to pay closer attention to Rudi. Not only had he read philosophy, but also literature, finding his own favourite writers and poets. In a conversation, Rudi would suddenly, just at the right moment, quote Schiller or Gottfried Benn, Goethe, Shakespeare, Thomas Wolfe or Nietzsche.

I was amazed, for he refuted all my conceptions of "typical" mathematicians. He told me more about himself. His parents owned a large farm in northern Germany. Born as the second son, he realized early in life that, much as he love the land with its wide open spaces, hills and cliffs and lush forests, he would have to leave it, as the farm would not be able to support two families. This realization was painful, tinged with bitterness. It forced him, at a very early age, to learn to create his own future. "God bless the child that's got his own", is a line from a Billie Holliday song. Rudi was such a blessed child—he had his own. He found his new world at school—his home became the world of books, the world of learning. And his aptitude in mathematics became apparent. At the age of ten he left his parents' home and lived with another family in the nearby larger town, where he could attend the Gymnasium, the secondary school which would prepare him for a university education. Later, at Gymnasium, he often felt excluded because of his background as a farmer's child. Some of his fellow students let him feel, very clearly, that he was lacking in social graces, that he came from an inferior social background. I think he never quite got over the pain of this discrimination. Learning became his passion. And this path led him from his humble elementary school in Dielmissen to the greatest universities in the world, to membership in the Russian Academy of Science. He had a fire in his mind, and this made conversations with him scintillating. This was the terrain where our minds met, and where I fell in love.

Many evenings, watching him sit in the turmoil of Larry's bar, he exuded a quality of tranquillity. He was above the fray, either focused on his chess game, or "in communion" with his own thoughts, which he would occasionally add to the paper lying before him. He clearly had something which very few others in the room had: a world of his own. He seemed incredibly strong and rooted in himself. Occasionally he would sit up, take notice of the life teeming around him, and then return again to this other, inner space.

This fascination with the world of mathematics became particularly evident one evening in the Spring of 1970. Richard Nixon had just announced the invasion of Cambodia. At universities around the country, massive strikes as a form of resistance took place. Soon the campus at Ohio State became a small battleground. Tanks rolled through the streets, students erected barricades and threw bricks and Molotov cocktails. Helicopters flew overhead, spraying the demonstrators with tear gas. Rudi and I sought refuge in the McDonalds on High Street, where we found Rudi's colleague, Bogdan Baishanski, also seeking shelter. Demonstrators ran into the McDonalds, followed by night-stick brandishing police. We fled back onto the streets. In front of me, I saw Rudi and Bogdan running from the police, jumping over barricades, clearly illuminated by the searchlights of the helicopters flying over our heads, throwing more tear gas in our direction. Stumbling blindedly behind them, I noticed that, as they ran, they were deep in conversation—about the (at that time still unsolved) four color conjecture!

A short time later, Rudi had been stopped in the middle of the night while driving home, for making a right turn without a full stop. Because of an outstanding traffic violation, he was arrested and led off in handcuffs. I scrambled to find two hundred dollars with which to bail him out. When I arrived at the jail the next morning, Rudi emerged smiling. He told me about the "interesting" evening he had spent, stuffed in a holding cell with his fellow inmates. And, he told me proudly, he had gotten a new idea in jail which led to a significant break-through in the paper he was currently writing!

Years later I read in book written by someone who was researching happiness, that the happiest people are those who have something in their lives which so absorbs them that it permits them to completely forget themselves and the world around them. This process of forgetting oneself is called flow. I think Rudi spent much of his life in this state. But of course this obliviousness to his surroundings left him vulnerable. Many times a date began with long searches in the parking lots around the Mathematics Department—Rudi simply could not remember where he had left the car that morning. Between us this was of course often a cause of exasperation on my part. One day, in a store, I noticed two young salesgirls giggling about Rudi, who was lost in space, smoking, and running his hands through his hair. A fierce determination to protect him in this vulnerability was born in me at that moment.

In this way, Rudi was like no one I had ever met. Years later, after we had moved to Germany, listening to my son and his friends recount funny anecdotes about Rudi, I realized that they were fascinated by precisely his way of being different from others, his eccentricity, to use another word. The word eccentric comes from the Greek words ek kentros, meaning not having the same center. Years later, after we had married, I stood in a market square with Rudi in Sicily, in Syracusa, the town where the great Archimedes had lived. He was killed when a Roman soldier accosted him in the market place, where he sat, drawing designs in the sand. Awed by Archimedes' fame, the soldier asked if there was anything he could do for him. Archimedes is said to have answered: "Don't disturb my circles". This story impressed me greatly, for I was sure that Rudi would have given the same answer, and I recognized that he was a kindred spirit.

Shortly after we met, Rudi returned to Germany for a few weeks. He wrote to me that he was reading a book by Giordano Bruno, entitled "Heroic Passions". It seemed so fitting. Years later, when we lived in Rome, we spent many an hour at the Campo dei Fiori, where Bruno was burned at the stake for refusing to renounce his scientific ideas. I had no doubt that Rudi would have ended there too had he lived in this time. Rudi was never politically correct. He said what he thought and accepted the consequences. Rudi was incapable of inauthenticity. There was a wild, almost savage need in him to stay true to himself, a need which caused him much conflict and grief. But suppressing his beliefs in order to attain some goal was beyond him. He paid a huge price in his life for that' and, at the same time, this is what made him so strong.

Rudi was the freest person I have ever met.

I saw Rudi for the last time on his last birthday, September 15, 2010. We spent the evening together, drinking a bottle of wine and talking of our son, of mutual old friends. The years passed by before our inner eyes. He was, as always, excited about life, looking forward to the new research he had embarked upon, and which he told me about, as always, with sparkling eyes. But something was different about this evening. After he finished talking, he asked me about myself. Amazed, I found myself telling Rudi about my life, my plans. He listened with a care and an attention that was new. We sat, side by side, companions of a shared life. I went home elated, feeling blessed and rich from this evening with Rudi.

Standing at his coffin in the cemetery, looking at his dead body, I realized there was only one word left to say to him: Thank you.

Comments by Rüdiger Reischuk

This volume again considers secure information transmission, but in a stronger setting. Instead of random noise that may generate errors now there is an active adversary that tries to corrupt messages, the problem of authentication. Even more, messages should not only be secured against changes of contents or authorship, they also have to be protected against getting known to third parties that observe the channel, the secrecy requirement. Shannon's entropy put classical (symmetric) cryptography on formal grounds. But large information theoretic distance turned out to be a very high requirement for many practical applications.

Diffie and Hellman had a groundbreaking new idea: asymmetric systems for which the security should depend on computational complexity requirements. Computational complexity was not one of the main focus of Rudolf Ahlswede's research. Still I remember extensive discussions with him on topics like Boolean functions and communication complexity. I met Prof. Ahlswede the first time as a graduate student in 1976 shortly after he had moved from Ohio to the University of Bielefeld. It took a while to correct my first impression about this man who did not seem to look and behave as professors are expected—noticeable, for example, playing chess in his office quite often, but also playing cards in the mensa with students being quite noisy. After intensive discussions I became aware of his real worth, his brilliant analytical ideas, his extraordinary mathematical skills and his philosophical thoughts.

After arriving in Bielefeld Rudolf Ahlswede immediately took responsibility in developing the young mathematical faculty there. He wanted to build a strong group in applied mathematics by hiring further colleagues from abroad. This was not an easy task because at that time applied mathematics was not considered *real mathematical science* by pure mathematicians in Bielefeld. But here and also later in controversial decisions of the faculty Rudolf Ahlswede fought for his ideas—in most cases successfully.

One of these new colleagues was Wolfgang Paul from Cornell who was known for his recent work in complexity theory and whom Ahlswede wanted to help adding computer science to the mathematical spectrum in Bielefeld—at least the theoretical part of informatics. I chose Wolfgang Paul as my advisor. His office was next to that

A. Ahlswede et al. (eds.), *Hiding Data – Selected Topics*,
Foundations in Signal Processing, Communications
and Networking 12, DOI 10.1007/978-3-319-31515-7

of Ahlswede and they got into closer contact. Ingo Wegener, one of Ahlswede's first Ph.D. students and assistant professors and coauthor of his later book on searching, got interested in Wolfgang Paul's research area, the complexity of Boolean functions. The cooperation between the two research groups grew and I was lucky to be part of this. Some years later after Rudolf Ahlswede had also considered successfully problems in other areas of mathematics besides information theory, I remember a discussion between Paul and Ahlswede. Rudi claimed that he would be able to solve important problems in any area of mathematics. Wolfgang replied that proving nontrivial lower bounds for the complexity of Boolean functions seems quite difficult and he should try that. This seems to be one of the rare examples where Ahlswede's ingenious combinatorial skills did not suffice for a breakthrough. Today, more than 30 years later no substantial progress has been made on this question and it seems that more time and completely new techniques are necessary.

This lack of proofs for lower complexity bounds, which are essential for the security of modern data hiding systems, may have been motivation for Rudolf Ahlswede as an emeritus to start studying crypto systems and its algorithmic foundation in detail and to prepare these lectures. One clearly notices his information theoretic background and new insights gained from this. *Hiding data* did not become one of his most active research areas, but he intensively investigated the *dual* question *searching data* in the last years of his life. His extraordinary mathematical research effort did not decrease when passing the age of retirement. This makes him even more outstanding.

Rudolf Ahlswede provided important help for my own scientific career. After my Ph.D. advisor Wolfgang Paul had left Bielefeld Rudolf Ahlswede stepped in and supported my habilitation in the area Theoretische Informatik at the faculty of mathematics. Later when my time in Bielefeld came to an end in 1985 we met again at several scientific conferences organized by him, meetings in Oberwolfach and at the ZIF in Bielefeld. Discussions with him stand out by technical deepness and bringing up novel ideas. Rudolf Ahlswede, I like to thank you for your advise and the many beautiful theorems and proof techniques you have invented.

List of Notations

K	A (perfect) field
$\overline{K}$	Algebraic closure of K
K^*	Group of invertible elements in K
K^+	Group of additive elements in K
$\mathrm{char}(K)$	Characteristic of K
$\mu_n(K)$	Subgroup of n-th roots of unity in K
$\mathbb{N}$	Non-negative integers
$\mathbb{Z}$	Integers
$\mathbb{Q}$	Rational numbers
$\mathbb{C}$	Complex numbers
$\mathbb{F}_p$	Finite field of p elements
$\mathbb{Z}_n$	$= \mathbb{Z}/n\mathbb{Z}$
$\mathbb{Z}_p, \mathbb{Q}_p$	Ring of p-adic numbers
$\mathrm{Div}^g(X)$	Group of divisors on X of degree g
$\mathrm{Div}_p(X)$	Principal divisors
$\mathrm{Pic}^0(X)$	0-part of the divisor class group (Picard group)
$\mathrm{Pic}^0(X)_n$	n-torsion subgroup of $\mathrm{Pic}^0(X)$
$\deg(D)$	Degree of the divisor D
$\Omega(X)$ $(\Omega^1(X))$	K-dimensional space of (holomorphic) differentials
$\#E(K)$	Number of rational points on a curve defined over K
$E[n]$	n-torsion point group
$E[n](K)$	n-torsion point group of K-rational points
$\mathrm{End}(E)$	Endomorphism ring of E
$\mathbb{A}^n$ $(\mathbb{P}^n)$	n -dimensional affine (projective) plane
$\mathcal{M}$	Set of possible plaintext messages
$\mathcal{C}$	Set of possible ciphertext messages
$\mathcal{K}$	Set of possible keys
lcm	Least common multiple
gcd	Greatest common divisor

A. Ahlswede et al. (eds.), *Hiding Data – Selected Topics*,
Foundations in Signal Processing, Communications
and Networking 12, DOI 10.1007/978-3-319-31515-7

Author Index

A. Ahlswede et al. (eds.), *Hiding Data – Selected Topics*,
Foundations in Signal Processing, Communications
and Networking 12, DOI 10.1007/978-3-319-31515-7

Subject Index

A. Ahlswede et al. (eds.), *Hiding Data – Selected Topics*, Foundations in Signal Processing, Communications and Networking 12, DOI 10.1007/978-3-319-31515-7

MIX
Papier aus verantwortungsvollen Quellen
Paper from responsible sources
FSC® C105338
www.fsc.org

If you have any concerns about our products,
you can contact us on
ProductSafety@springernature.com

In case Publisher is established outside the EU,
the EU authorized representative is:
Springer Nature Customer Service Center GmbH
Europaplatz 3, 69115 Heidelberg, Germany

Printed by Libri Plureos GmbH
in Hamburg, Germany